The Mathematics of the Heavens and the Earth

The Mathematics of the Heavens and the Earth

The Early History of Trigonometry

GLEN VAN BRUMMELEN

PRINCETON UNIVERSITY PRESS Princeton and Oxford

Published by Princeton University Press, 41 William Street, Princeton, New Jersey 08540

In the United Kingdom: Princeton University Press, 6 Oxford Steet, Woodstock, Oxfordshire OX20 1TW

Library of Congress Cataloging-in-Publication Data

Van Brummelen, Glen.
The mathematics of the heavens and the earth : the early history of trigonometry / Glen Van Brummelen.
p. cm.
Includes bibliographical references and index.
ISBN 978-0-691-12973-0 (hardcover : alk. paper) 1. Trigonometry—History. I. Title.
QA24.V36 2009
516.2409—dc22 2008032521

British Library Cataloging-in-Publication Data is available

This book has been composed in Times Roman

Printed on acid-free paper. ∞

press.princeton.edu

Printed in the United States of America

10 9 8 7 6 5 4 3 2 1

To

Len Berggren, my mentor,

and

Heide, my partner

Contents

Preface

Imagine a subject that is so pervasive that almost all of us see some of it during our high school education, one whose story goes back well into ancient times. This subject crosses most major cultures and places; indeed, it is not easy to identify societies that contributed significantly to science without using it in some way. It might even be argued that this discipline was the link that brought geometric models of the cosmos together with numerical computation in a synthesis that allowed theory to be converted into prediction: the birth of the exact sciences. All this makes it hard to believe that trigonometry has never been given a proper book-length historical treatment in English.

Trigonometry is particularly interesting because it is rare among mathematical subjects: it was originally motivated almost entirely by outside concerns, in this case, astronomy and geography. Many theorems that we now call trigonometric had been known before in other mathematical contexts; others came to our awareness only through the need to solve particular scientific problems. Thus trigonometry often retooled existing mathematics for new purposes, but partly through its interface with the sciences it blossomed into an independent discipline, providing structures and ways of thinking that eventually returned to inspire mathematics itself.

The most recent effort to write the history of trigonometry in a Western language was over a century ago, with Anton von Braunmühl's two-volume classic *Vorlesungen über Geschichte der Mathematik.*[1] The *Vorlesungen* was an outstanding example of the best work of its time: rigorous and thorough, yet concise and clear. However, most of the professional discipline of the history of mathematics has happened since it was written, and after one hundred years it is well past time to revisit the topic. In no area is this more true than in the non-Western cultures most relevant to us, India and Islam. The great nineteenth-century historians of mathematics, I am sure, would be astounded and impressed to see what manuscripts and discoveries have come to light in these fields.

I am proud to say that this book owes much to the efforts of the historians of von Braunmühl's generation. It is fashionable to criticize them these days, and not without cause. Current historiographic trends emphasize

[1] [von Braunmühl 1900/1903]. If one considers Russian to be a Western language, then there is also the notable [Matvievskaya 1990].

cultural contextualization of mathematical history,[2] and looking back, it is clear that our predecessors too often brought their nineteenth-century contexts into their interpretations. It is part of the job of the historian to attempt to disengage from the modern context, and to try to live the past the way the past lived it. Surely this task is quixotic, but there is much to be gained in the perpetual struggle to shed our own skins. Nevertheless, our success in this is elusive, and really can only be evaluated after much time has passed. I wonder whether twenty-second-century historians will judge us as harshly as we judge the historians of one hundred years ago.

What I find admirable about the great historians who did so much to establish the foundation of our field is their unremitting attention to rigor and reliability, and their commitment to both the history and the mathematics. We may sometimes question their historicity, but we can trust their honest attempts to face up to the material. They did their best to place their findings in an appropriate context, but they did not shy away from the mathematics itself. One cannot genuinely practice the history of a scientific subject without also living and breathing the science itself. Von Braunmühl knew the mathematics of which he wrote, and I have done my best to live up to that heritage.

This is not to say that I will stick slavishly to the style of the past masters. The value of placing results in their original context is well taken; one of my efforts in this direction is to include a number of extracts from (translations of) the original texts in the narrative. This should give the reader the chance to experience directly what the texts say, and to judge how they should be read. Explanations are provided to help decode passages that would otherwise be obscure. These should also help to illustrate that the modern mathematical notation used throughout this book is a transformation of the original texts; hopefully the inclusion of excerpts will allow the reader to decide how great this transformation is.[3] I hope that some of the excerpts will prove useful in the classroom: most of them stand on their own, and

[2] The paper that is sometimes said to have sparked the current trend of searching for and destroying modern mathematical conceptions in Greek mathematics is Sabetei Unguru's "On the need to rewrite the history of Greek mathematics" [Unguru 1975/76]; see also [Saito 1998] for an update. The issue of appropriate contextualization is felt particularly strongly in the study of the most ancient mathematical cultures; see [Imhausen 2003b] on Egypt, and [Robson 1999] and [Robson 2001] on Mesopotamia.

[3] The degree to which the arguments in historical mathematical texts should be translated to modern notation is a matter of current debate. For instance, many ancient and medieval texts will refer to "the rectangle between *AB* and *CD*." The arguments often require merely the product of the lengths of the line segments, but writing it as such changes the mental process by removing the geometric content, possibly changing our perception of what the author thought in some significant way. In this book—a scholarly survey, but not a research paper—I have tried to respect the original sources as much as possible while making the arguments accessible to the modern reader.

some might help to enliven trigonometric or astronomical concepts by providing historical context.

So, the reader will find here some of the flavor of an "episodes" book; it is not necessary to read it from cover to cover. Nevertheless there is enough of a flow of ideas that the completist reader should also be satisfied.

The question arises who I intend to read this book. I would be delighted if the interested lay public finds it to be useful, and I hope that they are given the chance. However, my first loyalty is scholarly: I aim to provide a solid, reference-supported account of the best understanding of the history of trigonometry that the academic community can provide. Hence the hundreds of footnotes: for those readers who wish to delve into the issues in more depth, the notes contain appropriate links to the literature, much of the back story, and details on the mathematics. However, I have been careful to keep scholarly debates out of the main text. My goal is to give the historical characters the stage. Nevertheless, this is a mathematical rather than a social history. There are plenty of other places to go to find out about the lives and contexts of the scientists. Instead, I have chosen to emphasize the mathematical arguments. Many of them are beautiful; most are practical. They are the heart of trigonometry, and I will not apologize for asking the reader to experience their pleasures.

The practical nature of trigonometry is a distinctive feature compared to most other mathematical disciplines and is worth highlighting, especially in this first of two volumes. Through the sixteenth century, trigonometry existed mostly (often exclusively) to serve the needs of mathematical astronomers. Without astronomy, there wouldn't have been much point. Thus the reader should expect to encounter at least the basics of spherical and planetary astronomy within these pages. I have done my best to avoid requiring any outside astronomical reading, pleasurable though that might be. The brief opening chapter on the ancient heavens should convey almost all the spherical astronomy that is needed to make sense of the book.

The dependence of trigonometry on outside sources of inspiration informs other distinctive choices I have made that might surprise some. Following the theme of contextualization, I focus on topics that the historical practitioners—not necessarily the modern scholars—found interesting and useful. Since these practitioners were astronomers, they were often less interested in harmonious theory than in practical results. So, in addition to the trigonometric identities and theorems, I have gone into some depth for instance on the methods that were developed to compute trigonometric tables. And in addition to some beautiful direct solutions to astronomical problems, I also discuss the numerical methods that were applied when geometry was not up to the task. Some modern readers find a pleasing aesthetic quality in the geometry and consider table-making a thing of the past, but this is after all a work of history. The past is my business.

I do find occasion to apologize in one respect. When considering what to include in a survey of this scope, the author must eventually draw a line in the sand. This leads inevitably to regret that some valuable topics missed the cut. In my case this includes trigonometric traces that may be found outside of my major cultural groupings—in particular, Jewish, Byzantine, and Chinese. In the case of China the story is episodic early and much happens well after the end date of this volume, so it will be treated in the sequel. There are also certain topics that I wish had received a proper scholarly treatment before the writing of this book (the role of the analemma in Islam comes to mind). However, this will always be the case, and eventually one has to put pen to paper.

How to Read This Book

I will be grateful to the enterprising reader who starts on page one and works faithfully to the end; there is reward in following a tale to its conclusion. But I do not expect everyone to have the same zeal that I do for my subject. Many people will use this book as a reference, concentrating on one or two of the major chapters at a time, or even dipping into individual applications that look intriguing. I have done my best to accommodate this, but there is a problem: a certain depth of astronomical understanding is required for a number of the episodes in this book. I cannot repeat the same astronomical introduction again and again. So, the text itself contains only brief explanations of the astronomy involved in each particular problem. If you find the text too terse, I have included a preliminary chapter which orients newcomers to the ancient heavens. That chapter should give all the context that is needed.

Some formatting details to keep in mind while reading: throughout the time periods covered in this volume trigonometric functions were defined not in the unit circle, but in circles of different radii. These are notated by capitalizing the first letter of the function ("Sin" rather than "sin"); the value of the radius R should be indicated somewhere nearby. Secondly, many of the calculations reported here were performed in the sexagesimal (base 60) system. Numbers are represented as follows: for example,

$$7,8;12,34,56 = 7 \cdot 60 + 8 + \tfrac{12}{60} + \tfrac{34}{60^2} + \tfrac{56}{60^3}.$$

Sometimes the whole parts of numbers are given in decimal format; for instance, the number above can be written as 428;12,34,56. Finally, errors in trigonometric values are given in units of the last place. For instance, if the value given in the historical text is 12;34,56 while the correct value is 12;34,59, the error is displayed as [-3].

I originally intended to cover the entire history of trigonometry in one book, but soon realized that task to be overwhelming. This volume is book-ended, more or less, by the developments of the Earth-centered and the Sun-centered universe. Trigonometry proper began with the origins of the Ptolemaic geocentric system, and a nice breaking point is found with the beginning of Copernicus's heliocentric model. From this point trigonometry began to move with the rest of mathematics toward symbolic algebra and (eventually) analysis, losing some of its geometric character and picking up a more familiar modern flavor. As the medieval Islamic scientists often wrote in their conclusions: there is much more to say, and I hope some day with the will of God to write the necessary sequel. It took von Braunmühl three years to publish his second volume; I hope that readers will be more patient with me. It might take a little longer.

Reference Sources

This project would not have been possible without access to a number of literature databases, both modern and old-fashioned. The opportunity to acknowledge these sources does not present itself in the main text, so I would like to give them a nod here. I turned most frequently to the following:

- *MathSciNet* (the online version of *Mathematical Reviews*);
- *Zentralblatt für Mathematik*;
- Kenneth O. May's *Bibliography and Research Manual of the History of Mathematics* [May 1973];
- The *Dictionary of Scientific Biography* reference sections;
- Dauben and Lewis's *The History of Mathematics from Antiquity to the Present: A Selective Annotated Bibliography*, revised edition on CD-ROM [Dauben 2000];
- The Research Libraries Group's History of Science, Technology and Medicine Database;
- Rosenfeld and Ihsanoğlu's *Mathematicians, Astronomers & Other Scholars of Islamic Civilisation and Their Works (7th–19th c.)* [Rosenfeld/Ihsanoğlu 2003];
- A private literature database on Islamic astronomy maintained by Benno van Dalen.

Although I have found ways of coping with most languages relevant to this study, Russian was a problem and I wish I could have handled this literature more effectively. A good Russian reference source is the extensive bibliography in [Matvievskaya 1990].

Acknowledgments

I have often been astonished at the long list of names in many books' acknowledgments sections and wondered if I would ever have the good fortune to know so many people willing to give their efforts to another cause. I am truly fortunate to be able to answer that in the affirmative. Among them are a number of people who provided input, commentary and support outside of the formal refereeing process: Len Berggren, John H. Conway, David Friedman, Alexander Jones, David King, Duncan Melville, Kim Plofker, Nathan Sidoli, Benno van Dalen, and Venessa Wallsten. Those who participated in the formal reviewing process, a task often without sufficient public recognition, deserve special thanks: Helmer Aslaksen, Len Berggren (again), Dennis Duke, Victor Katz, Kim Plofker (again), and Benno van Dalen (again), and two anonymous reviewers. Vickie Kearn, the editor at Princeton University Press, and her assistants Sarah Pachner and Anna Pierrehumbert were amazingly supportive. Technical support at Princeton provided by Terri O'Prcy, Lor Gehret, and Dimitri Karetnikov was meticulous and exemplary. Bezaleel Tekka was a great help in preparing the line drawings for publication. Reid Ginoza, a student at Bennington College, spent five valuable weeks reading the manuscript for mathematical accuracy. Heide Klix-Van Brummelen provided much translation assistance and support. I would like to identify two people who may not expect to see their names here: Radha Charan Gupta, whose extensive writings on the history of Indian trigonometry were so useful that I sometimes felt uncomfortably that I was summarizing his research career—I only wish he had gotten around to publishing his dissertation on the subject; and Marie-Thérèse Debarnot, whose summary of Islamic trigonometry in [Debarnot 1996] set a standard that I have struggled here to reach. I have also had the marvelous fortune to teach at two of the most exciting institutions of higher learning I have ever encountered: Bennington College and Quest University. The former granted me a sabbatical leave to work on this project and was in my corner throughout; the latter has embraced my work with equal generosity. As always, the author assumes all responsibility for weaknesses in the final product.

The final word goes to the two institutes that provided the crucial hospitality, funds and administrative support for this project: the Dibner Institute for History of Science and Technology at Massachusetts Institute of Technology (directors George Smith and Bonnie Edwards, and staff), and the Institut für Geschichte der Naturwissenschaften at the University of Frankfurt (director David King, and colleagues). Neither of these treasured institutions deserved the fates that befell them shortly after my stays; I would gladly give up this book to see either one of them re-open its doors. The field will never be the same. I have already dedicated this book, but an extra tip of the hat is called for here, and a raised glass to their memory.

The Mathematics of the Heavens and the Earth

The Ancient Heavens

This short chapter is not part of the main flow of the book. It is intended only as a reference for those readers who wish to become acquainted with the basic concepts of spherical astronomy that will arise later. Those readers may choose to read it now, or to wait until the need arises. All others should skip forward to chapter 1.

A night spent watching the heavens with a clear sky reveals some basic facts about the nature of the universe. Firstly, it is clear that the sky is the top half of a sphere, the bottom half being below the ground and therefore not visible to us. As the night progresses, we begin to notice that the stars are not fixed in place, but revolve slowly around us. It does not take long to realize that the stars are in fact attached to the ***celestial sphere***, and that the entire sphere is spinning slowly, at a rate of one rotation per day (figure I.1).[1] The Sun is one of the stars attached to the sphere, and it is the rotation of the celestial sphere that carries the Sun through the heavens, producing day and night. When the Sun is above the horizon the sky is too bright to see the other stars, but they are there nonetheless and they make their appearance when the Sun is carried below the horizon.[2]

Before long the existence of the half of the celestial sphere below the ground becomes clear: stars rise from below to above the horizon in the east, and they set in the west. Over the course of a year, as the observer becomes familiar with the constellations, a map of most of the celestial sphere and the locations of the stars on it is gradually constructed.[3]

The point directly above the observer is called the ***zenith***, and the great circle from the north point of the horizon, straight up to the zenith, and down to the south point of the horizon is called the observer's ***meridian***. As can be seen from figure I.1, the Sun reaches its highest altitude in the sky when it crosses the meridian; this is local noon.

The rotation of the celestial sphere leaves one point fixed around which the other stars move; this point is the celestial ***North Pole***, which is currently very close to Polaris, the North Star. Locating this star allows us to quickly determine the direction of North, as well as our terrestrial latitude—it is

[1] Of course, the modern reality is that the Earth is rotating rather than the heavens.

[2] In the rare occasion of a solar eclipse, the stars may be seen above the horizon during the day.

[3] Some stars, only visible in the Earth's southern hemisphere, will always stay below the horizon for our northern observer.

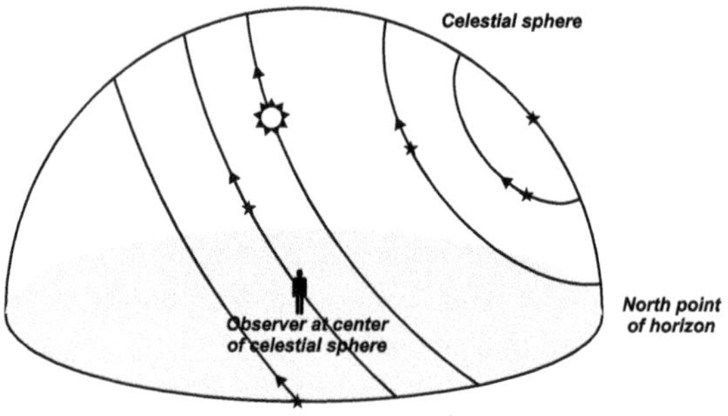

Figure I.1
The celestial sphere. The Sun and stars are on the surface of the sphere, and are carried in the sphere's daily rotation around the observer.

equal to the altitude of the North Star above the horizon.[4] The equator of the rotating celestial sphere, the ***celestial equator*** (see figure I.2), is one of the most important great circles in the sky. Stars that happen to lie on the equator will rise directly east, and will set directly west.

The celestial equator spins at the constant rate of one rotation per day, or 15° per hour. Positions on the celestial equator are measured with respect to one particular point on it, the ***vernal equinox*** ♈,[5] which we shall describe in more detail shortly. The positions can be measured in degrees, but due to the special nature of the equator, they are usually measured in hours. Thus the ***right ascension*** α of a point on the celestial equator will be anywhere from 0^{hr} to 24^{hr} starting from ♈. This suggests a coordinate system for the celestial sphere: a star that does not have the good fortune to lie on the equator can be imagined to drop perpendicularly onto it (see figure I.2). The angular distance that it falls is the star's ***declination*** δ;[6] the right ascension of the point where the star drops onto the equator is the star's right ascension α. The pair (α, δ) form the star's ***equatorial coordinates***.

At this point it is still conceivable for the observer to be standing on a flat, or a spherical, Earth. It is in fact quite easy to see that the Earth is a sphere. For instance, during a lunar eclipse the Earth's shadow cast on the Moon is always circular. Or, when a ship disappears over the horizon, the hull of the ship disappears before the mast does. There was never any question of the sphericity of the Earth for the astronomers in this book, and we shall take it for granted as well. Also, for them the center of the universe is obviously the center of the Earth. There are a few occasions (for example, in Archimedes'

[4] The North Pole gradually moves over the centuries; Polaris has not always been so conveniently located.

[5] At first it may seem peculiar to refer to the equinox as a point rather than a date. Here, we mean the Sun's place on the celestial sphere at the moment in time that is called the equinox.

[6] δ is positive if the star is north of the celestial equator, and negative if it is below. In ancient times, of course, δ was described simply as north or south of the equator.

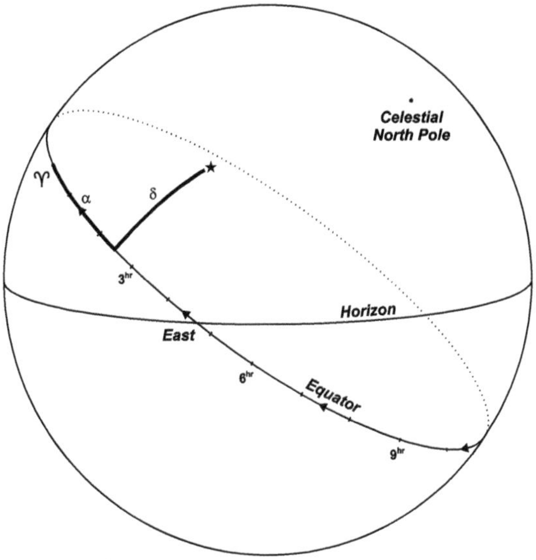

Figure I.2
Equatorial coordinates. For this star, $\alpha \approx 2\frac{1}{2}$ hrs and $\delta \approx +30°$.

Sand Reckoner) where there is an important distinction between the center of the Earth and the observer's location on its surface. But most of the time the Earth is an infinitesimal speck at the center of the celestial sphere.

If the Sun kept its place with respect to the rest of the stars in figure I.1, we would see the same star show every night: the same stars would rise as the Sun set, and at any given time of night the stars would be in the same place regardless of the time of year. However, the Sun does not keep its place. It wanders through the constellations, changing its place every day by about 1° of arc, making a complete circuit around the celestial sphere over the course of a year. Six other celestial bodies (the Moon, Mercury, Venus, Mars, Jupiter, and Saturn) also move through the background of the "fixed" stars, and collectively these seven objects are named ***planets*** after the Greek word for "wanderer."

But the Sun's motion is by far the most regular of the planets. Its path is a perfect great circle through the celestial sphere, known as the ***ecliptic*** (see figure I.3). The Sun does slow down and speed up slightly over the course of the year, but it does not stray at all from the ecliptic. The other six planets travel close to the ecliptic, but can be above or below the ecliptic by at most about 5° (depending on the object). This band around the ecliptic is called the ***zodiac***. The ecliptic is inclined to the celestial equator at an angle of $\varepsilon \approx 23.7°$;[7] this angle is called the ***obliquity of the ecliptic***.

[7] For the time period of this book, ε may be considered to be a constant. In fact it has been decreasing slightly over the millennia, and is now equal to about 23.4°.

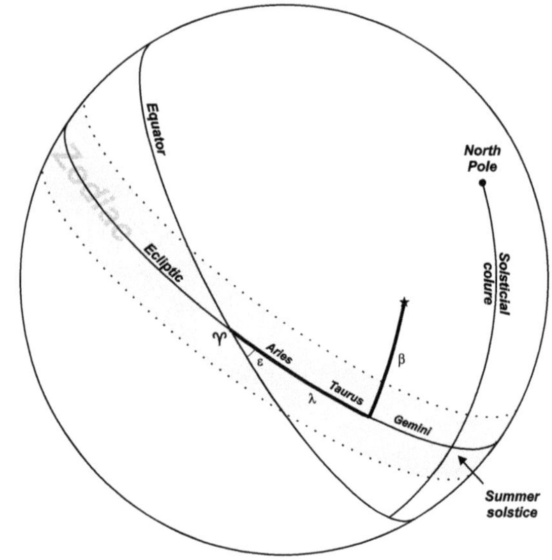

Figure I.3
The celestial sphere, highlighting the ecliptic and the zodiac. The ecliptic coordinates of the star shown in the diagram are $\lambda \approx 60°$ and $\beta \approx 30°$.

It is the obliquity that gives us our seasons. As the Sun changes its position over the course of the year, it spends varying amounts of time per day above the horizon. When it lies on one of the intersection points with the equator (the ***vernal equinox*** ♈ or the ***autumnal equinox*** on the other side) the Sun, like every other point on the equator, rises in the east, sets in the west, and spends 12 hours above the horizon in the interim. When the Sun reaches the point on the ecliptic furthest North from the equator (the ***summer solstice***), it rises and sets somewhat north of the east and west cardinal points and spends more than 12 hours per day above the horizon. This allows us to define the ***solsticial colure***, the great circle through the poles and the solstices (see figure I.3).

The existence of a new fundamental great circle on the celestial sphere, the ecliptic, suggests a new coordinate system based on it. An object's ***longitude*** λ is measured along the ecliptic from ♈, and its ***latitude*** β is the object's distance from the ecliptic along the sphere's surface (see figure I.3; these coordinates are unrelated to terrestrial longitudes and latitudes). This system is particularly useful when studying the planets: since they stay reasonably close to the ecliptic, their motions can be broken into their primary position λ and the smaller effect β. Most ancient planetary astronomers emphasized λ, but there was some concern for β as well.

Now, the 360° of the ecliptic/zodiac are divided into 30° segments, starting at ♈. These segments, the ***zodiacal signs***, are named after the constellations that (roughly) occupied these spaces in ancient times: Aries,

Taurus, Gemini, and so forth to Pisces. However, over the past two thousand years the constellations have drifted about 30°, or one zodiacal sign, away from their namesake segments of the zodiac. The effect of this gradual drift of the background stars from their positions on the ecliptic (or vice versa) is known as the ***precession of the equinoxes***, and measures about 1° every 72 years. This requires a decision to be made in terms of coordinates: is the fixed reference point from which the coordinates are measured going to be ♈ (which drifts slowly westward with respect to the background stars), or a fixed position within the map of fixed stars (which, conversely, drifts slowly eastward with respect to ♈)? In the former case we call them ***tropical coordinates***; in the latter, they are called ***sidereal coordinates***.

As the Sun travels along the ecliptic throughout the year, its declination varies from +23.7° (at the summer solstice) to −23.7° (at the winter solstice). On a given day we may assume that the Sun is effectively motionless along the ecliptic. It is thus carried in a circle by the rotation of the celestial sphere; this is called the Sun's ***day-circle*** (figure I.4). The ***hour-angle*** formed at the North Pole using the meridian and the arc that heads to the Sun determines the number of hours until local noon, since it corresponds to the equatorial arc between where the Sun is now and where the Sun will be at noon. The ***diurnal arc*** is the part of the day-circle above the horizon; in the summer this comprises more than half of it. So, in the summer there are more than 12 hours of daylight. However, some cultures measure hours as twelfths of the daylight period; i.e., for them hours are longer in the summer than in the winter. These variable hours are called ***seasonal day hours***; they occur often in astronomical timekeeping.

There is a hint of a third coordinate system in figure I.4, based on the horizon circle. The first coordinate, the ***azimuth***, is the position of the object along the horizon; the second coordinate, the ***altitude*** h, is the angular height of the object above the horizon. The altitude shall get some attention here, but we shall not refer much to the azimuth. Nevertheless a related quantity arises from time to time: the ***ortive amplitude*** ψ, the place on the horizon where the Sun rises (relative to the East point).

A problem that arises time and again in historical spherical trigonometry is the determination of ***rising times of arcs of the ecliptic***. It is related to timekeeping by the Sun, and we shall see it in several chapters of this book. We have seen already that the celestial equator rotates into itself over the course of a day and thus does not change its position; this is however not true of the ecliptic, which changes its position from hour to hour. Figure I.5 shows the celestial sphere at sunrise; the Sun's position on the ecliptic (its longitude λ) is assumed to be known by the time of year. We shall see that the Sun's

Figure I.4
Solar
timekeeping

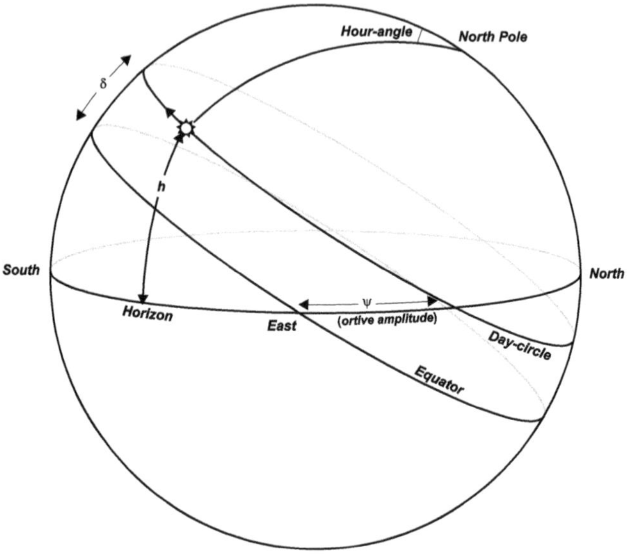

Figure I.5
The oblique
ascension θ

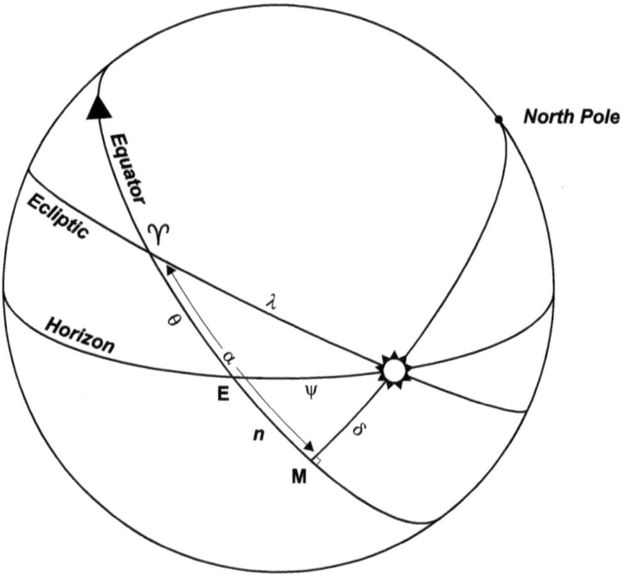

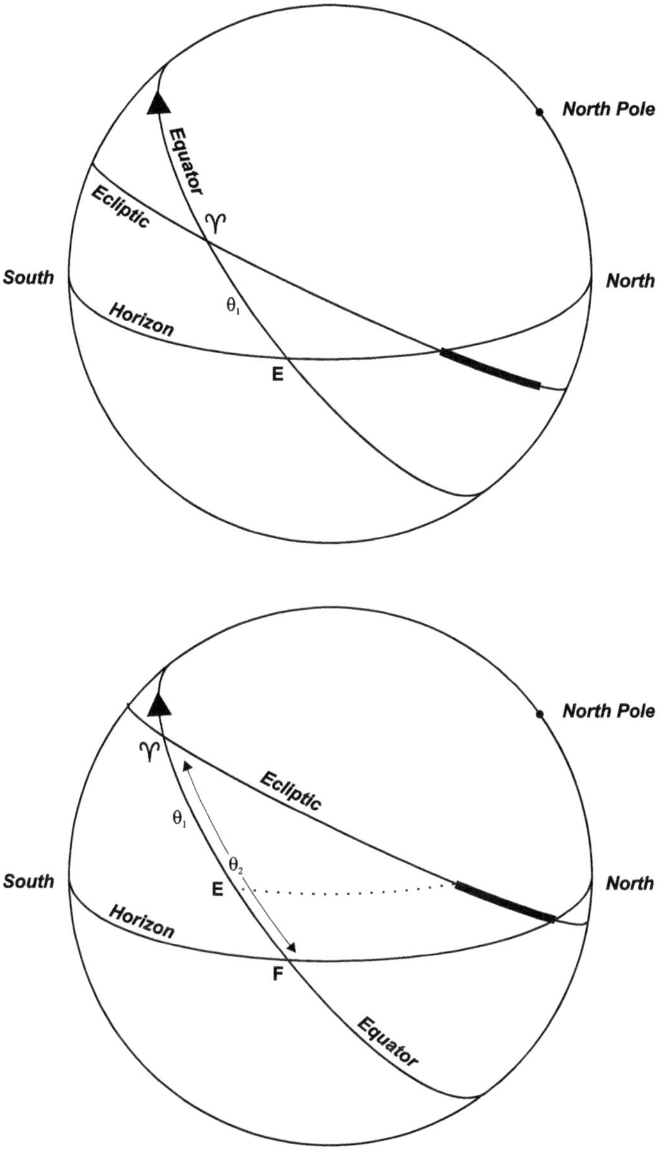

Figure I.6 The rising time $\widehat{EF} = \theta_2 - \theta_1$ of a given arc of the ecliptic (drawn in bold)

a) first contact

b) last contact

equatorial coordinates—its right ascension α and declination δ—may be found from λ. The ortive amplitude is also prominent in the diagram as $\psi = \overset{\frown}{E☼}$; it often enters into rising time computations.

Given the direction of the equator's rotation, it is clear from the situation in figure I.5 that the vernal equinox ♈ rose somewhat recently. Exactly how recently may be determined by finding the Sun's ***oblique ascension*** $\theta = \overset{\frown}{♈E}$: since θ is an arc on the equator, it can be converted to time by the relation 360° = 24 hours. Incidentally, if the observer is on the Earth's equator, the celestial equator will rise at right angles to the horizon and the right ascension and oblique ascension will coincide. But when the observer is elsewhere on the Earth's surface, the equator is oblique to the horizon, and right and oblique ascension differ by the amount $n = \overset{\frown}{EM}$. This quantity is called the ***ascensional difference***, or (in the Arab world) the ***equation of daylight***.

Given a particular arc of the ecliptic (usually one of the zodiacal signs), the ***rising time*** of that arc is the length of time it takes for it to rise above the horizon. In figure I.6(a) the top of the arc has just reached the horizon, and the oblique ascension is $\theta_1 = \overset{\frown}{♈E}$. A couple of hours later, as shown in figure I.6(b), the arc has crossed the horizon completely and has become entirely visible; the oblique ascension is now $\theta_2 = \overset{\frown}{♈F}$. The rising time is thus $\theta_2 - \theta_1 = \overset{\frown}{EF}$, the difference between the two oblique ascensions.

There is a great deal more to ancient mathematical astronomy; in particular, we have skipped over the theory of the motions of the planets along the ecliptic. However, this brief introduction should give the reader enough background to make sense of the astronomy in this book.

1 Precursors

What Is Trigonometry?

This deceptively difficult question will shape the opening chapter. The notion that sines, cosines, and other modern functions define what we mean by "trigonometry" may be laid to rest instantly; these functions did not reach their modern forms, as ratios of sides in right-angled triangles, until relatively recently. Even in their historical forms they did not appear until medieval India; the Greeks used the length of the chord of an arc of a circle as their only trigonometric function.[1] The word itself, meaning "triangle measurement," provides little help: it is a sixteenth-century term, and much ancient and medieval trigonometry used circles and their arcs rather than triangles as their reference figures.

If one were to define trigonometry as a science, two necessary conditions would arise immediately:

- a standard quantitative measure of the inclination of one line to another;[2] and
- the capacity for, and interest in, calculating the lengths of line segments.

We shall encounter sciences existing in the absence of one or the other of these; for instance, pyramid slope measurements from the Egyptian Rhind papyrus fail the first condition, while trigonometric propositions demonstrated in Euclid's *Elements* (the Pythagorean Theorem, the Law of Cosines) fail the second.

What made trigonometry a discipline in its own right was the *systematic* ability to convert back and forth between measures of angles[3] and of lengths. Occasional computations of such conversions might be signs of something better to come, but what really made trigonometry a new entity was the ability to take a given value of an angle and determine a corresponding length. Hipparchus's work with chords in a circle is the first genuine instance of this, and we shall begin with him in chapter 2. However, episodes that come

[1] The chord is the length of a segment cut off across a circle, usually considered as a function of the angle it subtends at the center.

[2] Today this is done with angles, although in certain contexts such as spherical trigonometry, circular arcs (treating the given lines as radii of the circle) are more convenient.

[3] For lack of a more convenient term, we shall use "angular measurement" to refer to any direct measure of inclination, whether angles or arcs are measured.

close—a prehistory—do exist in various forms before Hipparchus, and we shall mention some of them in this preliminary chapter.

There is a case to be made that definitions are unwise in a historical account, especially one covering several cultures and a vast time span. Any idea that produces fruit for as long a time as trigonometry is likely to be altered to the point of unrecognizability after a while—consider, for instance, the tortured history of the mathematical meaning of "analysis"—and trigonometry is no exception. It is hard to reconcile the use of sines and cosines in the analytical work of men like Euler, Fourier and Cantor with the geometric spirit of trigonometry's Greek progenitors. So, a definition that fits one part of the history may well be cumbersome elsewhere. Fortunately, our definition coincides (mostly) with the realization and growth of a powerful single idea, but in cases where they diverge, we will stay with the flow rather than the definition.

The *Seqed* in Ancient Egypt

The pyramids are among the most recognized structures in the world; the Great Pyramid at Giza is the oldest of the seven ancient wonders of the world, and the only one still standing. They have become the center of speculative theories of all sorts, purportedly revealing advanced age, advanced religious significance, advanced astronomy, advanced mathematics, and advanced engineering. Many of these theories have been shown to be little more than advanced silliness; nevertheless, the pyramids are undoubtedly some of the most remarkable buildings ever constructed. The precision in both their North-South alignments and their slopes on such a staggering scale are an achievement regardless of the level of sophistication required to produce them.

Part of the reason for the vast literature on the construction of the pyramids is, no doubt, the paucity of real evidence. In fact, much of what we know about Egyptian mathematics in general comes from only a few papyri that somehow survived the millennia. The most famous of these texts is the Rhind Mathematical Papyrus (RMP),[4] named after the nineteenth-century Scotsman A. Henry Rhind, who obtained it in Luxor in 1858. This document, over 5 m wide and about 32 cm high, consists mostly of the famous

[4] Much modern scholarship on the RMP is based on the edition [Peet 1923] and the two-volume edition, translation, and commentary by A. B. Chace [Chace 1927/1929]. A new translation has recently appeared within volume 3 of Marshall Clagett's *Ancient Egyptian Science: A Source Book* [Clagett 1999]. [Gillings 1972] contains substantial commentaries on the RMP. See also [Robins and Shute 1987] for photographic plates of the entire papyrus and an explanation of its mathematics.

2/*n* table used to convert fractional quantities into sums of unit fractions, a table of division of integers by 10, and a set of 84 mathematical problems. The problems are mostly arithmetical or involve the distribution of food, but some deal with the geometry of circles, rectangles, triangles, and pyramids. Problems 56–60 in particular contain calculations regarding the slope of a pyramid given its horizontal and vertical dimensions, and some have posited these problems as a kind of proto-trigonometry.[5]

The cause of these claims is the notion of the *seqed*, a term referring to the slope of an inclined side in Egyptian architecture. Used in the RMP only with respect to pyramids, there is evidence to suggest that the *seqed* was also used for the inclinations of temple gateways.[6] As a measure of "slope" it inverts our use of the word: the *seqed* is the amount of horizontal displacement, measured in palms, for every seven palms of vertical displacement (thus $s = 7/m$, where m is the modern definition of slope). Seven palms corresponded to one architectural royal cubit. Curiously, analysis of inclinations in Egyptian art suggests that something similar to the *seqed* may have been at work in that field as well, but using the small cubit of six palms instead. In the case of the temple of Ramesses II at Luxor, the use of a different cubit leads to a *seqed* of 1 (per 7) for the gateway itself, but an inclination of 1 (per 6) in a relief of the same gateway on a wall inside the temple!

Text 1.1
Finding the Slope of a Pyramid
(Rhind Mathematical Papyrus, Problem 58)

In a pyramid whose altitude is 93⅓, make known the seqed of it when its base-side is 140 [cubits].

Take ½ of 140, which is 70. Multiply 93⅓ so as to get 70. ½ of 93⅓ is 46⅔. ¼ of it is 23⅓. Take ½ ¼ of a cubit. Operate on 7: ½ of it is 3½; ¼ of it is 1½ ¼; the total is 5 palms 1 finger. This is the seqed.

[5] From [Robins and Shute 1985]: "We know from the pyramid exercises in the Rhind mathematical papyrus . . . that the ancient Egyptians used a simple trigonometry for determining architectural inclinations . . ." See also [Vetter 1925] and [Vogel 1959, 72–73]. The image on the cover of the recent *Trigonometric Delights* by Eli Maor ([Maor 1998]) contains both the Sphinx and the Great Pyramid.

[6] See [Robins and Shute 1985, 113]. For the use of the notion of slope in water clocks, see [Clagett 1995, 76]. For a description of how the *seqed* would have been implemented in pyramid construction, see [Rossi 2004, 192–196]. The extent to which the *seqed* was actually used is controversial; see the chapter on *seqed* theory in [Herz-Fischler 2000, 30–45] for a careful analysis, including a summary of various nineteenth-century opinions. Finally, [Imhausen 2003a, 162–168] contains an algorithmic analysis of the RMP *seqed* calculations.

Working out:

1 93⅓
\½ 46⅔
\¼ 23⅓

Total ½ ¼
Produce ½ ¼ of a cubit, the cubit being 7 palms.

1 7
½ 3½
¼ 1[½]¼

Total: 5 palms 1 finger, which is the seqed.[7]

Explanation: In figure 1.1, the base is bisected to give $BC = 70$ cubits. The calculations that follow establish that $BC / AB = 70 / (93\frac{1}{3}) = \frac{3}{4}$. To convert this ratio to a *seqed* measurement, multiply this result by 7, which gives a value of 5¼ (commonly used in pyramids of the late Old Kingdom).

The trigonometric connection resides in the fact that the *seqed* is simply the ratio BC / AB (scaled by the unit conversion from palms to cubits), and is thus equal to the cotangent of the angle at C. Is this legitimate trigonometry? The text does not refer explicitly to $\angle C$, but as we shall see, Hellenistic astronomers often used arcs rather than angles for computations that are clearly trigonometric. Here, however, there is no notion of the measurement of $\angle C$ in any form. The purpose of the *seqed* is architectural, and to read it as a cotangent is anachronistic. For this reason, and also since the *seqed* did not transmit to the Greeks, we shall say no more about it here.[8]

Babylonian Astronomy, Arc Measurement, and the 360° Circle

Both angles and lengths are continuous quantities; to be able to move easily between them requires a good system of numeration for fractional quantities. The ancient Egyptians used unit fractions exclusively, and forbade repeated use of the same unit fraction when representing a number. Clearly this would have made converting arbitrary angles to lengths cumbersome. The first powerful number system incorporating fractions arose in the third

[7] This translation is taken from [Clagett 1999, 167].
[8] Similar concepts of slope are found in other ancient cultures; for instance, see [Robson 1999, 222] for examples from Old Babylon. We shall ignore them here for the same reason that we pass over the *seqed*.

Figure 1.1
RMP Problem 58, illustrating the *seqed*

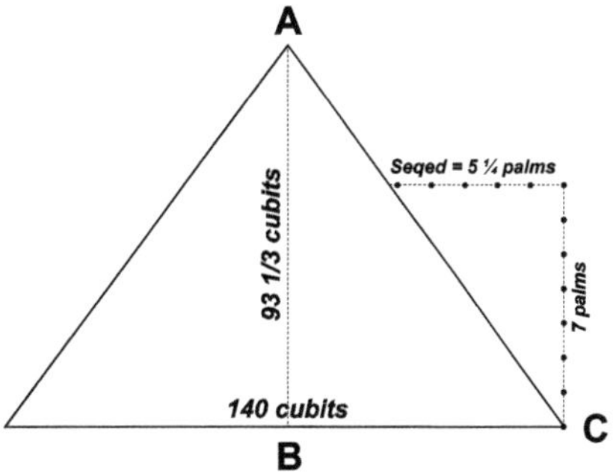

millennium BC in Sumer, out of a system of metrological units, particularly measurements of weight.[9] A quantity of ten of one unit became the next unit up, then six of that unit became the next, and so on. Eventually pairs of values in the alternating bases of 6 and 10 came to be thought of as single values in a number system with a base of 60. Since 60 has many divisors, sexagesimal calculation (as it is now called) is not as difficult as it may sound.[10] An administrative reform in the twenty-first century BC, requiring the keeping of meticulous records, may have instigated its birth.[11] Whatever the cause, it was to become the standard system of numeration in astronomy and trigonometry, at least for the fractional parts of numbers, for millennia.[12]

We shall represent sexagesimal numbers in the standard modern transcription; that is,

$$23;51,20 = 23 + \frac{51}{60} + \frac{20}{60^2}.$$

Since sexagesimals had arisen from metrology there was nothing particularly special about any given unit; the measurement could always be subdivided into something smaller. Hence the sexagesimal point, here indicated by a semicolon, was not as significant as the decimal point is to us, and was not written. The lack of a symbol for zero would also have caused some confusion until placeholder symbols for the absence of a quantity in a sexagesimal digit were invented.

[9] See [Thureau-Dangin 1921], [Neugebauer 1927], [Thureau-Dangin 1928a], and the nice survey article [Thureau-Dangin 1939].

[10] It is often said erroneously in textbooks that the base of 60 was chosen specifically because of its many divisors.

[11] [Høyrup 1992, 604]. The case for this dating is made in [Powell 1976].

[12] Indeed, [Nissen/Damerow/Englund 1993, 151] asserts that sexagesimal arithmetic was used initially only in arithmetic texts, and if not for the birth of astronomy, it may have "fallen into oblivion."

The sexagesimal system is still seen today, for instance, in the division of the hour into minutes and the minute into seconds, and of angular degrees into minutes and seconds. Place value numeration was helpful then, as now, for complicated arithmetic computations; the Babylonians were adept at addition, subtraction, multiplication, and division (although certain complications arise for division[13]). Tables of square and cube roots reveal that the Babylonians also had methods for these operations.[14]

By far the most famous computational tablet of the early Babylonian period is Plimpton 322, composed around 1700 BC. Excavated during the 1920s, it contains a collection of Pythagorean triples—that is, sets of three integers that satisfy the Pythagorean theorem $a^2 + b^2 = c^2$, thereby forming a right triangle. The smallest examples of such triples are 3-4-5 (a version of which appears on the tablet) and 5-12-13. One way to interpret this tablet is by trigonometry: after all, the triples appear to compare lengths of sides in a triangle, and they are arranged within the table in order of decreasing ratio of the two sides touching the right angle. However, there are a host of competing explanations, including an interpretation in terms of reciprocal pairs.[15] With no explicit system of angle measurement in this tablet or indeed in any early Babylonian mathematics,[16] nor any tradition of similar tablets of triples to follow it, Plimpton 322 fits neither our definition of trigonometry nor our narrative; hence we move on.[17]

However, angular measurement did begin with the Babylonians; but it was over a millennium later and in the context of astronomy. Observations of celestial phenomena date back almost as far as human history, but the earliest documents that record these events systematically and develop a quantitative science to accompany them began around the eighth century BC.[18]

[13] The Babylonian procedure for division is simply to find the reciprocal of the divisor and multiply by the numerator. Tables of reciprocals existed to aid the process, but usually gave only reciprocals whose sexagesimal representation terminated after a finite number of places. See [Aaboe 1964, 10–16 and 20–22], among many other sources, for more details.

[14] This is well known; indeed, [Neugebauer/Sachs 1945, 33–35] lists some tables, and p. 42 analyzes a procedure text for cube roots. A consideration of an algorithm that may have been used for evaluating square roots may be found in [Fowler/Robson 1998].

[15] The literature on Plimpton 322 is extensive, perhaps well beyond its significance for Babylonian mathematics. See [Neugebauer/Sachs 1945, 38–41] for its original publication. Most recently, see [Robson 2001] and [Robson 2002] for a revival of the reciprocal pairs theory; the former contains descriptions of competing theories and an extensive bibliography.

[16] [Robson 2001, 179–183] describes the Old Babylonian approach to circles as conceptualized "from the outside in"; even the radius was a subsidiary idea. See also [Gandz 1929, 458–459] for a refutation of the notion of angle in either Babylonian or Egyptian mathematics.

[17] The use of Pythagorean triples is also controversial in Egyptian architecture and mathematics. See [Rossi 2004, 60–65 and 216–221] for a good summary of the evidence.

[18] See [Neugebauer 1975, part I, pp. 351–353] for a short summary of our sources for Babylonian astronomy.

The driving forces behind these endeavors would have been partly calendric (for civil and agricultural purposes), but astrological needs were dominant. Perhaps as an outgrowth of this horoscopic heritage Babylonian astronomy was computational and predictive, rather than geometrical and explanatory. Rather than predict locations of celestial bodies at given times, astronomers predicted the times and places of given celestial events. There was no underlying geometric theory of how the heavens should work, only computational schemes to say when and where a periodic event would next occur.

Nevertheless, being able to answer the "where" question requires a means of pointing to a particular place in the sky, and to grasp the Babylonian solution we shall need some basic astronomy—although we should be clear that the Babylonians did not necessarily share our need for the geometric pictures to follow.[19] The dome of the sky is called the ***celestial sphere*** (see figure 1.2). Over the course of a day the celestial sphere performs a single rotation around us from east to west, carrying the stars and the Sun with it, thereby causing day and night. The stars remain in the same place with respect to each other, hence the name ***fixed stars***. To any observer the ***celestial equator*** remains in the same place in the sky, turning into itself with its daily rotation. Seven bodies (the Sun, Moon, and the five visible planets) also participate in this motion, but they also move slowly with respect to the fixed stars. For instance, the Sun travels on its own great circle in the celestial sphere, known as the ***ecliptic***, completing one revolution per year in the direction opposite to that of the celestial sphere's daily motion, or roughly 1° per day. The ecliptic does not remain fixed in place; it is carried by the daily rotation of the celestial sphere. Twice per year the Sun crosses the celestial equator, and at these times day and night are of the same length; these points are called the ***vernal and autumnal equinoxes***. (A fuller description of these astronomical concepts may be found in the chapter on the ancient heavens.)

For the simple reason that the shape of the solar system is close to a disc, most interesting celestial events—those involving the seven moving bodies—take place within a few degrees of the ecliptic. This band, known as the ***zodiac*** (see figure 1.2), was therefore the astronomer's arena. The MUL.APIN, a well-known text from the seventh century BC, contains an early version of the zodiac before the appearance of the twelve constellations that became the zodiacal signs, mentioning these twelve and another six constellations

[19] There is a debate over the extent to which the Babylonians visualized their astronomy; even the celestial sphere may not have been necessary. See [Brack-Bernsen 2003, esp. 23–24].

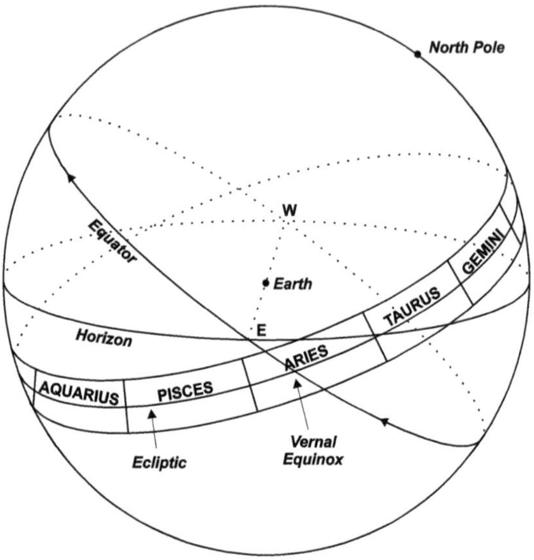

Figure 1.2
The celestial sphere, the equator, the ecliptic, and the zodiac

"who stand in the path of the Moon."[20] The Sun's motion was described as follows:

> From XII 1 to II 30 the Sun is in the path of Anu:
> Wind and storm.
> From III 1 to V 30 the Sun is in the path of Enlil:
> Harvest and heat.
> From VI 1 to VIII 30 the Sun is in the path of Anu:
> Wind and storm.
> From IX 1 to XI 30 the Sun is in the path of Ea:
> Cold.[21]

Hence the Sun moves on an inclined circle divided into four regions, spending three months in each.[22] From here it would be an easy step to split each region into three parts, one corresponding to each month; this division happened by the early fifth century BC. Each of the twelve regions was identified

[20] [Van der Waerden 1974, 79–80]. *Science Awakening II: The Birth of Astronomy* gives a fairly thorough account of the emergence of the zodiac in scattered places, and is partly the basis of the following discussion. See also [Evans 1998, 5–8] for a description of MUL.APIN, and [van der Waerden 1953] and [Brack-Bernsen 2003] on the development of the zodiac and ecliptic in Babylon and Greece.

[21] Quoted from [van der Waerden 1974, 80].

[22] "Month" refers to the time it takes for the moon to orbit the Earth.

with a constellation on or near the zodiac, giving rise to the ***zodiacal signs*** Aries, Taurus, and so forth.[23]

The signs, eventually made equal in size, were each further subdivided into 30 *uš* (meaning length). This division is convenient for a couple of reasons: the Sun travels roughly one uš per day, and 30 is a handy number in sexagesimal calculations. In fact the uš was used only for computation; for recording observations, another measurement called the *kùš* (equal to about 2.4 uš) was standard.[24] Nevertheless the zodiac was now divided into 360 units, later to be called degrees, and a unified system of arc measurement was in place,[25] at least for the zodiac.

Now, Babylonian astronomy was not designed with underlying geometric models in mind, and while it might seem natural to us to consider a planet's path as a geometric object in the heavens and work toward astronomical predictions from that, this was not necessarily so for the Babylonians. Consider, for instance, their approach to the motion of the Sun. It travels about 1° per day in its annual trip around the ecliptic, but its speed is not constant; in ancient times it was slower in the spring, and faster in the fall.[26] The Babylonians used two methods to represent the Sun's changing speed: System A, which alternates the speed between two fixed values; and System B, which varies the speed linearly between a maximum and a minimum (see figure 1.3).[27] It may be tempting, but is anachronistic, to consider these methods as approximations to trigonometric curves;[28] functions and graphs are modern inventions. In any case, these systems were numeric constructs,

[23] This zodiac differed from the Greek one, in that the vernal equinox happened to fall either 8° or 10° into Aries, and was fixed with respect to the background stars. (For research on this topic, see for instance [van der Waerden 1953] and [Huber 1958].) This becomes important with the later Greek discovery of the ***precession of the equinoxes***, a very slow movement of the vernal equinox with respect to the background stars (1° every 72 years). The Greeks fixed the beginning of Aries to the vernal equinox rather than the stars. Thus, today the zodiacal signs have moved about 30° away from the constellations for which they were named.

[24] See [Swerdlow 1998, 34–37].

[25] For more detail, including the relation to time measurement, see [Thureau-Dangin 1928b], [Sidersky 1929], [Thureau-Dangin 1930], [Thureau-Dangin 1931], [Neugebauer 1938], and [Neugebauer 1983, 8]. A candidate for the earliest text with a 360° zodiac is one published in [Sachs 1952, 54–57], probably dated 410 BC. Alongside ecliptic coordinates, another system was used for certain types of texts ("Diaries," "Goal-Year texts," and "Normal-Star Almanacs,") whereby locations were identified with respect to 31 "normal stars" scattered irregularly around the zodiac. See [Neugebauer 1975, 545–547].

[26] In modern times the summer is the longest season in the northern hemisphere.

[27] Fuller descriptions of Babylonian models of the motions of the planets may be found in any number of places; for instance, a survey in [Neugebauer 1957/1969, 97–144] and thorough coverage in [Neugebauer 1975, 347–540] and [Brown 2000], and conjectures on the theory's construction in [Aaboe 1965], [Aaboe 1980], and [Swerdlow 1998].

[28] It has been tried, for example, in [Dittrich 1934].

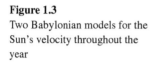

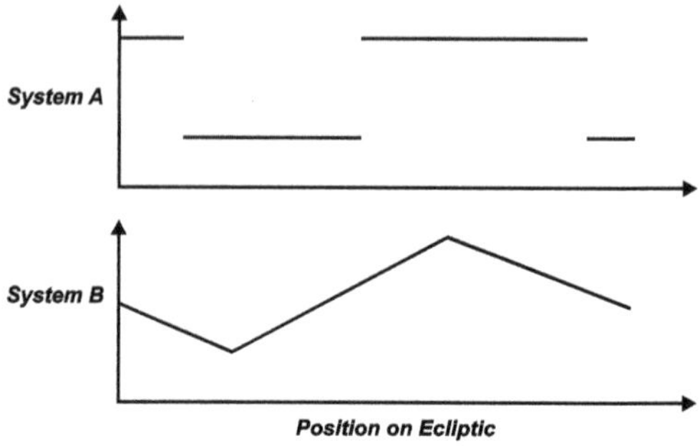

Figure 1.3
Two Babylonian models for the Sun's velocity throughout the year

allowing the direct prediction of solar positions without the intervening presence of a geometric model.

The Geometric Heavens: Spherics in Ancient Greece

Meanwhile, Greek astronomy took a very different turn. The earliest astronomical interests in Greece were likely similar to other cultures: firstly calendrical, concerning the coordination of dates and weather events with the risings and settings of certain stars and constellations. The belief that the heavens and the Earth are concentric spheres is very old; the bringing together of this cosmological statement with these astronomical problems may have come with Eudoxus of Cnidus in the mid-fourth century BC.[29] The next decades produced several texts on the mathematics of spheres with an eye to their use in the heavens. Noteworthy among these were Autolycus of Pitane's *On a Moving Sphere* and *On Risings and Settings*,[30] and Euclid's *Phaenomena*.[31] These works were among the treatises called the "Little Astronomy," used by later writers as an approach to the study of Ptolemy's *Almagest*.[32]

[29] [Goldstein/Bowen 1983] argues that Eudoxus's two-sphere model marks the beginning of a new phase of Greek astronomy, linking calendar problems with cosmology and allowing mathematical reasoning to operate on both. One scholar believes that Eudoxus performed trigonometric calculations to determine terrestrial latitudes from the lengths of shadows cast by the gnomon (a simple astronomical instrument to be discussed in the next chapter); see [Szabó 1985a].

[30] The Greek text and an English translation of these works may be found in [Autolycus (Bruin/Vondjidis) 1971]; Greek text and French translation are in [Autolycus (Mogenet) 1950]; and German translations of Autolycus and Theodosius are in [Czwalina 1931].

[31] See [Berggren/Thomas 1996] for a translation and study of Euclid's *Phaenomena*.

[32] [Neugebauer 1975, 768–769] doubts that the "Little Astronomy" actually existed.

The problems dealt with in these treatises had much to do with the earlier astronomical and geographical concerns;[33] for instance, the varying lengths of daylight throughout the year for a given location, the division of the Earth's surface into climate zones, or the angles between important great circles in the celestial sphere. Euclid's *Phaenomena* is geared particularly to the times required for various arcs of the ecliptic to rise above the horizon. Since the Sun is a point on the ecliptic, rising times were crucial for the reckoning of the passage of time throughout the day, and we shall return to this topic several times. One typical theorem reads as follows:

> **[Proposition 9b]**. Semicircles of the ecliptic that do not begin on the same parallel rise entirely in unequal times, [with] that following Cancer in the greatest, those following, each in turn, in lesser [times], and that following Capricorn in the least. And those [semicircles] beginning on the same parallel rise in equal times.[34]

To see what this is saying consider figure 1.4, which positions the celestial sphere so that the equinoxes (the places where the equator and ecliptic intersect) are due north and south. The semicircle "following Cancer" is drawn in bold;[35] dashes or dots are on the back surface of the sphere from our perspective. The equator rotates at a constant speed, carrying the ecliptic along with it. The point on the equator that is now at E on the horizon will take twelve hours to rise above the horizon, travel through the sky, and reach the horizon again at W. On the other hand the bolded part of the ecliptic has already begun to rise above the horizon, and twelve hours from now it will have as far to go before setting as it has risen in the figure. So the bolded semicircle takes more than twelve hours to rise. Conversely, the part of the ecliptic not drawn in bold will take less than twelve hours to rise.

Euclid's proof of this proposition is entirely geometric. Even so, it has been suggested that he may have been aware of Babylonian arithmetic schemes dealing with length of daylight (similar to the solar schemes of figure 1.3), and he may have been trying to put the symmetries used in these schemes on firm geometrical ground.[36]

[33] See [Aujac 1976] for a discussion of the interaction between astronomy and geography, and the latter's use of both spherics and "spheropoeia" (which included the use of instruments).

[34] [Berggren/Thomas 1996, 71].

[35] By now the signs of the ecliptic were positioned so that Aries began at the vernal equinox, as opposed to the Babylonian practice of figure 1.2.

[36] [Berggren/Thomas 1996, 2] and [Berggren 1991b, especially 237–238].

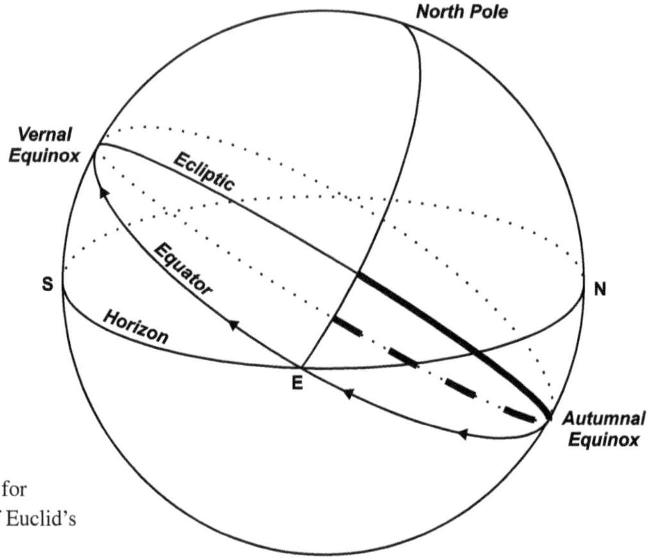

Figure 1.4
Simplified figure for Proposition 9b of Euclid's *Phaenomena*

A Trigonometry of Small Angles? Aristarchus and Archimedes on Astronomical Dimensions

Although early spherics was able to say a great deal about the qualitative behavior of the celestial sphere, it was not the basis for a predictive science.[37] For instance, one could not use it to infer precisely where a particular body would be at a given time. In this case the geometers would have been of little help. Our notion of geometry, using coordinates, leads us to think easily of lengths as quantities, measured by numbers; and locations as pairs (or triples) of numbers referred to coordinate axes. Indeed, we often think of a curve as an algebraic equation: for instance, a parabola *is* an equation of the form $y = ax^2 + bx + c$. By contrast, Greek geometry operated on its own and had no need for such a numeric foundation.[38]

Indeed, if we were to impose a numerical understanding of magnitudes on Greek geometry, then we might say that many trigonometric results were

[37] See [Knorr 1989] for a discussion of four Greek scientific disciplines that do not employ numeric techniques, including early spherics. Knorr argues that their purpose was to explain, not predict, phenomena.

[38] The case is made in detail in [Fowler 1987] that Greek mathematics and astronomy, up to and including Archimedes, were not "arithmetized"; see also [Fowler 1992].

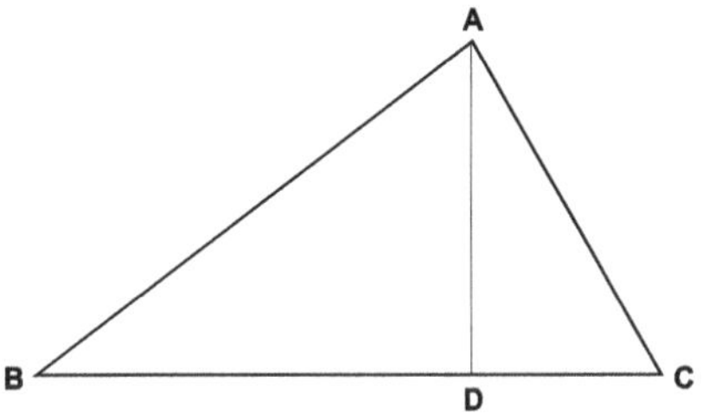

Figure 1.5
Diagram for *Elements* II.13

known well before the birth of trigonometry. The most obvious is the Pythagorean Theorem, which (for a unit circle) is equivalent to the identity $\sin^2\theta + \cos^2\theta = 1$. More dramatically, Proposition II.13 of Euclid's *Elements* reads:

> In acute-angled triangles the square on the side subtending the acute angle is less than the squares on the sides containing the acute angle by twice the rectangle contained by one of the sides about the acute angle, namely that on which the perpendicular falls, and the straight line cut off within by the perpendicular towards the acute angle.[39]

In figure 1.5, if $a = BC$, $b = AC$, and $c = AB$, then $BD = c \cos\angle B$, and we have the Law of Cosines:

$$b^2 = a^2 + c^2 - 2ac \cos\angle B.$$

But these sorts of statements are not much closer to trigonometry than was the Egyptian *seqed*. We have a working definition of angle, but no quantitative measurement or systematic conversion to lengths.[40]

The first real glimmer of a proper trigonometry came with the earliest stages of a quantitative astronomy in the third century BC. As so often in the history of science, it is very difficult to assert a precise beginning; a blurry gradual awakening does not lend itself to the identification of a single blessed moment. The inevitable result is that the birth of mathematical astronomy has been attributed to several different people.[41] One of the earliest figures in this

[39] [Euclid 1925, 406].

[40] See Chapter 4, on John of Muris, for an example of II.13 being used more legitimately as the Law of Cosines to solve a triangle.

[41] [Neugebauer 1959], for instance, argues that "all the evidence points to Apollonius as the founder of Greek mathematical astronomy" based on his work on planetary motion. [Goldstein/Bowen 1983], on the other hand, claim that Eudoxus was "[t]he person largely responsible for turning astronomy into a mathematical science."

story is Aristarchus of Samos (ca. 310 BC–230 BC),[42] who is most famous for being the first to propose a Sun-centered solar system, about 1,700 years before Copernicus. The little we know of "the mathematician," as he was called, includes that he was a student of the philosopher Strato of Lampsacus; Vitruvius credits him with the invention of the scaphe, a type of sundial.[43] Unfortunately most of his works are lost, including what he wrote on his heliocentric system, and a book on optics. Our knowledge of Aristarchus's solar system therefore must be rebuilt from references in writings of later authors, especially Archimedes' *Sand Reckoner*, of which we shall speak shortly.[44]

Aristarchus's only surviving work, *On the Sizes and Distances of the Sun and Moon* (see figure 1.6), has nothing to do with his heliocentric theory; it sets out to measure the relative and absolute sizes of the Earth, Moon, and Sun.[45] It was one of the treatises in the "Little Astronomy," and it is probably due to its inclusion in this collection that we have it today.[46] It reads in the classical Greek style, reminiscent of Euclid, and its theorems are worthy of one known as "the mathematician."

The key datum for Aristarchus's determination of the relative distances of the Moon and the Sun comes from the moment when precisely half of the Moon's face is visible from the Earth. In figure 1.7, when the observer sees the half Moon, the Sun–Moon–Earth angle is a right angle; the Moon–Earth–Sun angle is asserted to be "less than a quadrant by one-thirtieth of a quadrant" $\approx 3^\circ$ i.e., 87°.[47] (The correct value is close to $89^\circ 50'$.) This hopelessly poor value, in addition to some other incorrect observations and assumptions, lead one to question whether Aristarchus's purpose was less quantitative than proof of concept, "a purely mathematical exercise"[48]—albeit an ingenious one.

To convert his angular data to ratios of distances, Aristarchus is forced into approximations of trigonometric quantities. As a sample of his technique we shall follow the first half of Proposition 7, in which he demonstrates that

[42] The source on which most twentieth-century work on Aristarchus is based is [Heath 1913], which contains a translation of *On the Sizes and Distances of the Sun and Moon*. See also the recent detailed textual and mathematical discussion of *On the Sizes* in [Berggren/Sidoli 2007]. For a biographical essay, see [Stahl 1970].

[43] [Vitruvius 1914, 273] or [Vitruvius 1934, vol. 2, 255].

[44] Aristarchus's heliocentric hypothesis has attracted a fair bit of scholarly attention, especially the extent to which Copernicus may have been indebted to him. Among others, see [von Erhardt/von Erhardt-Siebold 1942 and 1943], [Neugebauer 1942], [Africa 1961], [Derenzini 1974], [Wall 1975], [Rosen 1978], [Gingerich 1985], [Zhitomirskiĭ 1986], and [Christianides et al. 2002]. [Wall 1975] contains a summary of all extant information on Aristarchus.

[45] [Tannery 1883, 243–244] argues that Eudoxus may have originated the method that follows.

[46] [Heath 1913, 317–318].

[47] [Heath 1913, 353].

[48] Neugebauer makes this case in [Neugebauer 1975, 634–643], especially p. 643.

ARISTARCHI
DE MAGNITVDINIBVS,
ET DISTANTIIS SOLIS,
ET LVNAE, LIBER
CVM PAPPI ALEXANDRINI
explicationibus quibuſdam.
A' FEDERICO COMMANDINO
Vrbinate in latinum conuerſus, ac
commentarijs illuſtratus.
Cum Priuilegio Pont. Max. In annos X.

PISAVRI, Apud Camillum Franciſchinum.
M D L X X I I.

Figure 1.6
Title page of Commandino's 1572 edition of Aristarchus of Samos's *On the Sizes* (courtesy of the Burndy Library)

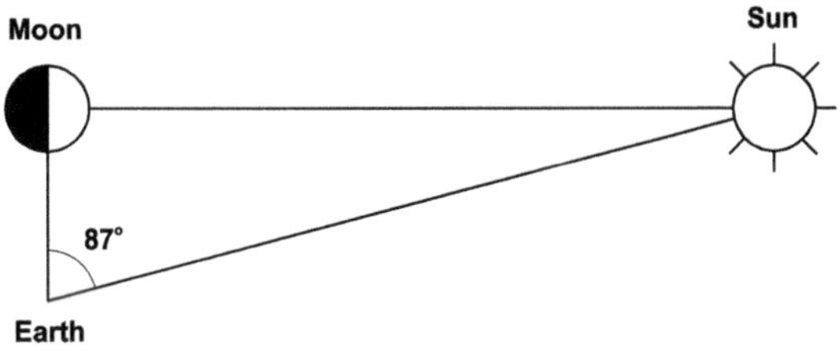

Figure 1.7
Earth, Sun, and Moon at half moon, from Aristarchus's *On the Sizes*

the ratio of the Sun's distance to the Moon's distance, is greater than 18 but less than 20.

Text 1.2
Aristarchus, the Ratio of the Distances of the Sun and Moon
(Proposition 7a from *On the Sizes*)

[See figure 1.8.] *The distance of the Sun from the Earth is greater than eighteen times, but less than twenty times, the distance of the Moon from the Earth.*

For let *A* be the center of the Sun, *B* that of the Earth.

Let *AB* be joined and produced.

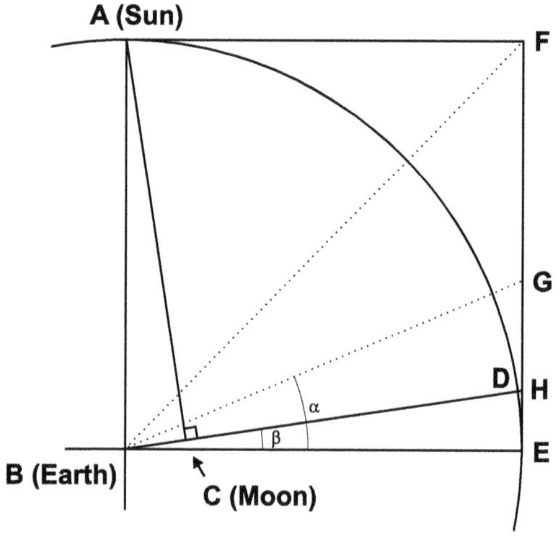

Figure 1.8
Diagram for Proposition 7 of *On the Sizes*

Let *C* be the center of the Moon when halved; let a plane be carried through *AB* and *C*, and let the section made by it in the sphere on which the center of the Sun moves be the great circle *ADE*.

Let *AC*, *CB* be joined, and let *BC* be produced to *D*.

Then, because the point *C* is the center of the Moon when halved, the angle *ACB* will be right.

Let *BE* be drawn from *B* at right angles to *BA*; then the circumference *ED* will be one-thirtieth of the circumference *EDA*; for, by hypothesis, when the Moon appears to us halved, its distance from the sun is less than a quadrant by one-thirtieth of a quadrant.

Thus the angle *EBC* is also one-thirtieth of a right angle.

Let the parallelogram *AE* be completed, and let *BF* be joined.

Then the angle *FBE* will be half a right angle.

Let the angle *FBE* be bisected by the straight line *BG*; therefore the angle *GBE* is one fourth part of a right angle.

But the angle *DBE* is also one thirtieth part of a right angle; therefore the ratio of the angle *GBE* to the angle *DBE* is that which 15 has to 2: for, if a right angle be regarded as divided into 60 equal parts, the angle *GBE* contains 15 of such parts, and the angle *DBE* contains 2.

Now, since *GE* has to *EH* a ratio greater than that which the angle *GBE* has to the angle *DBE*, therefore *GE* has to *EH* a ratio greater than that which 15 has to 2.

Next, since *BE* is equal to *EF*, and the angle at *E* is right, therefore the square on *FB* is double of the square on *BE*.

But, as the square on *FB* is to the square on *BE*, so is the square on *FG* to the square on *GE*; therefore the square on *FG* is double of the square on *GE*.

Now 49 is less than double of 25, so that the square on *FG* has to the square on *GE* a ratio greater than that which 49 has to 25; therefore *FG* also has to *GE* a ratio greater than that which 7 has to 5.

Therefore, *componendo*, *FE* has to *EG* a ratio greater than that which 12 has to 5, that is, that which 36 has to 15.

But it was also proved that *GE* has to *EH* a ratio greater than that which 15 has to 2; therefore, *ex aequali*, *FE* has to *EH* a ratio greater than that which 36 has to 2, that is, than that which 18 has to 1; therefore *FE* is greater than 18 times *EH*.[49]

Explanation: By hypothesis $\angle ABC = 87°$ therefore $\angle DBE = 3°$. Since angles are measured by fractions of right angles, we shall write $\angle ABC = \frac{29}{30}R$ and $\angle DBE = \frac{1}{30}R$, where $R = 90°$.

By construction, $\angle FBE = \frac{1}{2}R$ and $\angle GBE = \frac{1}{4}R$; therefore $\dfrac{\angle GBE}{\angle DBE} = \dfrac{15}{2}$.

Since $\dfrac{GE}{EH} > \dfrac{\angle GBE}{\angle DBE}$, we know that $\dfrac{GE}{EH} > \dfrac{15}{2}$. This key step, which applies the inequality without comment or proof (see below), allows Aristarchus to convert from an equality of *angles* to an inequality of *lengths*.

Now ΔFBE is a right isosceles triangle, so $FB^2 = BE^2$; but (by *Elements* VI.3) $\dfrac{FB^2}{BE^2} = \dfrac{FG^2}{GE^2}$, so $FG^2 = 2GE^2$.

Next, since $\sqrt{2} > 7/5$, we have $\dfrac{FG}{GE} > \dfrac{7}{5}$, so $\dfrac{FE}{GE} > 1 + \dfrac{7}{5} = \dfrac{12}{5}$.

Finally, $\dfrac{FE}{EH} = \dfrac{FE}{GE} \cdot \dfrac{GE}{EH} > \dfrac{12}{5} \cdot \dfrac{15}{2} = 18$.

The lemma applied by Aristarchus in the argument above is stated by Archimedes as follows:

[49] Excerpted from [Heath 1913, 376–381].

> If of two right-angled triangles, (one each of) the sides about the right angle are equal (to each other), while the other sides are unequal, the greater angle of those toward [next to] the unequal sides has to the lesser (angle) a greater ratio than the greater line of those subtending the right angle to the lesser, but a lesser (ratio) than the greater line of those about the right angle to the lesser.[50]

Thus in figure 1.9 where $AC = DF$ and $\alpha > \beta$,

$$\frac{EF}{BC} < \frac{\alpha}{\beta} < \frac{DE}{AB}. \tag{1.1}$$

Aristarchus applies the second inequality to ΔBEH and ΔBEG in figure 1.8; elsewhere in *On the Sizes* he uses both inequalities. We may understand this result in modern terms as follows: since $AC/BC = \sin\alpha$, $DF/EF = \sin\beta$, $AC/AB = \tan\alpha$, and $DF/DE = \tan\beta$,

$$\frac{\sin\alpha}{\sin\beta} < \frac{\alpha}{\beta} < \frac{\tan\alpha}{\tan\beta}.^{51} \tag{1.2}$$

These facts, seen easily from the graphs of the sine and tangent functions, may also be proved geometrically.[52]

We find similar uses of trigonometry in one of the works of deservedly the most famous of ancient scientists, Archimedes (287–212 BC). One of the most creative mathematicians of all time, Archimedes is renowned for his ability to approach difficult geometric problems, including his determinations of areas of curved figures. His "method of exhaustion" works by gradually inscribing, and simultaneously circumscribing, the required figure with polygons that successively approximate the curves more closely. Archimedes also wrote a number of works in topics that would now be classified under physics, including mechanics, optics, and hydrostatics. His feared war machines and other instruments spread his fame far beyond the confines of science.[53]

[50] From the *Sand Reckoner*, translated by Wilbur Knorr in [Knorr 1985]. The origin of this lemma is lost, but it has been speculated that it was part of a pre-Euclidean collection of lemmas for use in the study of spherics [Hultsch 1883].

[51] See, for instance, [Heath 1921, vol. 2, 5], and similar statements in [Zeller 1944], 3–4.

[52] One such proof is given in [Heath 1913, 366].

[53] See [Clagett 1970] for an extensive biographical essay on Archimedes. The two leading modern sources for Archimedes are the translations by Thomas Little Heath in [Archimedes 1897] (including most of the *Sand Reckoner*) and the accounts of [Dijksterhuis 1956], especially the Princeton 1987 reprint which contains an essay by Wilbur Knorr bringing the state of Archimedes scholarship up to date.

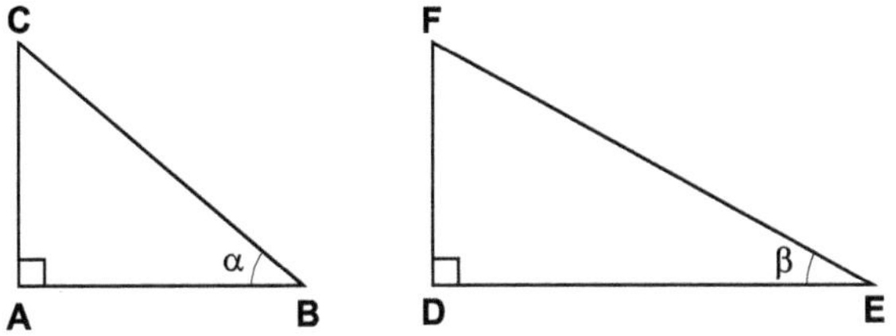

Figure 1.9
An ancient trigonometric lemma

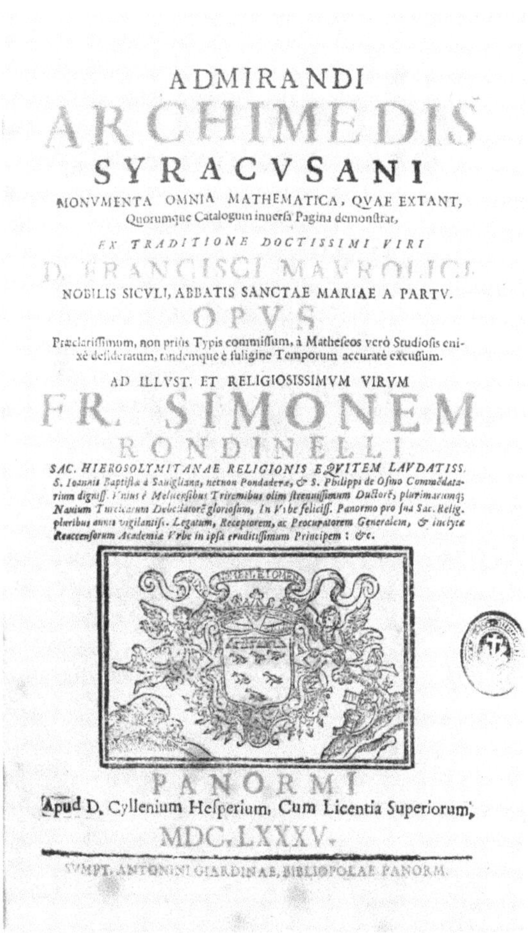
ADMIRANDI
ARCHIMEDIS
SYRACVSANI
MONVMENTA OMNIA MATHEMATICA, QVAE EXTANT,
Quorumque Catalogum inuersa Pagina demonstrat,
EX TRADITIONE DOCTISSIMI VIRI
D. FRANCISCI MAVROLICI
NOBILIS SICVLI, ABBATIS SANCTAE MARIAE A PARTV.
OPVS
Præclarissimum, non priùs Typis commissum, à Matheseos verò Studiosis enixè desideratum, tandemque è fuligine Temporum accuratè excussum.
AD ILLVST. ET RELIGIOSISSIMVM VIRVM
FR. SIMONEM
RONDINELLI
SAC. HIEROSOLYMITANAE RELIGIONIS EQVITEM LAVDATISS. S. Ioannis Baptistæ à Sauigliana, necnon Pondaderæ, & S. Philippi de Osimo Commēdatarium digniss. Vnius è Melitensibus Triremibus olim strenuissimum Ductorē, plurimarumq; Nauium Turcicarum Debellatorē gloriosum, In Vrbe feliciss. Panormo pro sua Sac. Relig. pluribus annis vigilantiss. Legatum, Receptorem, ac Procuratorem Generalem, & inclytæ Reaccensorum Academiæ Vrbe in ipsa eruditissimum Principem: &c.
PANORMI
Apud D. Cyllenium Hesperium, Cum Licentia Superiorum.
MDC.LXXXV.
SVMPT. ANTONINI GIARDINAE, BIBLIOPOLAE PANORM.

Figure 1.10
Title page of Maurolico's 1685 edition of Archimedes' collected works, including the *Sand Reckoner* (courtesy of the Burndy Library)

The *Sand Reckoner* (see figure 1.10), one of several of Archimedes's works intended for a wider audience, was written to show the difference between a large finite quantity and the infinite. He illustrates this amply by calculating the number of grains of sand that would be required to fill the universe. His result, an upper bound on the order of 10^{63}, was astronomically large (and required the invention of an ingenious number system to express), but not infinite.[54] To make sure that he had truly capped the number with an upper bound Archimedes needed to use the largest available estimate for the size of the universe, hence his use of Aristarchus's heliocentric system. If the Earth travels around the Sun, then the vantage point from which we see the stars is moving, so they should appear to move over the course of the year. In fact they do, but by such a tiny amount that the effect is nowhere near visible. Thus Aristarchus had to move the fixed stars far enough away that this ***stellar parallax*** was negligible, resulting in the largest universe known to Archimedes.

The *Sand Reckoner* begins with a measurement of the angular magnitude of the diameter of the Sun, d_s, which Archimedes reports to be

$$\frac{1}{200}R < d_s < \frac{1}{164}R \tag{1.3}$$

(where $R = 90°$). (It had been thought that Archimedes obtained these bounds by drawing lines of sight on his observational instrument and measuring their angles physically. Recently, a procedure has been reconstructed that uses mathematical methods similar to what follows, including an application of (1.1).[55]) Using (1.3), Archimedes needs to find a lower bound on the length of the Sun's diameter. Since he already has Aristarchus's estimate of the size of the circle in which the Sun orbits, Archimedes needs only to find a bound relative to that circle.

In figure 1.11 (see also figure 1.12), E is the observer at a moment after the Sun has risen, $\angle PEQ$ is the Sun's angular diameter seen from E, and $\angle FCG$ its angular diameter seen from the center of the Earth C. A and B are the extensions of the lines of sight from C. From (1.3) we know that $\angle PEQ < \frac{1}{164}R$. Since E is closer to the Sun than C, the angular diameter seen from C is smaller, so $\angle FCG < \frac{1}{164}R$. Thus, the arc $\widehat{AOB}$ in the corresponding circle is less than $\frac{1}{164\times 4} = \frac{1}{656}$ of a circle, and AB is less than one side of a 656-gon. Hence $AB / CO < \frac{1}{656} \cdot 2\pi < \frac{1}{100}$.[56] But AB is the Sun's diameter,[57] and since the Earth is assumed to be smaller than the Sun, the sum of the Earth's and Sun's radii is less than $\frac{1}{100}CO$. Therefore $\frac{100}{99} > CO / HK > CF / EQ$.[58]

[54] For an extended but accessible discussion of Archimedes' purpose in this work, see [Netz 2003].

[55] See [Delambre 1817, vol. 1, 104–105], [Hultsch 1899, 197], and [Lejeune 1947, 38–41] for the original theories, and [Shapiro 1975] for his rebuttal and reconstruction of Archimedes' method.

[56] Here Archimedes applies his lower bound for π, 22/7, from *Measurement of a Circle*.

[57] Since $CA = CO$, AM is perpendicular to CO, and OF is perpendicular to AC, by symmetry we have $AM = OF$, and the latter is the Sun's radius.

[58] The first of these inequalities is true because HK is what is left over when the two radii are taken away from CO. The second inequality is true because $CO > CF$ and $HK < EQ$.

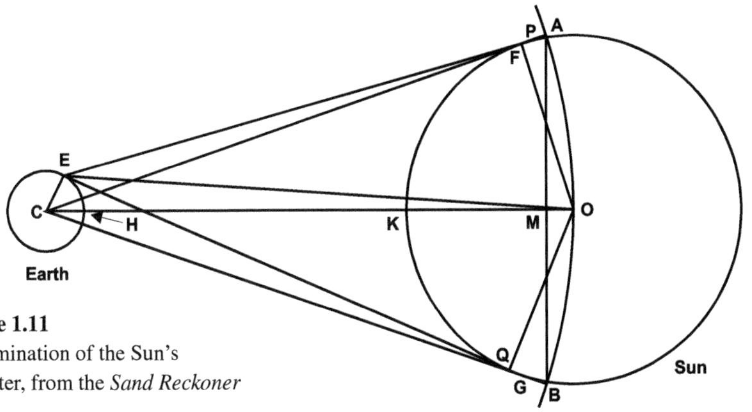

Figure 1.11
Determination of the Sun's diameter, from the *Sand Reckoner*

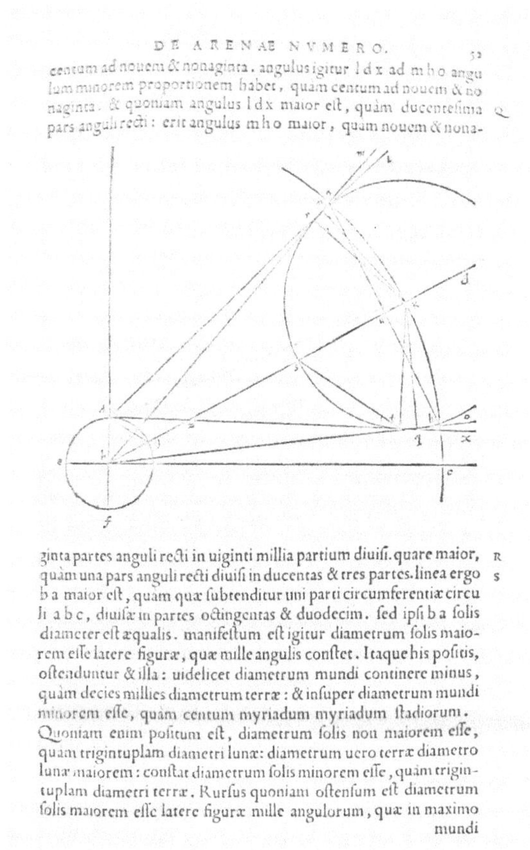
DE ARENAE NVMERO. 52

centum ad nouem & nonaginta. angulus igitur l d x ad m h o angulum minorem proportionem habet, quam centum ad nouem & nonaginta. & quoniam angulus l d x maior est, quàm ducentesima pars anguli recti: erit angulus m h o maior, quam nouem & nonaginta partes anguli recti in uiginti millia partium diuisi. quare maior, quàm una pars anguli recti diuisi in ducentas & tres partes. linea ergo b a maior est, quam quæ subtenditur uni parti circumferentiæ circuli a b c, diuisæ in partes octingentas & duodecim. sed ipsi b a solis diameter est æqualis. manifestum est igitur diametrum solis maiorem esse latere figuræ, quæ mille angulis constet. Itaque his positis, ostenduntur & illa: uidelicet diametrum mundi continere minus, quàm decies millies diametrum terræ: & insuper diametrum mundi minorem esse, quàm centum myriadum myriadum stadiorum. Quoniam enim positum est, diametrum solis non maiorem esse, quam trigintuplam diametri lunæ: diametrum uero terræ diametro lunæ maiorem: constat diametrum solis minorem esse, quàm trigintuplam diametri terræ. Rursus quoniam ostensum est diametrum solis maiorem esse latere figuræ mille angulorum, quæ in maximo mundi

Q
R
S

Figure 1.12
The equivalent to figure 1.11 in Commandino's edition of Archimedes' *Sand Reckoner* (courtesy of the Burndy Library)

Now Archimedes applies (1.1) to ΔOEQ and ΔOCF (since $OF = OQ$ but $EQ<CF$, since E is closer to the Sun than C), which gives

$$\frac{CO}{EO} < \frac{\angle OEQ}{\angle OCF} < \frac{CF}{EQ}.$$

This result allows him to convert from his inequality of lengths to one of angles. Doubling both angles and recalling that we already know that $\angle PEQ > \frac{1}{200}R$, it is relatively easy sailing (using the second inequality above) to arrive at $\angle ACB > \frac{99}{20000}R > \frac{1}{203}R$, so that $\widehat{AB}$ is greater than $\frac{1}{203\times 4} = \frac{1}{812}$ of a circle, and finally that AB is greater than one side of an 812-gon. Archimedes rounds downward to the side of a chiliagon (1000-gon), and has a convenient lower bound for the Sun's diameter.

That both Aristarchus and Archimedes use (1.1) with no comment or proof suggests that the lemma would have been familiar to their readers. In fact there are nine proofs of one or the other inequality in this lemma in antiquity, but the earliest proof of the first inequality does not appear until Ptolemy's construction of his chord table in the *Almagest* four centuries later.[59] How the lemma might have arisen may only be speculated. Perhaps it was recorded in a mathematical work alongside and adjunct to the study of spherics,[60] or perhaps it was more closely bound with the emergence of practical astronomical problems.[61] The lemma is useful, for instance, in the determination of rising times of arcs of the ecliptic, or in the measurement of time with sundials.[62]

Whether or not the lemma allows us to say that a more extensive trigonometry existed as early as the third century BC is a matter of debate.[63] Clearly its existence alone does not qualify as sufficient evidence. Nor does it provide us with a useful general trigonometric tool; although it is perfectly plausible to measure angles as fractions of a right angle, the lemma gives reasonable and useful bounds only for very small arcs.

[59] [Knorr 1985] gives a detailed comparison of these nine sources, noting their remarkable similarities. An equivalent to the tangent half of the lemma appears in Euclid's *Optics*, in a different context and format. Opinions on which is the earliest text to contain this proof have varied; see [Knorr 1991, 195], [Knorr 1994], and [Jones 1994].

[60] As suggested in [Hultsch 1883], especially pp. 415–420.

[61] This view is favored by [Björnbo 1902].

[62] [Knorr 1985, 383–384].

[63] Views on this matter go back at least to the nineteenth century; see for instance [Tannery 1893, 60–68]. More recently, Toomer suggests that lemma (1.1) may have been "of fundamental importance in evaluating triangles before the development of trigonometry" [Diocles 1976, 162], a view that Knorr is inclined to support in [Knorr 1985, 385].

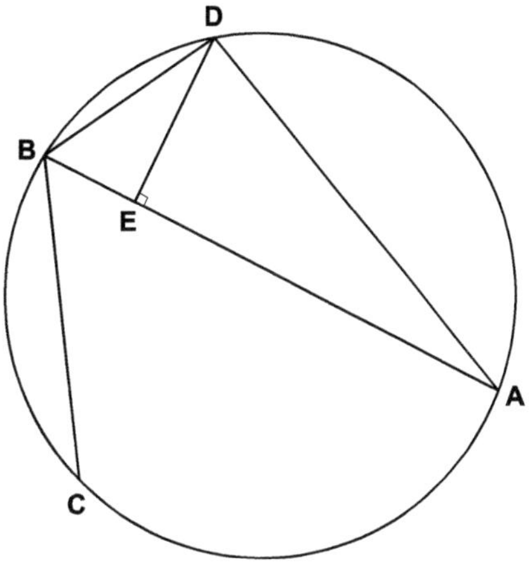

Figure 1.13
Archimedes' Theorem of the Broken Chord

Given an arbitrary arc of sufficient magnitude, the lemma cannot produce bounds of any use to an astronomer in a reasonable time. On its own, then, (1.1) is unable to provide a systematic conversion between arcs and lengths.

We are not yet finished with Archimedes. In his treatise *Book on the Derivation of Chords in a Circle*, the early eleventh-century Muslim scientist al-Bīrūnī attributed the following result to Archimedes, sometimes called the "Theorem of the Broken Chord": in figure 1.13, given unequal arcs $\overset{\frown}{AB} > \overset{\frown}{BC}$ in a circle, let D be the midpoint of $\overset{\frown}{AC}$ and drop perpendicular DE onto AB. Then $AE = EB + BC$.[64] From here it is possible to derive results equivalent to the sine or chord angle sum and difference identities, among others. This fact led one scholar to claim that Archimedes had a more sophisticated trigonometry than what we have witnessed thus far, one that would have given him the capacity to construct a table of chords as powerful as Ptolemy's.[65] However, the trigonometric identities do not drop entirely without effort from this theorem,[66]

[64] [Al-Bīrūnī 1948, 7–8], a German translation along with commentary and the appearance of the same theorem in works of other medieval Arabic authors is available in [Suter 1910–11]. Al-Bīrūnī also mentions this theorem in his *Masʿudic Canon* [Al-Bīrūnī 1954, 273–274]; the relevant section is translated into German in [Schoy 1927, 2–7, esp. 3–4]. See also the exposition of al-Bīrūnī's text in [Dold-Samplonius 1973, 34–39].

[65] [Tropfke 1928]; announced in [Miller 1928] and repeated elsewhere; see for instance [Karpinski 1945, 269].

[66] Toomer derives the chord difference formula, for instance, as follows: let O be the center of the circle, F the point diametrically opposite from D, $\alpha = \angle AOD$, and $\beta = \angle DOB$. Now, $\Delta DEB \sim \Delta DFA$ (since $\angle DBE = \angle AFD$ subtend the same chord DA—thanks to Alina Kononov and Eric Gorlin for this) and $\Delta AED \sim \Delta FBD$ ($\angle BAD = \angle BFD$ since they subtend the same chord BD). So $BE = (BD \cdot AF) / DF$ and $AE = (AD \cdot BF) / DF$. Then, from Archimedes' theorem,

and, just as seriously, we have no idea how Archimedes (as opposed to his Muslim successors) might have intended to use it.[67] It can only be said that, had Archimedes the idea and numerical system needed to build a systematic chord table, he would have had the mathematics at his disposal to pull it off.[68]

$BC = \text{Crd}(\alpha - \beta) = AE - BE = (AD \cdot BF - BD \cdot AF) / DF$. Since $AD = \text{Crd}\ \alpha$, $BD = \text{Crd}\ \beta$, DF is the diameter, and BF and AF may be found by applying the Pythagorean Theorem with known quantities, the chord difference law results.

[67] This argument is made convincingly in [Toomer 1973a, 20–23].

[68] See [Schneider 1979, 149–151] for an argument to this effect. Other cases have been made for more extensive calculations with chords in the third century BC than are found in the extant literature; see for instance [Tannery 1893, 60–68], [van der Waerden 1970, 7], and especially [van der Waerden 1986, esp. 400–401] for their views on Apollonius, and [Szabó 1985a] on Eudoxus. Since these arguments must remain speculative (barring further discoveries), we simply refer the interested reader to these articles. With respect to Archimedes, we have also his *Measurement of a Circle*. In his computation of bounds for the value of π he uses the equivalent of a chord half-angle formula, which we shall see is a vital tool for the construction of a chord table. Once again, we have no evidence that Archimedes used it outside of the context of computing π.

2 Alexandrian Greece

Convergence

The arrival of Babylonian astronomy in Greece in the second or third century BC transformed that science utterly.[1] Although the details of this transmission may never be known, eventually the geometry of earlier Greek science was wedded to Babylonian arithmetic practice. The introduction of the sexagesimal (base 60) number system simplified calculations with arbitrary lengths; the use of a 360° circle made it natural to apply sexagesimal arithmetic to arcs. Finally, Babylonian observational records and predictive schemes for positions of celestial events gave Greek scientists the tools they needed to build geometric astronomical models that were also quantitatively accurate.[2]

The sexagesimal system first appeared in mid-third century BC Greece, in the geographical work of Eratosthenes. His division of the Earth into five climate zones (one tropical, two temperate, two frigid) was defined by breaking the meridian circle of the Earth into sixty parts, and allocating certain numbers of sixtieths to each zone.[3] The practice of dividing the circle into sixty parts did not stop with Eratosthenes; it may be found in scattered later writers as late as the seventh century AD.[4]

The 360° circle is first mentioned on a second-century BC astronomical inscription found in 1893 in Keskintos, on Rhodes. Within a list of astronomical parameters we find the following phrase:

> A circle comprises 360 degrees or 9720 *stigmai* [points]; a degree comprises [27] points.[5]

The unit "points" appears nowhere else; until very recently it was thought that the tablet read 720 points rather than 9720, implying that a point is $\frac{1}{2}^\circ$,[6] but new research suggests that 9720 is in fact correct. This curious unit

[1] This is the establishment view; several scholars are (to put it mildly) skeptical of the role of Babylonian arithmetical astronomy in Greece. See especially many articles published in the journal *DIO*.

[2] The earliest precise astronomical observations made in Greece were in the third century BC, but probably not in degree units. See [Goldstein/Bowen 1991, esp. 103–105].

[3] Strabo reports this in his *Geography*, [Strabo 1917, vol. 1, 436–439].

[4] The list includes Geminus, Plutarch, Galen, and Severus Sebokht. See [Neugebauer 1975, 590, n.2] for references.

[5] [Jones 2006, 13].

[6] See [Tannery 1895, 51–52] and [Neugebauer 1975, 698–699]. As we shall see, the demi-degree did have some currency in later astronomy.

might have arisen by dividing the diameters of the Sun and Moon into 15 equal parts.[7]

Also in the second century BC we find the following passage in *On the Rising of the Twelve Signs of the Zodiac* by Hypsicles of Alexandria (best known as the author of Book XIV of Euclid's *Elements*):

> The circumference of the zodiac circle having been divided into 360 equal arcs, let each of the arcs be called a *degree in space*, and similarly, if the time in which the zodiac circle returns to any position it has left be divided into 360 equal times, let each of the times be called a *degree in time*.[8]

Hypsicles's purpose was to solve the following problem: given the ratio of the longest to the shortest day at a given location, determine the times that it takes for the various signs of the zodiac to rise above the horizon.[9] We have seen rising times treated geometrically in spherics (for instance by Euclid); Hypsicles's approach was completely different. He assumes that the values of the rising times for successive zodiacal signs form an arithmetical progression, and he arrives at a numerical sequence strikingly similar to the Babylonian arithmetical schemes for astronomical phenomena.[10] Without trigonometry, this method was perhaps the best option. However, at least in Greece, the end of this approach was in sight. The convergence of structured numeration and mensuration with the geometrical style of astronomy was about to pay rich dividends.

Hipparchus

A figure as foundational as Hipparchus of Rhodes deserves better preservation than history has granted him; his works are almost completely lost. All that survives is a relatively minor treatise, *Commentary on the Phaenomena of Aratus and Eudoxus* (figure 2.1), and probably it is with us only because of the fame of Aratus's astronomical poem. Born in Nicaea early in the second century BC, Hipparchus probably spent most of his career in Rhodes.

[7] [Jones 2006, 14–16].

[8] [Thomas 1941, vol. 2, 394–397].

[9] Each sign of the zodiac is a particular segment of the ecliptic 30° in length. As the celestial sphere rotates over the course of a day, a given segment will at some point broach the horizon; after a while it will completely clear the horizon. This time interval is the zodiacal sign's ***rising time***.

[10] Hypsicles's work has been edited and translated into German in [Hypsicles (DeFalco/Krause) 1966]. A commentary by Otto Neugebauer in the same volume (pp. 5–22) explains Hypsicles's arithmetical approach to rising times; see also [Neugebauer 1975, 715–718] for a summary. [Tannery 1912, vol. 2, 261–262] suggests that this method came directly from Babylon.

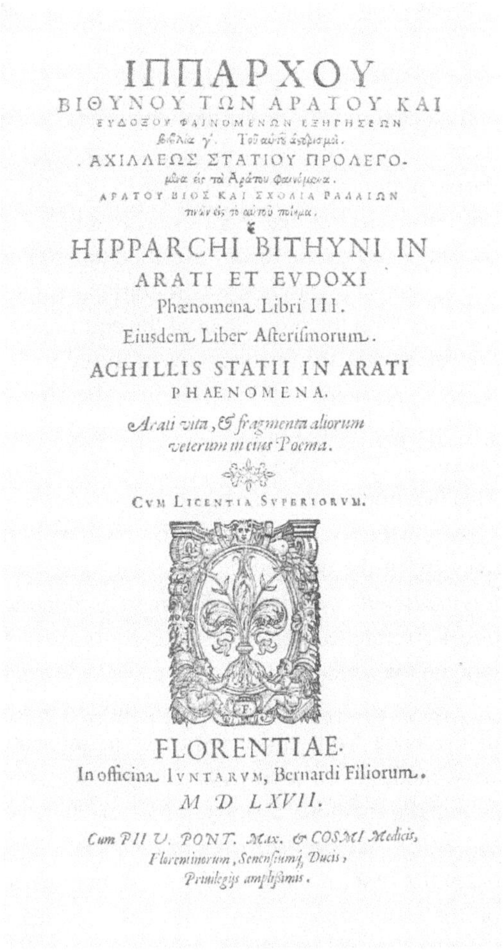
ΙΠΠΑΡΧΟΥ
ΒΙΘΥΝΟΥ ΤΩΝ ΑΡΑΤΟΥ ΚΑΙ
ΕΥΔΟΞΟΥ ΦΑΙΝΟΜΕΝΩΝ ΕΞΗΓΗΣΕΩΝ
βιβλία γ'. Τοῦ αὐτοῦ ἀστερισμοί.
ΑΧΙΛΛΕΩΣ ΣΤΑΤΙΟΥ ΠΡΟΛΕΓΟ-
μένα εἰς τὰ Ἀράτου φαινόμενα.
ΑΡΑΤΟΥ ΒΙΟΣ ΚΑΙ ΣΧΟΛΙΑ ΠΑΛΑΙΩΝ
τινῶν εἰς τὸ αὐτοῦ ποίημα.

HIPPARCHI BITHYNI IN
ARATI ET EVDOXI
Phænomena Libri III.
Eiusdem Liber Asterismorum.
ACHILLIS STATII IN ARATI
PHAENOMENA.
Arati vita, & fragmenta aliorum
veterum in eius Poema.

CVM LICENTIA SVPERIORVM.

FLORENTIAE.
In officina IVNTARVM, Bernardi Filiorum.
M D LXVII.

Cum PII V. PONT. Max. & COSMI Medicis,
Florentinorum, Senensiumq; Ducis,
Priuilegijs amplissimis.

Figure 2.1
Title page from a 1567 edition of Hipparchus of Rhodes' *Commentary on Aratus and Eudoxus* (courtesy of the Burndy Library)

We know from many references to him by later authors that he wrote on a variety of scientific topics, especially astronomy, but also optics, combinatorial mathematics, astrology, and geography (an attack on Eratosthenes).

Most of what we know of Hipparchus's astronomy comes from reports in Ptolemy's *Almagest*, written almost three centuries later.[11] Our understanding of the authorship of the *Almagest*'s innovations has changed dramatically. Many were once thought to be mere records of Hipparchus's

[11] The best places to start with Hipparchus's astronomy are [Neugebauer 1975, 274–343] and the somewhat shorter [Toomer 1978], which relies partly on Neugebauer but also includes some of Toomer's reconstructions of Hipparchus's methods. [Linton 2004, 51–60] is a good capsule review.

achievements.[12] But today, the *Almagest* is taken by most (although not all) scholars to be close to its word when attributing discoveries to Hipparchus or claiming them for Ptolemy himself. There is no doubt that Hipparchus constructed models of the motions of the Sun and Moon, since Ptolemy tells us as much. However, it is usually accepted today that Hipparchus did not derive a theory of the motions of the planets, and that he restricted himself only to observing that theories current at his time were deficient (especially the simple epicyclic theory ascribed to Apollonius in the *Almagest*).[13]

Among Hipparchus's astronomical achievements, few rank as high as his work with the precession of the equinoxes. As we described in chapter 1, the ***equinoxes***, the intersections of the ecliptic with the celestial equator (see figure 1.2), were within Aries and the sign directly opposite, Libra. Hipparchus discovered that the equinoxes do not stay fixed with respect to the background stars, but move very slowly, at a rate of at least 1° per century.[14] This fact implies that one must choose whether to fix one's ecliptic coordinate system to the fixed stars, or to the slowly moving equinoxes. The Babylonians, unaware of the difference, chose the fixed stars as their reference, hence the term ***sidereal coordinates***. For reasons relating to the theory of the Sun, the Greeks' reference points were the equinoxes (***tropical coordinates***), and they laid out the zodiacal signs so that Aries always begins at the vernal equinox.

Since we have so little of Hipparchus's writings, we must rely on scraps of evidence from various sources to reconstruct what he contributed and what methods he used. Like his predecessors, Hipparchus may have used arithmetic methods to solve putatively trigonometric problems. It has been suggested that *On the Rising of the Twelve Signs of the Zodiac* was one such example; if so, it must have evoked memories of Hypsicles's earlier treatise on the same topic[15]—but it has been argued recently that Hipparchus may have had access to spherical trigonometry for this problem.[16] Certain fragments of mathematical methods in Hipparchus's geography also suggest the use of arithmetic schemes,[17] but again there are hints of more sophisticated methods elsewhere.[18] We shall return to these debates shortly.

[12] The classic and influential *Histoire de l'Astronomie Ancienne* [Delambre 1817] was perhaps the most prominent work to take this view.

[13] This is based primarily on Ptolemy's report in *Almagest* IX.2, [Ptolemy (Toomer) 1984, 421–422].

[14] The correct figure is about 1° every 72 years.

[15] See [Neugebauer 1975, 301].

[16] [Sidoli 2006].

[17] For instance, [Strabo 1917, 283] seems to attribute to Hipparchus an arithmetic relation between the length of longest daylight and the Sun's maximum altitude.

[18] The other problem is the determination of one's local latitude from the length of the longest day. See the upcoming section on spherical trigonometry before Menelaus.

Other of Hipparchus's lost works are more certain to have incorporated trigonometry. For instance, recent reconstructions of Hipparchus's method to find the distances of the Sun and Moon require the knowledge of chords of certain arcs arising within the problem, for which Aristarchus's methods would have been insufficient.[19] It is tempting but probably facile to assign to Hipparchus both a pre-trigonometry and a post-trigonometry phase; scientific research does not usually proceed so linearly.

A Model for the Motion of the Sun

The Sun's annual journey through the celestial sphere would have been one of the most obvious signals that objects in the heavens cannot move strictly according to the Aristotelean principle of uniform motion about a circle with the Earth at its center. Certainly the Sun travels around the ecliptic, a great circle centered at the Earth (see figure 1.4). But, according to Hipparchus, it takes $94\frac{1}{2}$ days to traverse the 90° arc from the vernal equinox to the summer solstice (its most northerly point above the equator), and only $92\frac{1}{2}$ days to proceed from there to the autumnal equinox. Thus spring is actually longer than summer.[20] The lengths of fall and winter can be inferred from these measurements to be a mere $88\frac{1}{8}$ and $90\frac{1}{8}$ days respectively.[21]

Geometric models to account for anomalies in the motions of the planets had been advanced before Hipparchus. While the Sun and Moon accelerate and decelerate, the other planets move more dramatically, generally traveling close to the ecliptic but occasionally reversing their path before continuing on their way. (Figure 2.2 illustrates one of Mars's ***retrograde motions*** from 2003. Although the Greeks also considered motion above and below the ecliptic, we shall focus on motion along the ecliptic.) Eudoxus (fourth century BC) had proposed that the planets move according to a system of linked, nested spheres. His ingenious design moved the planets in curves known as ***hippopedes***, reproducing qualitatively, but unfortunately not quantitatively, the planets' retrograde motions. Apollonius, in the late third century BC, is the first author we know to discuss the simple theory of epicycles.[22] In figure 2.3, which represents the plane containing the ecliptic,

[19] See [Toomer 1974], building on [Swerdlow 1969]. The chords that arise are considerably larger than in Aristarchus, and lower/upper bound methods would have been ineffective.

[20] This assumes the convention that seasons begin at the equinoxes and solstices.

[21] These numbers, relying on a year length of 365¼ days, are found in *Almagest* III.4 [Ptolemy (Toomer) 1984, 155].

[22] Apollonius may be the originator of the epicyclic model. The only evidence we have comes from *Almagest* XII.1 [Ptolemy (Toomer) 1984, 555–562], where a couple of Apollonius's mathematical results are presented.

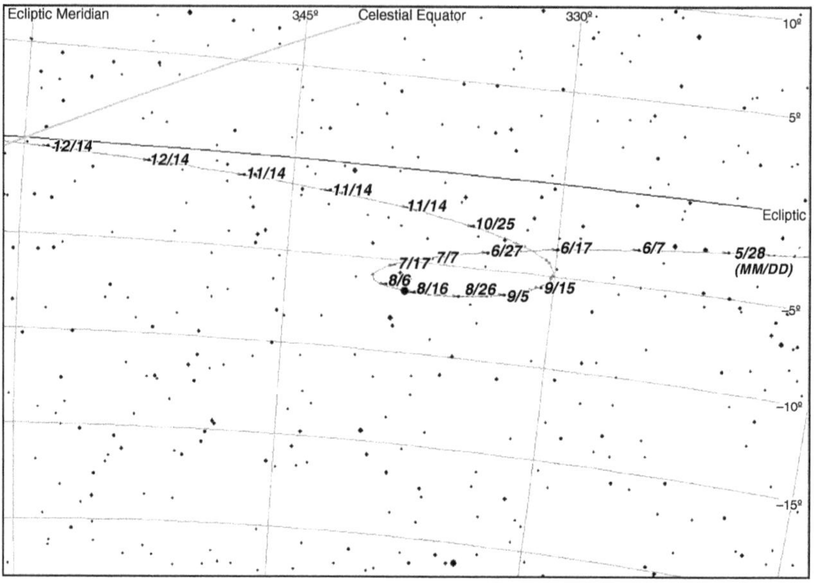

Figure 2.2 The path of Mars through the fixed stars from May to December 2003

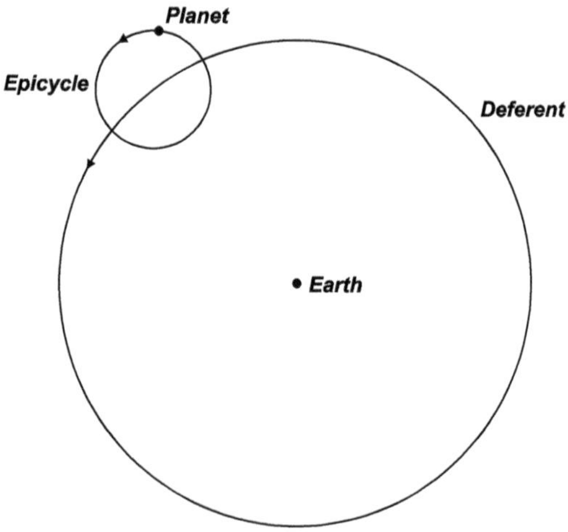

Figure 2.3
The simple epicyclic model of planetary motion

the center of a small circle (the ***epicycle***) travels at a uniform rate around the edge of a large circle (the ***deferent***). At the same time the planet travels uniformly around the epicycle. Varying the sizes of the circles and the ratios of the two velocities appropriately can produce accelerations, decelerations, or retrograde motions, as desired.

As far as we know Apollonius's epicycles remained geometric; he did not attempt to fit the model to observations by quantifying the parameters—which would have furnished astronomy with its first precise theory. Hipparchus seems to have realized that this simple model is unable to reproduce the planets' actual retrograde arcs: each of the loops is the same size, disagreeing with what we see. He thus stayed away from the planets.[23] However, he showed no such reluctance with the Sun and Moon, and his work on these models may have led him to bring together Babylonian arithmetic and Greek geometry.[24]

The epicyclic model turns out to be identical to the ***eccentric*** model, where the Sun revolves uniformly on a circle not centered at the Earth. Knowing precisely where to place the Earth within this circle to match the perceived changes in the Sun's velocity is one of Hipparchus's major goals, and the problem can be solved from the data he provides: the lengths of the spring and summer ($94\frac{1}{2}$ and $92\frac{1}{2}$ days) and the year's length ($365\frac{1}{4}$ days). Hipparchus's solution is lost, but we have a solution in Ptolemy's *Almagest* III.4.[25]

Text 2.1
Deriving the Eccentricity of the Sun's Orbit
(from Ptolemy's *Almagest* III.4)

[See figure 2.4.] Now since the Sun traverses circle *ΘKLM* with uniform motion, it will traverse $\widehat{\Theta K}$ in $94\frac{1}{2}$ days, and $\widehat{KL}$ in $92\frac{1}{2}$ days. In $94\frac{1}{2}$ days its mean motion is approximately 93;9°, and in $92\frac{1}{2}$ days 91;11°. Therefore

$$\widehat{\Theta KL} = 184;20^\circ$$

and, by subtraction of the semi-circle *NPO* [from arc *ΘKL*],

$$\widehat{N\Theta} + \widehat{LO}[= 184;20^\circ - 180^\circ] = 4;20^\circ.$$

So $\widehat{\Theta NY} = 2\widehat{\Theta N} = 4;20^\circ$ also,

[23] *Almagest* IX.2, [Ptolemy (Toomer) 1984, 421–422].

[24] The possibility has been raised that Hipparchus used Babylonian predictive schema while building his models for the Sun and Moon. This was first noticed in [Kugler 1900]; among many other sources see also [Aaboe 1955] and [Jones 1991b]. [Toomer 1998] is a summary of the evidence in favor of Babylonian methods in Hipparchus's astronomical work.

[25] One cannot always assume that what is attributed to Hipparchus in the *Almagest* is an accurate reflection of Hipparchus's work. In fact, [Jones 1991a] argues that much of Ptolemy's account of the Sun does not reflect Hipparchus's approach, particularly the tables of solar motion, the epoch position, and the equation of time.

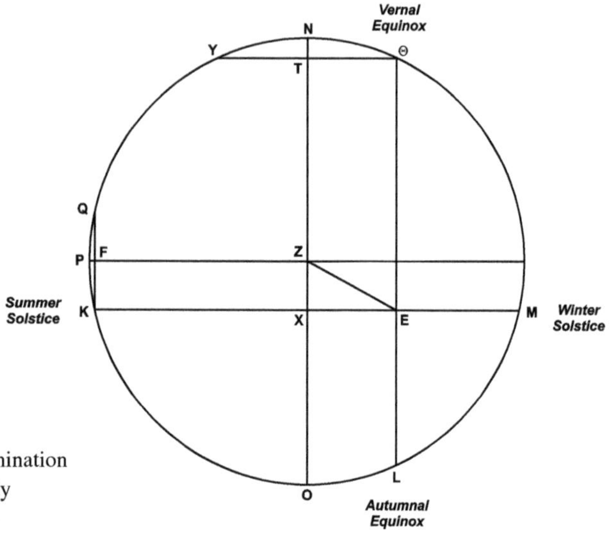

Figure 2.4
Ptolemy's determination of the eccentricity of the Sun's orbit

$$\left.\begin{aligned} \therefore \Theta Y &= \text{Crd}\,\widehat{\Theta NY} \approx 4;32^p \\ \text{and } EX = \Theta T &= \tfrac{1}{2}\Theta Y = 2;16^p \end{aligned}\right\} \text{where the diameter of the eccentre} = 120^p.$$

Now since $\widehat{\Theta NPK} = 93;9°$, and $\widehat{\Theta N} = 2;10°$ and quadrant $NP = 90°$, by subtraction, $\widehat{PK} = 0;59°$, and $\widehat{KPQ} = 2\widehat{PK} = 1;58°$.

$$\left.\begin{aligned} \therefore KFQ &= \text{Crd}\,\widehat{KPQ} = 2;4^p, \\ \text{and } ZX = KF &= \tfrac{1}{2}KFQ = 1;2^p \end{aligned}\right\} \text{where the diameter of the eccentre} = 120^p.$$

And we have shown that $EX = 2;16^p$ in the same units.

Now since $EZ^2 = ZX^2 + EX^2, EZ \approx 2;29\frac{1}{2}^p$ where the radius of the eccentre $= 60^p$.

Therefore the radius of the eccentre is approximately 24 times the distance between the centers of the eccentre and the ecliptic.[26]

Explanation: Since this is Ptolemy's and not Hipparchus's work, the conventions used here do not necessarily apply to Hipparchus. Ptolemy sets the radius of the eccentre, the circle of the Sun's orbit, to 60 units (an arbitrary but convenient choice, since he uses sexagesimal numeration). The notation ^{p} stands for parts in this unit of distance. Ptolemy's chord function, here denoted "Crd," assumes a radius $R = 60^p$ (see figure 2.5).

[26] Excerpted from [Ptolemy (Toomer) 1984, 155].

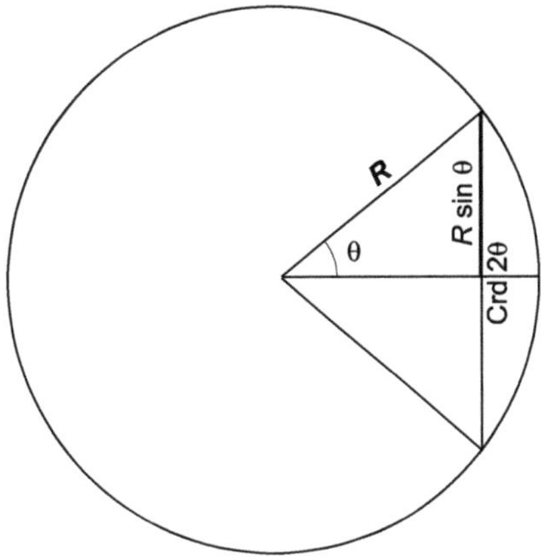

Figure 2.5
The relation between the chord function and the modern sine

In figure 2.4 *FZ*, *KM* and *YΘ* are parallel, and are perpendicular to *NO*, *ΘL*, and *QK*. The Sun travels from the vernal equinox Θ to the summer solstice *K* in $94\frac{1}{2}$ days. Since Ptolemy supposes that it does so at constant speed, this arc (seen from the Earth to be 90°) is $\overset{\frown}{\Theta K} = 94\frac{1}{2}\text{ days}\cdot\dfrac{360^\circ}{365\frac{1}{4}\text{days}} \approx 93;9^\circ$. Likewise, since the summer is $92\frac{1}{2}$ days, $\overset{\frown}{KL} \approx 91;11^\circ$.

From here it is easy to determine $\overset{\frown}{\Theta NY} = 4;20^\circ$. Using his table of chords Ptolemy finds $\Theta TY = \text{Crd}\,\overset{\frown}{\Theta NY} = 4;32^p$; similarly $KFQ = \text{Crd}\,\overset{\frown}{KPQ} = 2;4^p$. Cut the lengths in half and transfer them to *EX* and *ZX*; then by the Pythagorean Theorem we get the ***eccentricity*** of the Sun's orbit $ZE = 2;29\frac{1}{2}^p$. Rounded to $2;30^p$, this is $\frac{1}{24}$th of the radius of the eccentric circle.

Hipparchus's Chord Table

If (as seems likely) Hipparchus proceeded in the above manner, he would have been faced with the task of finding the chords of $\overset{\frown}{\Theta NY}$ and $\overset{\frown}{KPQ}$. Although they correspond to somewhat small arcs, these chords do not have "nice" lengths like the sides of the regular polygons of Aristarchus; their values derive from observational data and are not likely to be convenient numbers. Methods involving bounds would not have given the accuracy that Hipparchus seems to have achieved, so he must have had a means of producing chords for arbitrary arcs.

Indeed, we have a report from Theon of Alexandria that Hipparchus wrote a treatise on chords in 12 books.[27] It is difficult to take Theon at his word here: even taking into account the intricacies of building a chord table (see *Ptolemy's Chord Table* later in this chapter), it is hard to imagine why the topic would have required 12 books. Although the length of the treatise is disputed, its existence is not—especially since we also have a report by the later Greek astrologer Vettius Valens that he used a table constructed by Hipparchus to determine the Sun's position (although not the Moon's).[28] Such a table would surely have been built with the help of a chord table.

However, it is quite possible that Hipparchus's chord table looked very different from tables that came later. For instance, for our modern trigonometric functions we use the unit circle ($R=1$). Ptolemy chose $R=60^p$, which scales his chord function up by that factor (although since his number base was 60, this was hardly an inconvenience). Figure 2.5 illustrates the simple relation between the modern sine function and a chord in a circle of radius R,

$$\text{Crd}\, 2\theta = 2R \sin\theta. \tag{2.1}$$

One might decide instead that the circumference of the circle should be divided into a fixed number of parts, say, $360^\circ \cdot 60 = 21{,}600'$. In this case R would have length $21600/2\pi \approx 3438$ parts. We shall see later that this system was actually used in ancient Indian astronomy, which probably was transmitted from pre-Ptolemaic Greece. One advantage to this system is that a small arc (measured in minutes) and its chord are almost equal, since they are measured in the same units of length.[29]

There may be countless ways in which Hipparchus's table differed from others, including the spacing of the arcs for which the chords are tabulated, the numeration system, the level of rounding, and so on. One might despair of ever getting a clear idea of what this first trigonometric table may have looked like, except for a clue buried in Ptolemy's exposition of the lunar theory in the *Almagest*. Following Hipparchus, Ptolemy takes observational records of three lunar eclipses, separated by just under 180 days each, and applies them to the simple epicyclic model of figure 2.3. A somewhat involved trigonometric calculation eventually produces the ratio of r, the radius of the epicycle, to $\mathfrak{R}$, the radius of the deferent.

[27] See [Rome 1931/1936/1943, vol. 2, 451].

[28] [Vettius Valens (Pingree) 1986, 339]. [Jones 1991b, 445–447] argues that Hipparchus's solar tables, if they existed, came late in his career. The solar and lunar positions on which his models depend thus would have been calculated using Babylonian arithmetic schemes (see [Jones 1991a])!

[29] The reasoning here is similar to the modern use of radian measure, which has the desirable property that $\sin x \approx x$ for small values of x.

Ptolemy gets $r / \mathfrak{R} = 5;14 / 60$, but he also reports Hipparchus's result, $r / \mathfrak{R} = 247\frac{1}{2} / 3122\frac{1}{2}$. When this matter is considered using a simple eccentric circle for the Moon's motion instead of an epicycle and a different set of data, Hipparchus gets $e / \mathfrak{R} = 327\frac{2}{3} / 3144$ (where e is the circle's eccentricity).

The numbers in these ratios are peculiar indeed. Attempts to explain them by reconstructing Hipparchus's calculations have led several researchers to the conclusion that they arose from the use of a chord table with $R = 3438$: if one follows the computation through as closely to Hipparchus as one could hope for, the numerators and denominators are close to his numbers (although the match is not perfect).[30] If this theory is correct, the first trigonometric table would have looked something like figure 2.6.

Might Hipparchus, then, have been the source for later Indian sine tables, which also use $R = 3438$? The Indian tables list sines for every $3\frac{3}{4}°$, which corresponds to chords for every $7\frac{1}{2}°$; this structure matches with a tradition in Greek geography, which divides the circle into 48 parts of $7\frac{1}{2}°$ each.[31] To complete the circle, one of those scientists who used the 48-part division was Eratosthenes, whose geographical work we know to have been criticized by Hipparchus.

All of this evidence is convincing, yet circumstantial, and cannot be proved definitively unless new documents come to light. In the next chapter we shall visit alternate theories concerning the construction of the Indian tables.

[30] Early researchers had assumed that Hipparchus's chord table looked much like Ptolemy's. [Biot 1859, 408–410], for instance, had already raised the possibility that Hipparchus's chord table was similar enough to Ptolemy's that it might be the source of the Indian sine tables; and [Tannery 1893, 63–68] speculated that Indian tables derived from a Greek table even earlier than Hipparchus. (See [Szabó 1985b] for a claim that Hipparchus must have had the *Almagest* chord table itself, based on results of certain gnomon shadow length calculations.) The reconstruction of a chord table with $R = 3438$ from the lunar parameters was first explored in [Toomer 1973a] and asserted too strongly as "conclusively established." However, an incorrect dating for one of the eclipse observations flawed the analysis, and it was abandoned when the correct date failed to confirm that $R = 3438$ ([Swerdlow 1979b], Toomer in [Ptolemy (Toomer) 1984, 215 n. 75]). Further, an analysis of Toomer's success in matching the *Almagest* parameters [Klintberg 2005] recently concluded that Toomer's evidence for $R = 3438$ as opposed to, say, $R = 3600$ is not as strong as it seems. However, [Duke 2005a] revisited the issue. A new path through the computations results in an extremely good match with Hipparchus. Finally, [van der Waerden 1986] speculates that Indian chord tables are based not on Hipparchus (who was "not a great geometer"), but on a table that van der Waerden believed to have been composed by Apollonius. He reconstructed this hypothetical table based on the *Almagest* declination table in [van der Waerden 1988], but this analysis was undermined in [Van Brummelen 1993, 90–101].

[31] [Neugebauer 1972, 250–251]; this article's suggestion of a link between Hipparchus and Indian sine tables instigated the trail of research by Toomer and Duke that led to the reconstructions of a chord table with $R = 3438$.

θ	**Crd θ**
0	0
$7\frac{1}{2}°$	450
15°	897
$22\frac{1}{2}°$	1341
30°	1780
$37\frac{1}{2}°$	2210
45°	2631
$52\frac{1}{2}°$	3041
60°	3438
$67\frac{1}{2}°$	3820
75°	4186
$82\frac{1}{2}°$	4533
90°	4862
$97\frac{1}{2}°$	5169
105°	5455
$112\frac{1}{2}°$	5717
120°	5954
$127\frac{1}{2}°$	6166
135°	6352
$142\frac{1}{2}°$	6511
150°	6641
$157\frac{1}{2}°$	6743
165°	6817
$172\frac{1}{2}°$	6861
180°	6875

Figure 2.6
A reconstruction of Hipparchus's table of chords

If we go further to ask how Hipparchus's chord table was calculated with only the barest traces of evidence even of its existence, we risk taking speculation to an extreme. However, it is worth noting that it would take very little geometry to do it. The chord of 60° is just R (the radius of the base circle), and the chord of 90° is $R\sqrt{2}$ (see figure 2.7 (a) and (b)). The Pythagorean Theorem provides an easy calculation of the chord of the supplement of a given arc:

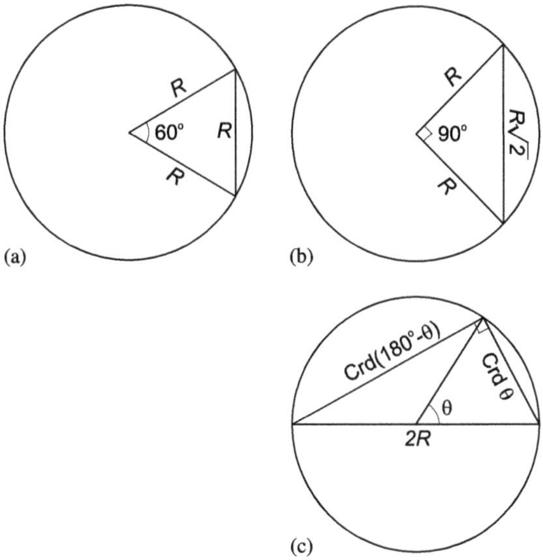

Figure 2.7
Elementary chord relations

$$\mathrm{Crd}(180° - \theta) = \sqrt{(2R)^2 - (\mathrm{Crd}\,\theta)^2} \tag{2.2}$$

(figure 2.7(c)[32]). Just one more theorem is needed to complete the table, and it had been known already to Archimedes a century earlier.

In *Measurement of the Circle*, Archimedes had found that $3\frac{10}{71} < \pi < 3\frac{1}{7}$ by approximating the circle with inscribed and circumscribed hexagons, then successively bisecting the sides and calculating their lengths. As one proceeds to dodecagons, 24-gons, 48-gons, and finally to 96-gons, the polygons approximate the circle more and more closely.[33] Archimedes's side bisection process also provides a ready means to determine Crd $\theta/2$ given Crd θ, for any θ. In figure 2.8, let $CB = \mathrm{Crd}\,\theta$ and $DB = \mathrm{Crd}\,\theta/2$; given CB, the goal is to find DB. The angles at A are half the angles at O,[34] so AD bisects $\angle CAB$. Now $\Delta ADB \sim \Delta ACE$ (since the angles at C and D are right, and the angles at A are equal). So

[32] It was well known that any triangle inscribed in a circle, with one side equal to the diameter, is right-angled.

[33] The medieval astronomer Jamshīd al-Kāshī (c. AD 1400) extended Archimedes's method to a polygon of 805,306,368 sides, obtaining a value of π accurate to the equivalent of about sixteen decimal places. See a German translation of the text in [Luckey 1953], or [Van Brummelen 1998] for a popular account.

[34] By *Elements* III.20.

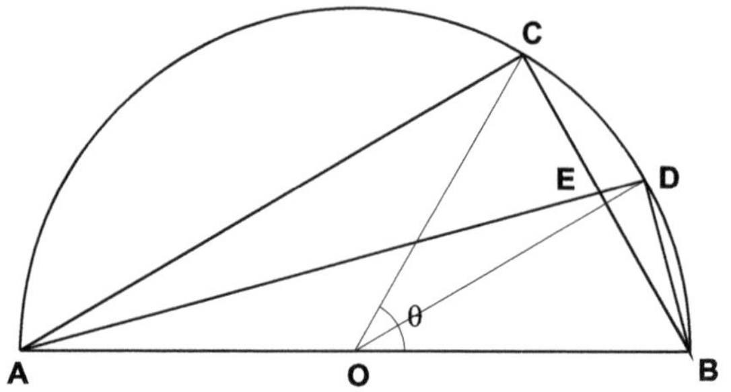

Figure 2.8
Archimedes's half-chord calculation

$$\frac{AD}{DB} = \frac{AC}{CE} \underset{(Elts\,\text{VI.3})}{=} \frac{AB}{BE} = \frac{AB+AC}{BE+CE} = \frac{AB+AC}{BC}. \tag{2.3}$$

The quantities on the right-hand side of this equality are all known (since AC can be found using (2.2) above). Since $AD = \sqrt{AB^2 - DB^2}$, we may solve for the only remaining unknown DB, which is the sought Crd $\theta/2$.[35] This process turns out to be mathematically equivalent to the well-known half-angle formula for sines,

$$\sin\frac{\alpha}{2} = \sqrt{\tfrac{1}{2}(1-\cos\alpha)}. \tag{2.4}$$

From (2.2), (2.4), and the starting points of Crd 60° and Crd 90°, it is a simple (if tedious) exercise to build the chord table of figure 2.6.[36] If one needs to find chords of arcs between entries in the table, linear interpolation is an option. This method is what Ptolemy uses in the *Almagest*, and the errors that result are not large enough to affect any of the above arguments. Thus for the first time the Greeks had the ability to answer the question: given an arbitrary arc, how long is its chord?

The Emergence of Spherical Trigonometry

Whatever it might have looked like, Hipparchus's table of chords likely served his treatments of the motions of the Sun and Moon well. However,

[35] The proof in the *Almagest* is slightly different from this [Ptolemy 1984, 52–53], but also goes back to Archimedes, in a work on the heptagon known only from Arabic sources [Schoy 1927, 81–82].

[36] [Hughes 2000] suggests that Hipparchus also might have had a value for Crd 72° (as was later used by Ptolemy) to build a chord table with many more entries and a more complicated construction, perhaps explaining Theon's assertion that Hipparchus's work on chords consisted of twelve sections. This would force one to conclude that the early Indian sine tables derived from an earlier, more primitive Greek chord table, or were computed independently of the Greeks.

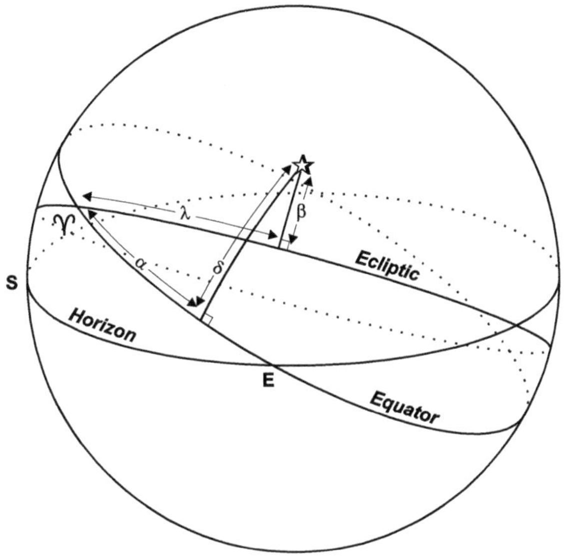

Figure 2.9
Celestial coordinate systems

many astronomical problems arise on the surface of a sphere, not on a flat plane, and this requires mathematics beyond what we have seen so far. For instance, knowing where the Sun will be on the ecliptic does not tell us where it will be in the sky. Over the course of the day the position of the ecliptic changes as the celestial sphere performs its daily rotation. Knowing the Sun's location instead with respect to the celestial equator (see figure 1.4) would be much more convenient, since the equator stays in place, rotating into itself at a constant rate of 15° per hour.

Positions on both the ecliptic and the equator are measured eastward with respect to the vernal equinox ♈ (figure 2.9). The Sun's place on the ecliptic is its ***longitude*** λ. If we are interested in some other object such as a star or a planet, its arc distance above or below the ecliptic is its ***latitude*** β. Similarly, measuring along the equator gives the ***right ascension*** α, while the arc distance above or below the equator is the ***declination*** δ. Hipparchus may have used an equatorial coordinate system for the fixed stars,[37] but the Sun, of course, moves in the ecliptic. Converting from ecliptic to equatorial coordinates thus would have been crucial for time-keeping, and this requirement is clearly a problem on the surface of a sphere, not a plane.

But in order to find out, say, the length of daylight on a given day at a given place, converting the Sun's position to equatorial coordinates is only

[37] The case for this has been revived recently in [Duke 2002], countering the view of [Neugebauer 1975, 277–280] that the *Almagest* contains the first appearance of an orthogonal coordinate system on the celestial sphere.

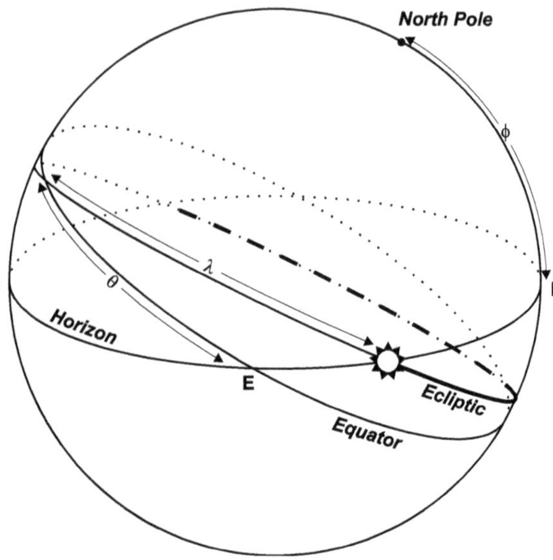

Figure 2.10
The oblique ascension θ and rising times of arcs of the ecliptic

part of the solution. Arcs on the equator act as a sort of clock; if two objects on the equator are separated by 15°, then they will rise above the horizon one hour apart. In fact, the right ascension α is still measured today in units of 24^h of time rather than 360°. However, most objects (including the Sun) are not on the equator (figure 2.10). Now, as the Sun rises, so does some point E on the equator; the distance along the equator from the vernal equinox to E is the ***oblique ascension*** θ.[38] So to find the length of daylight, we need to know the oblique ascension of the Sun's place on the ecliptic (at sunrise), and the oblique ascension of the point opposite the Sun (at sunset). The difference between these two is known as the ***rising time*** of the arc of the ecliptic drawn in bold. From the rising time, the 15°/hour conversion factor gives the length of daylight.

Finding the oblique ascension was thus a fundamental problem for the ancient astronomer. Simple twists of this problem, such as finding one's terrestrial latitude from the length of the longest day, had geographical significance as well. Clearly a unification of trigonometry with the science of spherics had the potential to produce a big payoff.

When and how this merger happened is entirely unclear. We have texts from two authors that illustrate the "before" and "after" stages, but as we shall see, even they cannot be trusted to reflect their own times.

[38] If the observer is on the Earth's equator, the celestial equator is perpendicular to the horizon, and the oblique and right ascensions are equal—hence the names. This special case, known as *sphaera recta*, is a commonly discussed situation in ancient astronomy, to be distinguished from *sphaera obliqua*.

Theodosius of Bithynia

The first text in spherics after Hipparchus came quite quickly, before the end of the second century BC. Theodosius of Bithynia's[39] *Spherics*[40] was one of three of his works which have been preserved as part of the "Little Astronomy"; the others, *On Habitations* and *On Days and Nights*, also deal with topics of spherical geography and astronomy.[41] Their survival has more to do with their inclusion in the "Little Astronomy" than with their own merits; Theodosius is seen as more a compiler of existing work than an original scholar.[42] Even so, he has done us a favor by showing the extent of Greek knowledge of spherics just before it was transformed by trigonometry.

Theodosius's *Spherics* (see figure 2.11) is similar in style to its predecessors; in fact, so many of its propositions are pre-Euclidean that this book is sometimes described as more representative of the late fourth century BC than of its own time.[43] The *Spherics* is divided into three books. Book I and the first half of Book II do for the geometry of the sphere roughly what Euclid's *Elements* does for the circle. We reproduce a couple of the theorems here to give a taste:

> **Proposition I.9:** Given a circle in a sphere, if from one of the poles of this circle a line is drawn perpendicular to the circle, it will fall on the center of the circle, and if extended, it will fall on the other pole of the circle.[44]

> **Proposition II.3:** If in a sphere two circles intersect the circumference of a great circle at the same point, and they both have their poles on this great circle, then they will be tangent to each other.[45]

[39]There is some confusion over his dates, and whether he hailed from Bithynia or Tripolis; see [Tannery 1893, 36–37].

[40]The *Spherics* is available in a Greek edition in [Theodosius (Heiberg) 1927b], in German translation in [Theodosius (Nizze) 1826], and in French translation in [Theodosius (ver Eecke) 1927c]. Our account of some of the theorems in the *Spherics* is based on ver Eecke's translation. See also [Malpangotto 2003], which deals especially with Pappus's generalizations of some of Theodosius's theorems.

[41]*On Habitations* and *On Days and Nights* is available in a Greek edition and Latin translation in [Theodosius 1927a].

[42]Information on Theodosius is scarce. See [Heath 1921, vol. 2, 245–252], [Theodosius (Fecht) 1927a, 1–12], and [Neugebauer 1975, 748–767], especially on the *Spherics*.

[43][Bulmer-Thomas 1976, n. 6]. [Berggren 1991b], however, argues that certain aspects of Theosodius's methods go beyond Autolycus and Euclid.

[44]Translated from [Theodosius (ver Eecke) 1927c, 14].

[45]Translated from [Theodosius (ver Eecke) 1927c, 35].

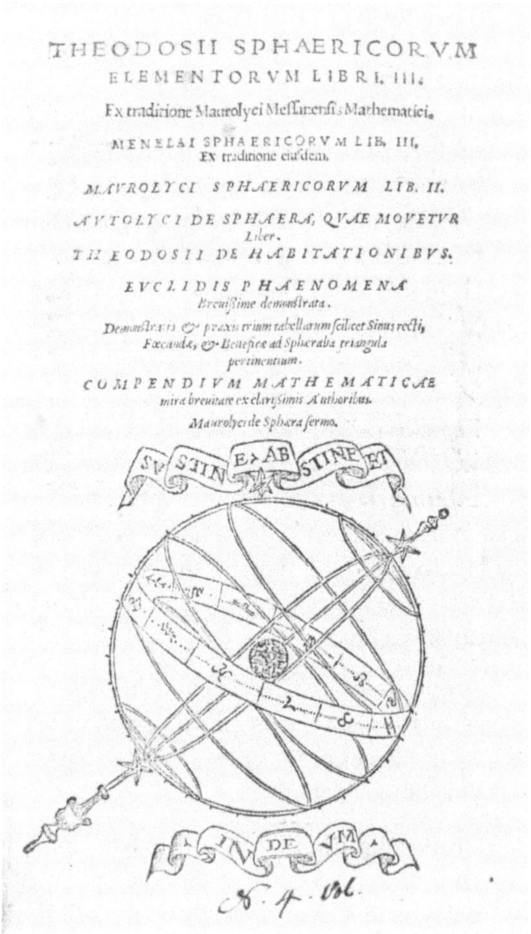
THEODOSII SPHAERICORVM
ELEMENTORVM LIBRI. III.
Ex traditione Maurolyci Messanensis Mathematici.
MENELAI SPHAERICORVM LIB. III.
Ex traditione eiusdem.
MAVROLYCI SPHAERICORVM LIB. II.
AVTOLYCI DE SPHAERA, QVAE MOVETVR Liber.
THEODOSII DE HABITATIONIBVS.
EVCLIDIS PHAENOMENA Breuissime demonstrata.
Demonstratio & praxis trium tabellarum scilicet Sinus recti, Fœcundæ, & Beneficæ ad Sphæralia triangula pertinentium.
COMPENDIVM MATHEMATICAE mira breuitate ex clarissimis Autoribus.
Maurolyci de Sphæra sermo.

Figure 2.11
Theodosius's *Spherics*, Francesco Maurolico's edition, 1558 (courtesy of the Burndy Library)

About halfway through Book II, the theorems become astronomical—although at no time is any astronomy mentioned explicitly. Most of these theorems give various inequalities, which are of course of little practical use. For instance:

> **Proposition III.6:** [See figure 2.12.] If the pole of parallels [A] is on the circumference of a great circle [$ADBEC$]; if two great circles, of which one is one of the parallels [BC] and the other is oblique to the parallels [DE] intersect this great circle at right angles; if we cut equal consecutive arcs [$\overset{\frown}{HQ}$ and $\overset{\frown}{QF}$] on the oblique circle on the same side of the largest of these parallels; and if we describe great circles [joining] the points thus determined with the pole [AHL, AQM, and AFN]: they will cut, in their intervals, unequal arcs on the

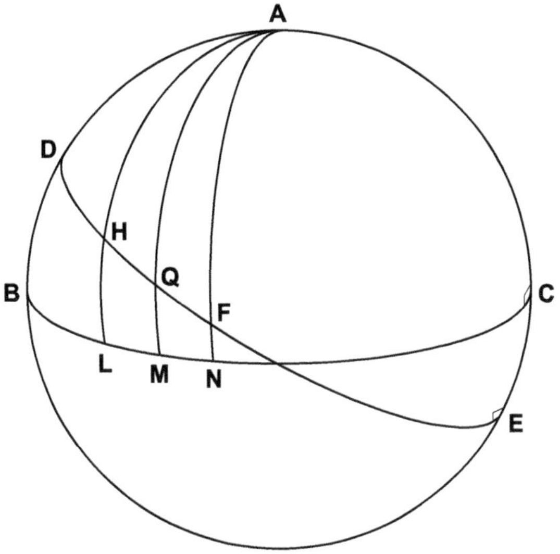

Figure 2.12
Diagram for Theodosius's *Spherics* III.6 (simplified). The "parallels" are the circles of constant latitude.

largest of the parallels; and the arc closest to the original great circle will always be larger than that which is further.[46]

This obscure theorem states the following: draw two great circles *BC* (with pole *A*) and *DE* on the sphere. Cut equal contiguous arcs $\widehat{HQ}$ and $\widehat{QF}$ on *DE*, both above *BC*, and draw great circle arcs from *A* through *H*, *Q*, and *F* to *L*, *M*, and *N*. Then $\widehat{LM} > \widehat{MN}$. This apparently useless result becomes interesting only when one interprets *BC* as the celestial equator and *DE* as the ecliptic. Then we conclude that equal arcs of the ecliptic rise in unequal times, and that ecliptic arcs further from the equator (closer to *ADB*, the ***solsticial colure***) take longer to rise than arcs nearer to the equator.[47] In modern terms, the right ascension α of an ecliptic arc λ increases more and more rapidly as λ increases from 0° to 90°; i.e., the second derivative of α with respect to λ is positive.

Another important result, reproduced here, contains an interesting proof.

Proposition III.11: [See figure 2.13.] If the pole of parallels [*A*] is on the circumference of a great circle [$\widehat{ADB}$] which intersects two great circles at right angles, of which one is a parallel [$\widehat{BQC}$] and the

[46] Translated from [Theodosius (ver Eecke) 1927c, 96–97].

[47] This statement is similar in intent and meaning to Euclid's *Phaenomena* Proposition 9b, quoted in the previous chapter. The solsticial colure is the great circle through the pole and crossing the celestial equator 90 degrees away from the equinoxes; it thus has right ascension 6^{hr} and 18^{hr}.

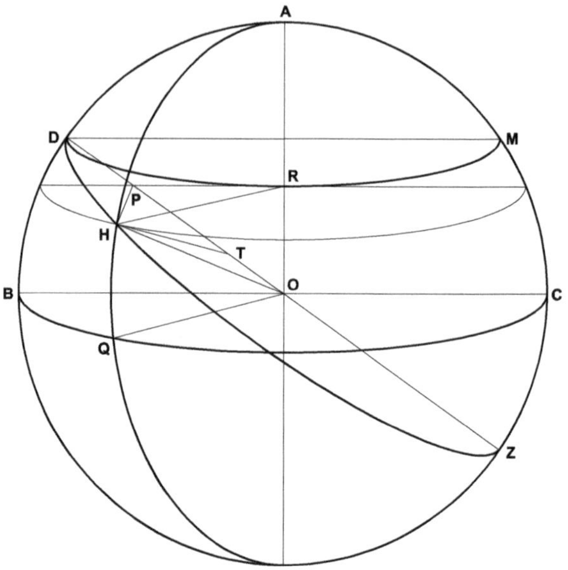

Figure 2.13 Diagram for Theodosius's *Spherics* III.11 (simplified slightly)

> other is oblique to the parallels [$\widehat{DHZ}$]; and if another great circle passing through the poles of the parallels [$\widehat{AHQ}$] cuts the oblique circle between the largest of parallels and that which touches the oblique circle [i.e., between parallel circles $\widehat{BC}$ and $\widehat{DM}$], the ratio of the diameter of the sphere to the diameter of the circle that touches the oblique circle [DM] is greater than that of the arc on the largest of parallels [$\widehat{BQ}$] situated between the original great circle and the great circle passing through the poles [$\widehat{AQ}$], to the arc of the oblique circle situated between these last circles [$\widehat{DH}$].[48]

Understood astronomically, $\widehat{BC}$ is the equator and $\widehat{DZ}$ is the ecliptic. The theorem states that the ratio of the ecliptic arc $\widehat{DH}$ to its corresponding rising time arc $\widehat{BQ}$ is greater than the ratio of the diameter of the tropic circle DM to the diameter of the sphere.

The full proof begins by establishing many facts about the figure which are obvious at a glance. We will trust the reader to fill these in, and paraphrase the rest:

[48] Translated from [Theodosius (ver Eecke) 1927c, 111–112].

Proof: Connect various lines as shown in figure 2.13. Drop a perpendicular from H to DZ, defining P. Transfer length PR to PT, defining T. Then $\Delta PTH = \Delta PRH$, since both angles at P are right, $PR = PT$, and PH is a side in both triangles.

Now, from the second inequality in lemma (1.1) applied to configuration $HPTO$, we have $\frac{OP}{PT} > \frac{\angle PTH}{\angle POH}$. Since $PT = PR$ and $\angle PTH = \angle PRH = \angle BOQ$, we have $\frac{OP}{PR} > \frac{\angle BOQ}{\angle POH}$. But $\frac{DZ}{DM} = \frac{OP}{PR}$,[49] and $\frac{\overset{\frown}{BQ}}{\overset{\frown}{DH}} = \frac{\angle BOQ}{\angle POH}$. Therefore the ratio of the diameters DZ/DM is greater than the ratio of the chords $\overset{\frown}{BQ} / \overset{\frown}{DH}$.

The use of lemma (1.1) in this spherical context leaves Theodosius in the same situation as Aristarchus and Archimedes had been in plane geometry: without trigonometry, he is unable to turn his result into a useful equality.

Menelaus of Alexandria

Our next *Spherics* is over two centuries later, and by then the subject had been completely transformed. The author, Menelaus of Alexandria, lived near the end of the first century AD. As is true of many Greek scientists, we know virtually nothing of his life. Aside from the *Spherics*, he authored several other works. Among them was a treatise on chords, which Theon tells us was in six books;[50] one would expect, then, that Menelaus built his own chord table, but no one knows for certain. Two passages in Heron's *Metrica* (first century AD) quote unspecified "works in chords of the circle" stating that the ratio of the chord of a side of a regular nonagon to the diameter of its circumscribed circle is about 1/3, and that the ratio for an 11-gon is

[49] This equality has been taken more than once to be equivalent to the relation $\tan a = \sin b \tan A$ in a spherical triangle with a right angle at vertex C ([Heath 1921, vol. 2, 251] and [Bulmer-Thomas 1976, 320]). In Figure 2.13 consider spherical triangle AHD with a right angle at D; then $DZ/DM = 1/\sin h$ (where $h = \angle ROP$ is the side of the triangle opposite H). Now $\angle A = \angle PRH = \angle PTH$, so $\tan A = PH/PT$; also $\angle a = \overset{\frown}{DH} = \angle HOP$, so $\tan a = PH/PO$. The result follows by substituting into $DZ/DM = OP/PR$. It seems to me that the context is too narrow to assert from this equality of ratios that Theodosius had access to a spherical trigonometric theorem in his astronomical work.

[50] [Rome 1931/1936/1943, vol. 2, 451]. This is the same quote from which we learn that Hipparchus wrote a treatise on chords in twelve books.

about 7/25.[51] Possibly these were found in the lost treatises of either Menelaus or Hipparchus.[52]

Menelaus's other treatises include the geometrical *On the Triangle* and *Elements of Geometry*,[53] both of which exist only in fragments in Arabic. He composed at least part of a star catalogue and a work on the settings of zodiacal signs, but these books are lost.

The *Spherics* does not exist in the original Greek either, but is available in a number of Arabic translations and commentaries,[54] many of which supplement the text with its astronomical applications. For, just as Theodosius before him, Menelaus treated his subject as a work solely of geometry. However, as we shall see, as with Theodosius, the astronomical meaning is always just below the surface.

Menelaus's *Spherics* is divided into three books. The first is genuinely geometric in content, and deals systematically with the spherical triangle (a figure on the sphere enclosed by arcs of three great circles) for the first time. Menelaus mimics the theorems in Book I of Euclid's *Elements* on plane triangles, extending them or altering them as appropriate. A couple of examples:

> **Proposition I.2:** The two angles on the base of any [spherical] triangle with two equal sides are themselves equal.[55]

> **Proposition I.5:** In each [spherical] triangle two of its sides, whichever they may be, are always larger than the remaining side.[56]

Some statements about plane triangles are not true of spherical triangles, and vice versa. For instance,

> **Proposition I.11:** The exterior angle in any [spherical] triangle is smaller than [the sum of] both its opposite interior angles.[57]

[51] [Heron (Schöne) 1903, 58–59, 62–63]. These values are equivalent to sin $20° \approx 0.333$ (as opposed to 0.342) and sin $16.36° \approx 0.28$ (as opposed to 0.282).

[52] This supposition is questioned in [Rome 1932], and especially [Rome 1933a]. A debate arose earlier whether or not Heron's lengths of sides of polygons may be considered to be trigonometric, [Cantor 1907, vol. 1, 399] in favor, and [Tannery 1882, 184–191] (implicitly) and [von Braunmühl 1900/1903, vol. 1, 9] against. [Schmidt 1900] argued that the derivations are purely geometric. Since they do not particularly involve arc or angle measurement, we are inclined to agree.

[53] [Hogendijk 1999/2000] restores some of the contents of *Elements of Geometry* from two Arabic treatises; a fragment of *On the Triangle* is referenced in [Steinschneider 1896, 199].

[54] Menelaus's *Spherics* is available in an Arabic edition and German translation in [Krause 1936]. See also [Björnbo 1902] for an extended study of the work.

[55] Translated from [Krause 1936, 122].

[56] Translated from [Krause 1936, 124].

[57] Translated from [Krause 1936, 129].

From this result we conclude immediately that the sum of the interior angles of a spherical triangle is greater than two right angles, as Menelaus points out.[58]

Book II resembles the astronomical theorems of Theodosius's *Spherics*; indeed, many of its results are extensions or generalizations of Theodosian analogues. For instance Proposition II.21 corresponds to Theodosius's III.6, except that Menelaus does not require the two given ecliptic arcs (*HQ* and *QF* in figure 2.12) to be contiguous. Later, in the middle of Book III, we find similar inequalities, this time relating ratios of arcs rather than the arcs themselves. Our example takes over a page and a half to state, let alone prove, so we paraphrase in modern notation:

> **Proposition III.14:** [See figure 2.14.] In spherical triangle *ABG* where $\widehat{BG} < 90°$, cut two arcs *GD* and *ZT* from *BG* and draw arcs *DE*, *ZH* and *TK* so that they form the same angle with $\widehat{AB}$ as $\widehat{AG}$ does. Then:
>
> - $\widehat{GD} = \widehat{ZT} \Rightarrow \widehat{GA} - \widehat{DE} < \widehat{ZH} - \widehat{TK}$;
> - $\widehat{GA} - \widehat{DE} = \widehat{ZH} - \widehat{TK} \Rightarrow \widehat{GD} > \widehat{ZT}$;
> - $\widehat{GD} + (\widehat{GA} - \widehat{DE}) = \widehat{ZT} + (\widehat{ZH} - \widehat{TK}) \Rightarrow \widehat{GD} < \widehat{ZT}$.

As medieval commentator Naṣīr al-Dīn al-Ṭūsī points out,[59] this is a cumbersome way of stating that

$$\frac{\widehat{GD}}{\widehat{ZT}} > \frac{\widehat{AG} - \widehat{DE}}{\widehat{ZH} - \widehat{TK}}. \tag{2.5}$$

In the case where the marked angles in the figure are right angles, *AB* is the equator, and *BG* is the ecliptic, this is equivalent to

$$\frac{\Delta\lambda_1}{\Delta\lambda_2} > \frac{\Delta\delta_1}{\Delta\delta_2}, \tag{2.6}$$

where $\Delta\lambda_1$ and $\Delta\delta_1$ are the differences in longitude and declination respectively for $\widehat{GD}$, and $\Delta\lambda_2$ and $\Delta\delta_2$ are the differences for $\widehat{ZT}$. This is

[58] Let the exterior angle be $180° - A$, and the two interior angles be *B* and *C*. The theorem states that $180° - A < B + C$, so $180° < A + B + C$. See [Krause 1936, 130].

[59] [Nadal/Taha/Pinel 2004, 420]. This paper illustrates many of the astronomical applications of propositions in Menelaus's *Spherics*, and the following discussion is indebted to it.

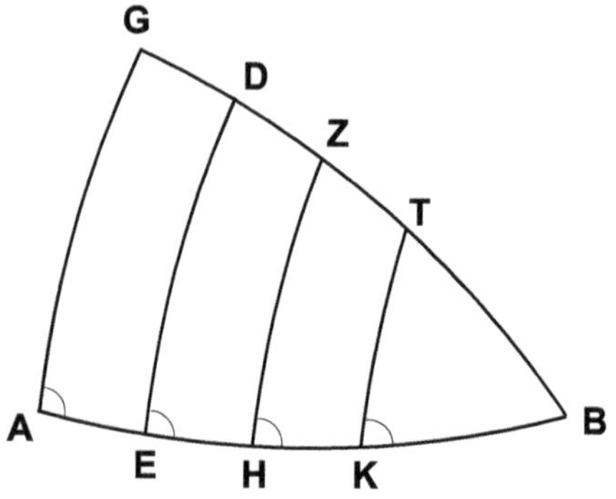

Figure 2.14
Figure for Menelaus's *Spherics* III.14

equivalent to the modern statement that the second derivative of δ with respect to λ is negative for $0 < \lambda < 90°$.

Supposing that Menelaus or other astronomers of his time were computing tables of functions like $\delta(\lambda)$,[60] relations of this sort would have been plainly visible in the gradually slowing increase in the values of δ as λ increased. Perhaps Menelaus included propositions such as this one to confirm the relationship, verifying the numeric patterns in the tables.[61]

The Foundations of Spherical Trigonometry: Book III of Menelaus's *Spherics*

The first several theorems of Book III establish the basis from which all of spherical trigonometry follows. In fact, as we shall see, the amazingly versatile Proposition III.1 (see figure 2.15) is the only theorem used by Ptolemy in the *Almagest* to solve all important problems of spherical astronomy. It is now called Menelaus's Theorem; it was named the Transversal Figure in medieval Islam after the shape of the diagram, where it remained the foundation of spherical trigonometry until around AD 1000.

Precisely how Menelaus's Theorem appeared in the *Spherics* is unclear, since it takes on different forms in the different Arabic editions. The following passage is translated from a manuscript that has been argued to be closest to the original.[62]

[60] This is not much of a stretch. Ptolemy's *Almagest* later in the century contains a very good table of declinations that was not computed directly from the *Almagest* chord table.

[61] This is suggested in [Nadal/Taha/Pinel 2004, 435]. One wonders about similar results in Theodosius, or even Euclid and Autolycus. The tables witnessed especially by these last two would have been Babylonian rather than Greek.

[62] [Lorch 2001, 327–335].

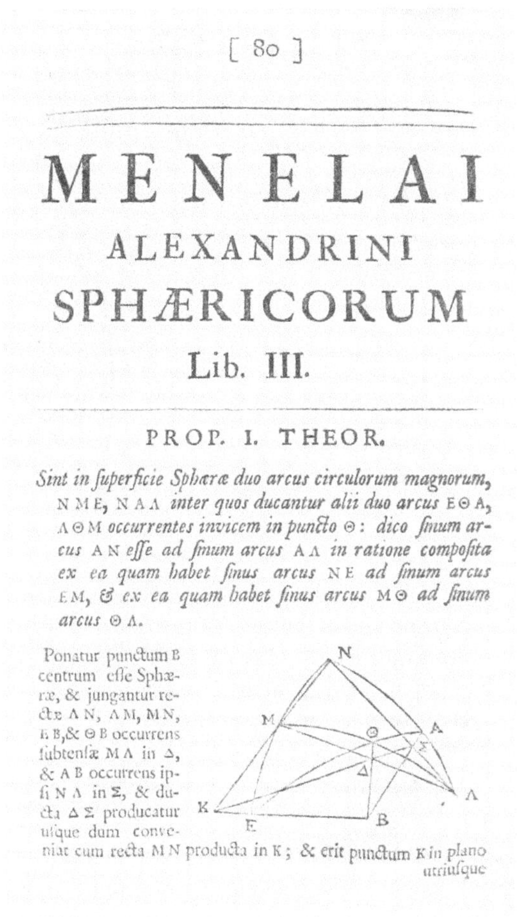

[80]

MENELAI

ALEXANDRINI

SPHÆRICORUM

Lib. III.

PROP. I. THEOR.

Sint in ſuperficie Sphæræ duo arcus circulorum magnorum, N M E, N A Λ *inter quos ducantur alii duo arcus* E Θ A, Λ Θ M *occurrentes invicem in puncto* Θ: *dico ſinum arcus* A N *eſſe ad ſinum arcus* A Λ *in ratione compoſita ex ea quam habet ſinus arcus* N E *ad ſinum arcus* E M, *& ex ea quam habet ſinus arcus* M Θ *ad ſinum arcus* Θ Λ.

Ponatur punctum B centrum eſſe Sphæræ, & jungantur rectæ Λ N, Λ M, M N, E B, & Θ B occurrens ſubtenſæ M Λ in Δ, & A B occurrens ipſi N Λ in Σ, & ducta Δ Σ producatur uſque dum conveniat cum recta M N producta in K; & erit punctum K in plano utriuſque

Figure 2.15
The first page of Book III of Menelaus's *Spherics* (Halley's edition, 1758) showing Menelaus's Theorem (courtesy of the Burndy Library)

Text 2.2
Menelaus, Demonstrating Menelaus's Theorem
(from the *Spherics* III.1)

[See figure 2.16.] Then, if arc *BED* intersects arc *GEZ* between the two arcs *BZA*, *GDA*, and each of the four arcs is less than a semicircle, I say that the ratio of the chord of twice *AZ* to the chord of twice *BZ* is compounded of the ratio of the chord of twice *AG* to the chord of twice *GD* and the ratio of the chord of twice *DE* to the chord of twice *EB*.

The proof of it is that we make point *H* the center of the sphere and we join *HZ*, *HE*, *HG* and we join *BD*, *BA*, *AD*. And let *AD*, *HG* meet in one of two directions at *T*. *BD* will intersect *HE* at *L*; *AB* will intersect *ZH* at *K*. Then points *K*, *L*, *T* are in the planes of both circle *ZEG* and triangle *ABD*; hence, line *KLT*

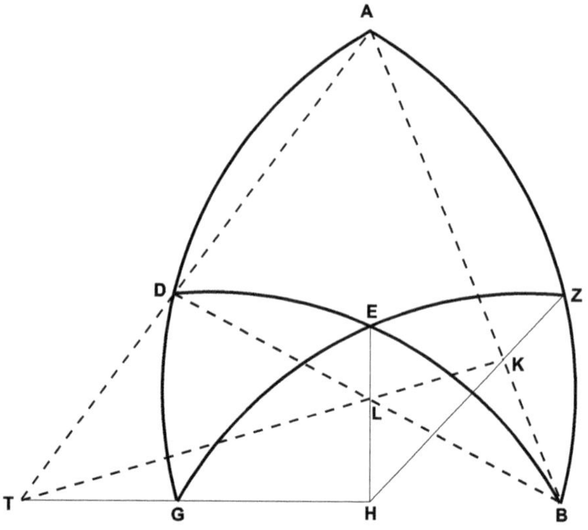

Figure 2.16 Menelaus's Theorem, III.1 of the *Spherics*

is straight. Then, according to what is in the diagram, the ratio *AK* to *KB* is compounded of the ratio *AT* to *TD* and the ratio *DL* to *LB*; and these are the ratios of the chords of twice the arcs attached to them. Hence, the ratio of the chord of twice *AZ* to the chord of twice *BZ* is compounded of the ratio of the chord of twice *AG* to the chord of twice *GD* and the ratio of the chord of twice *ED* to the chord of twice *EB*.[63]

[*Menelaus goes on to consider the case that AD is parallel to HG.*]

Explanation: The presentation is abbreviated and will require some unpacking; a couple of intermediate theorems are assumed without proof, which we shall fill in.

In modern notation, Menelaus asserts that

$$\frac{\mathrm{Crd}\,2\widehat{AZ}}{\mathrm{Crd}\,2\widehat{BZ}}=\frac{\mathrm{Crd}\,2\widehat{AG}}{\mathrm{Crd}\,2\widehat{GD}}\cdot\frac{\mathrm{Crd}\,2\widehat{DE}}{\mathrm{Crd}\,2\widehat{EB}}; \tag{2.7}$$

the equivalent statement in sines is simply

$$\frac{\sin\widehat{AZ}}{\sin\widehat{BZ}}=\frac{\sin\widehat{AG}}{\sin\widehat{GD}}\cdot\frac{\sin\widehat{DE}}{\sin\widehat{EB}}. \tag{2.8}$$

The first few sentences of the proof establish the geometry of the figure. The argument splits into cases when *HG* and *AD* are drawn, since they might

[63] Reprinted from [Sidoli 2006, 53].

meet AD on the side extended beyond D, or they might meet on the side extended beyond A, or they might be parallel. Our proof deals with the first and third cases, of which we shall examine only the first.

The key to the argument is the appeal to "a certain argument in the diagram," where Menelaus calls upon the planar equivalent of his theorem. In the shape in dotted lines in figure 2.16, he asserts (without proof) that

$$\frac{AK}{KB} = \frac{AT}{TD} \cdot \frac{DL}{LB}.^{64} \tag{2.9}$$

The phrase "and these are the ratios of the chords of twice the arcs attached to them" brings the proof home (subject to substitutions at the end). It relies implicitly on two related lemmas again taken for granted by Menelaus, but proved by Ptolemy a few decades later in the *Almagest* as follows:

Lemma A: [See figure 2.17(a).] In circle AZB, $AK/KB = \text{Crd}\,2\widehat{AZ} \,/\, \text{Crd}\,2\widehat{BZ}$.

Proof: $\Delta AKY \sim \Delta BKX$, so $AK/KB = AY/XB$. But $AY = \frac{1}{2}\text{Crd}\,2\widehat{AZ}$, and $XB = \frac{1}{2}\text{Crd}\,2\widehat{BZ}$.

Lemma B: [See figure 2.17(b).] In circle ADG, $AT \,/\, TD = \text{Crd}\,2\widehat{AG} \,/\, \text{Crd}\,2\widehat{GD}$.

Proof: $\Delta DTY \sim \Delta ATX$, so $AT \,/\, TD = AX \,/\, DY$. But $AX = \frac{1}{2}\text{Crd}\,2\widehat{AG}$ and $DY = \frac{1}{2}\text{Crd}\,2\widehat{DG}$.

Menelaus applies Lemma A twice and Lemma B once, to the three ratios in (2.9). Our case of Menelaus's Theorem (2.7) has now been established.[65]

Menelaus goes on to prove a related result: in figure 2.16,

$$\frac{\text{Crd}\,2\widehat{AB}}{\text{Crd}\,2\widehat{AZ}} = \frac{\text{Crd}\,2\widehat{BD}}{\text{Crd}\,2\widehat{DE}} \cdot \frac{\text{Crd}\,2\widehat{GE}}{\text{Crd}\,2\widehat{GZ}}. \tag{2.10}$$

The two statements (2.7) and (2.10) together would eventually become known as Menelaus's Theorem.

[64] Draw DX parallel to TK, meeting AB at X. Then $AK \,/\, KB = (AK \,/\, XK) \cdot (XK \,/\, KB)$. But $\Delta XAD \sim \Delta KAT$, so $AK/XK = AT/TD$; and $\Delta BKL \sim \Delta BXD$, so $XK/KB = DL/LB$. The result follows by substitution.

[65] It is possible, as argued in [Rome 1933b], that the proof in Menelaus's *Spherics* known to Ptolemy and Theon was missing the case that HD is parallel to AD in figure 2.16.

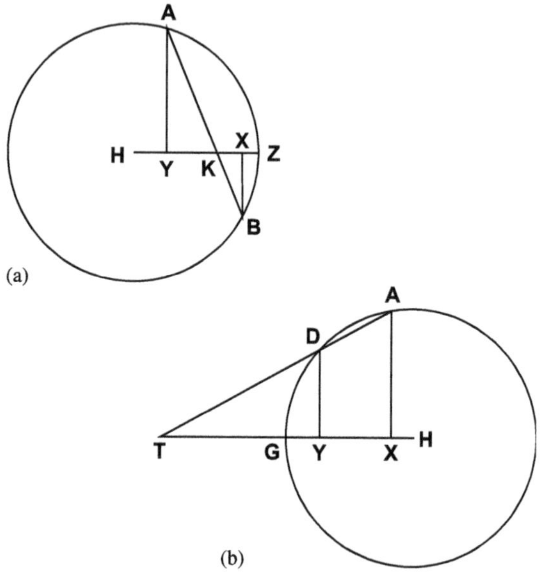

Figure 2.17
Lemmas for Menelaus's *Spherics*, III.1

Although the uses of this theorem may seem somewhat limited given the nature of the quadrilateral on which it operates, it is astonishingly versatile and solved many problems in spherical astronomy for 900 years. We shall glance at one application here, and again when we cover the *Almagest* and trigonometry in medieval Islam.

The problem of converting the Sun's longitude λ on the ecliptic to equatorial coordinates, i.e., determining α and δ, is handled by adding the appropriate arcs to the diagram (figure 2.18). Draw the solsticial colure *NCB* and a Menelaus configuration arises, namely, ♈*AB*☼*N*. Since *B* and *C* are both 90° removed from ♈ we have $\widehat{BC} = \varepsilon$, the fixed angle between the equator and the ecliptic called the ***obliquity of the ecliptic***. (Today $\varepsilon \approx 23.44°$; in the second century BC it was about 23.72°; the value given in the *Almagest* is about 23.86°.) Applying (2.10) to our figure gives

$$\frac{\mathrm{Crd}\,180°}{\mathrm{Crd}\,2\varepsilon} = \frac{\mathrm{Crd}\,180°}{\mathrm{Crd}\,2\delta} \cdot \frac{\mathrm{Crd}\,2\lambda}{\mathrm{Crd}\,180°};$$

simplified (letting $R = 60$ be the radius of the celestial sphere), we get

$$\mathrm{Crd}\,2\delta = \frac{\mathrm{Crd}\,2\varepsilon \cdot \mathrm{Crd}\,2\lambda}{2R}, \tag{2.11}$$

equivalent to the modern formula $\sin\delta = \sin\varepsilon \cdot \sin\lambda$. So we have the declination, one of the two sought quantities. We leave to the interested reader the

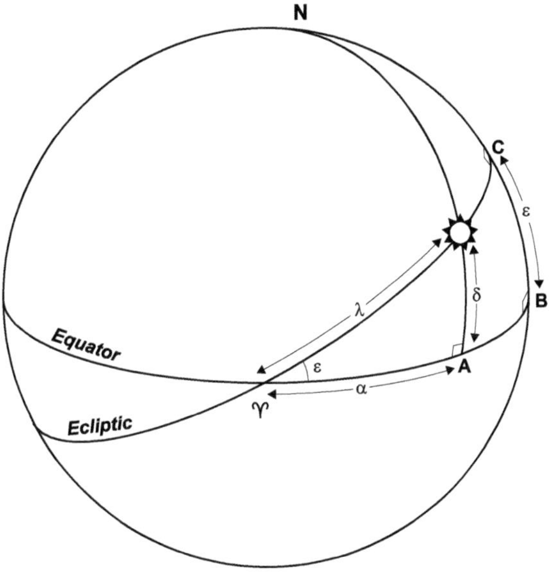

Figure 2.18
Using Menelaus's Theorem to calculate the declination of the Sun

task of applying (2.7) to the same figure to arrive at the equivalent of $\sin\alpha = \tan\delta \cdot \cot\varepsilon$, which gives the right ascension and completes the task of coordinate conversion.

The next few results in the *Spherics* derive from III.1, and are interesting because they eventually became foundations of medieval Islamic spherical astronomy (although not directly from Menelaus).

> **Proposition III.2:** [See figure 2.19.] If two spherical triangles share one angle (A and D), and another angle (G) is equal or supplementary to the corresponding angle (Z), then
>
> $$\frac{\sin\widehat{AB}}{\sin\widehat{BG}} = \frac{\sin\widehat{DE}}{\sin\widehat{EZ}}.[66] \tag{2.12}$$

The case where $\angle G = \angle Z = 90°$ would become known as the "Rule of Four Quantities." Its advantages are made clear by the derivation of the declination in figure 2.18: apply it to Δ♈☼A and Δ♈CB and the result follows immediately, without having to draw arcs up to the pole.

> **Proposition III.3:** [See figure 2.19.] If the angles at A and D are right and $\angle G = \angle Z$, then

[66] Of course Menelaus expressed his theorems using chords, not sines.

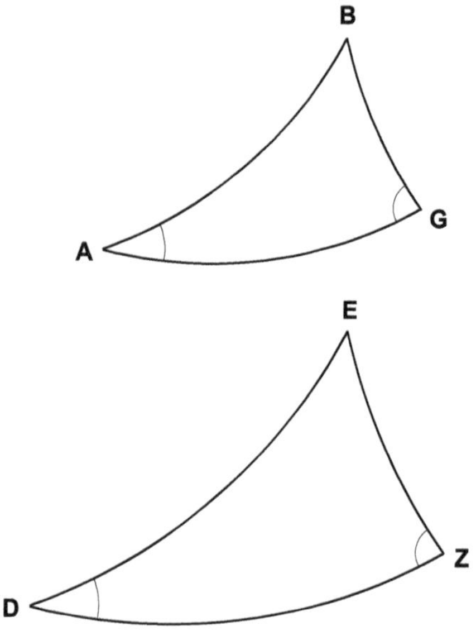

Figure 2.19
The Rule of Four Quantities

$$\frac{\sin AG}{\sin DZ} = \frac{\tan AB}{\tan DE}, \tag{2.13}$$[67]

the "Law of Tangents." A glance at figure 2.18 reveals that this leads to the formula for the right ascension $\sin\alpha = \tan\delta \cdot \cot\varepsilon$, to which we referred above.

The *Spherics* concludes with several theorems directly applicable to astronomy. One of these, III.22, corresponds to Theodosius's III.11 which we discussed earlier. It has two advantages over Theodosius: it is more general, and more importantly its use of spherical trigonometry allows Menelaus to achieve what Theodosius could not: an equality. In figure 2.20 $\widehat{DE}$ is an arbitrary arc on great circle *ADEB*, and the horizontal circles through *A*, *D*, and *E* are parallel to $\widehat{GHB}$. Then

$$\frac{\sin\widehat{GH}}{\sin\widehat{DE}} = \frac{d_A \cdot d_B}{d_D \cdot d_E}, \tag{2.14}$$

where the *d*'s are diameters of the parallel circles through the indicated points.

[67] The tangents arise because two of the quantities in the Menelaus expression are (equivalents of) sines of arcs complementary to *AB* and *DE*. They are thus cosines of *AB* and *DE* and form ratios with the sines, producing tangents.

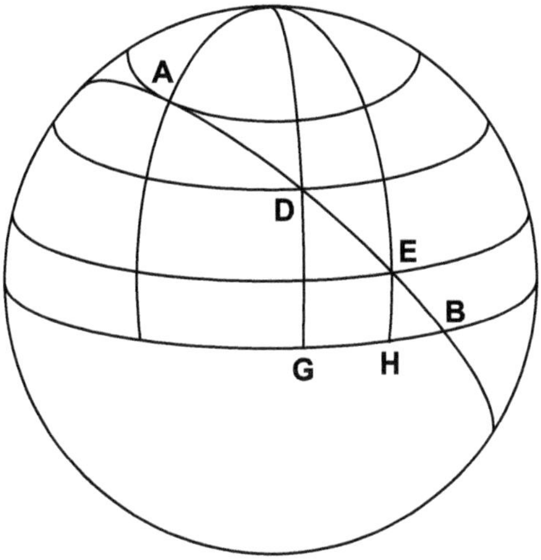

Figure 2.20
Menelaus's *Spherics*, Proposition III.22

This theorem may be put to several astronomical uses.[68] For instance, let $\widehat{GHB}$ be the equator, and *ADEB* be the ecliptic. If we let *D* coincide with *A* (so that $DB = GB = 90°$) and consider, the coordinates α, δ, and λ of point *E*, then

$$\frac{\sin \widehat{GH}}{\sin \widehat{DE}} = \frac{\cos \widehat{HB}}{\cos \widehat{EB}} = \frac{\cos\alpha}{\cos\lambda} = \frac{d_A \cdot d_B}{d_A \cdot d_E} = \frac{1}{\cos\delta}.$$

This gives the position of the Sun on the ecliptic directly from its equatorial coordinates.

Spherical Trigonometry before Menelaus?

Wilbur Knorr once made the following analogy concerning Greek mathematics:

> Imagine what it would be to piece together an account of contemporary mathematics on the basis of the books held in the undergraduate library of just one major university. Suppose further that these books had been scattered about into dozens of small holdings, and that before you could reassemble them, many had suffered damage, . . . Now project yourself far into the future, so that your entire enterprise would mean dealing with these twentieth century materials through an alien language and culture, and in an entirely different

[68] [Nadal/Taha/Pinel 2004, 412–414] lists some applications, including the one reported here.

mathematical idiom. . . . However desperate this scenario might seem, the situation with the historian of ancient mathematics is even worse. Directly or indirectly, every inquiry we make in this field is haunted by the fact of *loss*.[69]

We have done our best to reproduce the idiom of ancient mathematics but we cannot avoid the loss of part of the historical record, so we arrive at a gap here. Several clues suggest that spherical trigonometry was not born with Menelaus. His famous theorem (2.7)/(2.10) appears at the beginning of a book, where we seldom find new results—only the foundation from which others will be built.[70] Also curiously, in neither of the enunciations of Menelaus's Theorem does he refer to spherical triangles, so carefully studied in Book I and applied elsewhere—suggesting that the theorem was lifted from another context. Finally, the proof of Proposition III.5 of Menelaus's *Spherics* takes for granted the so-called ***anharmonic property*** of four great circles on a sphere, which states the (chord equivalent of the) following: in figure 2.21, with all arcs parts of great circles,

$$\frac{\sin \widehat{AD}}{\sin \widehat{DC}} \cdot \frac{\sin \widehat{BC}}{\sin \widehat{AB}} = \frac{\sin \widehat{WZ}}{\sin \widehat{YZ}} \cdot \frac{\sin \widehat{XY}}{\sin \widehat{WX}}. \tag{2.15}$$

Omitting its proof, Menelaus must have thought that his readers were aware of this important result.[71]

Evidence for the loss of a predecessor comes from the *Almagest* as well. Menelaus wrote his *Spherics* only a generation or two before the *Almagest*, yet Ptolemy never refers to him directly in connection with spherical trigonometry, only as an observer.[72] Finally, although Ptolemy relies heavily on (2.7) and (2.10), he never uses any of the more advanced results in Menelaus's *Spherics*.

All of this suggests that Menelaus borrowed his theorem from some previous book, and that Ptolemy also relied on some earlier work—perhaps the same one. The identity of this lost predecessor may be forever hidden, but the argument has been made more than once that the missing author was none other than Hipparchus. This identification has been speculated for many decades,[73] but recently some scraps of evidence have been brought to bear on the question.

[69] [Knorr 1990, 207].

[70] This case is made in [Sidoli 2004, 72].

[71] This case is made in [Heath 1921, vol. 2, 269–270].

[72] [Ptolemy (Toomer) 1984, n. 84].

[73] [Heath 1921, vol. 2, 270], for instance, speculated that both Menelaus's Theorem and the anharmonic property were known to Hipparchus, mostly on the basis of Ptolemy's reliance on

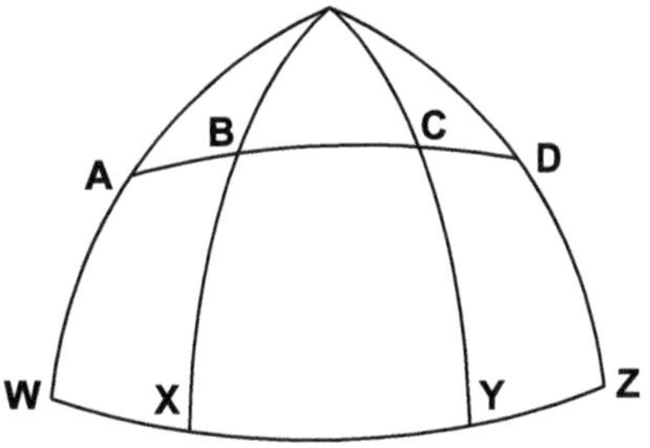

Figure 2.21
The anharmonic property

Strabo's *Geography* contains excerpts of Hipparchus's geographical writings, including some data on the calculation of an observer's terrestrial latitude, ϕ, from the length of the longest day of the year, M.[74] According to Strabo, Hipparchus divided zones of latitude on the Earth into *klimata*, defined by evenly-spaced values of M from 13 to 19 hours. He calculated the latitude for each *klima*; for instance, at Pontus ($M = 15\frac{1}{2}$ hours), he gave $\phi = 31{,}700$ stades (where a stade is $1/700^{\text{th}}$ of a degree, so $\phi \approx 45.3°$).

The question arises how Hipparchus was able to come up with his latitudes, which are generally quite accurate. A Babylonian-style arithmetic scheme has been fitted to his table of values of ϕ (M), but there is no explanation for how that scheme might have been generated.[75] Indeed, Hipparchus's numbers are so good that one suspects that they were found using the trigonometric formula

$$\tan\phi = \frac{-\cos(M/2)}{\tan\varepsilon}, \tag{2.16}$$

where ε is the obliquity of the ecliptic and M is measured by the standard $15° = 1$ hour. As one might expect, this correct formula matches Hipparchus's data better than does an arithmetic scheme.[76]

The question is, if Hipparchus used the trigonometric solution, how did he get his hands on it? The obvious answer is through spherical trigonometry,

---

Hipparchus in the *Almagest*. [Von Braunmühl 1897a, 5] even suggests the possibility of a Babylonian origin for the solution of spherical triangles.

[74] [Strabo (Jones) 1917, vol. 1, 507–521].

[75] [Neugebauer 1975, 304–306].

[76] First proposed in [Diller 1934], the hypothesis was revived and enhanced in [Rawlins 1982, 368]. A table in [Rawlins 1994, 56] compares the fit to Hipparchus of the arithmetic scheme with the trigonometric one. The comparison uses the value $\varepsilon = 23;40°$, better than Ptolemy's $\varepsilon = 23;51,20°$. Although the latitudes support Hipparchus's use of $\varepsilon = 23;40°$ (and see also [Nadal/Brunet 1984/1989] for further evidence that Hipparchus used this value), [Jones 2002] argues that a shadow ratio datum given in Strabo favors $\varepsilon = 23;51,20°$. The debate continues (see also *DIO* **4.2** (1994), p. 54 note C added 2002/03).

which is how Ptolemy does it, using Menelaus's Theorem.[77] If no other plausible historically-attested method were available, we would be forced to conclude from the numbers in Strabo's table that Hipparchus was capable of spherical trigonometry. However, there are other possibilities.

The ***analemma*** is an ancient geometrical device used to solve astronomical or geographical problems on a sphere. It consists essentially of taking a reference plane through the center of the sphere and rotating relevant arcs onto that plane, thereby reducing the original spherical problem to a planar one. The existence of analemma constructions in the theory of sundials suggests that they are very old.[78] There are only a handful of examples of Greek analemmas, but one of them—Heron of Alexandria's determination of the great circle distance between two cities based on simultaneous observations of a lunar eclipse—hints at prior, more extensive Greek work on the subject.[79] A passage in Hipparchus's *Commentary on Aratus* (which we shall visit in a moment) suggests that Hipparchus used them himself. A reconstructed analemma solution to the latitude problem would negate the necessity of the conclusion that spherical trigonometry goes back to Hipparchus.

Such a derivation exists;[80] we present it here to show how an analemma works. In figure 2.22 let *HOS* be the horizon, *OQ* the equator, and *OE* the ecliptic; ε is the obliquity of the ecliptic and ϕ is the local latitude. On the longest day of the year the Sun traces out a circle parallel to the equator through *E* that comes perpendicularly out of the page, so that the setting Sun ☼ is directly between *S* and the reader's eye. That circle *D*☼*E*, indicated by dotted lines, is rotated onto the plane of the diagram. Then, since *E* is the highest point on the ecliptic, $\overset{\frown}{☼E}$ corresponds to half the day; so where *M* is the length of daylight expressed in degrees, $M/2 = 90° + \theta$. Now

$$CE = \tfrac{1}{2}\mathrm{Crd}(180° - 2\varepsilon)\,[81] \text{ and } CO = \tfrac{1}{2}\mathrm{Crd}(2\varepsilon).$$

Therefore

$$SC = CE \cdot \tfrac{1}{2}\mathrm{Crd}(2\theta),$$

[77] [Ptolemy (Toomer) 1984, 76–78]. A complication arises if (2.16) is applied directly: since (as we shall see) the use of a chord table allows the computation of most trigonometric functions but not the arc tangent, ϕ may not be obtained directly from $\tan \phi$. [Neugebauer 1975, 37–38] describes how Ptolemy gets around this roadblock in this case.

[78] [Neugebauer 1975, 301].

[79] For a summary see [Neugebauer 1975, 845–848]; for the argument that Heron's work reflects an analemmatic tradition, see [Sidoli 2005].

[80] [Wilson 1997]; the following description is based on this paper.

[81] To see this join *OD*, and note that $\angle EOD = 180° - 2\varepsilon$.

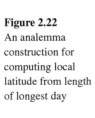

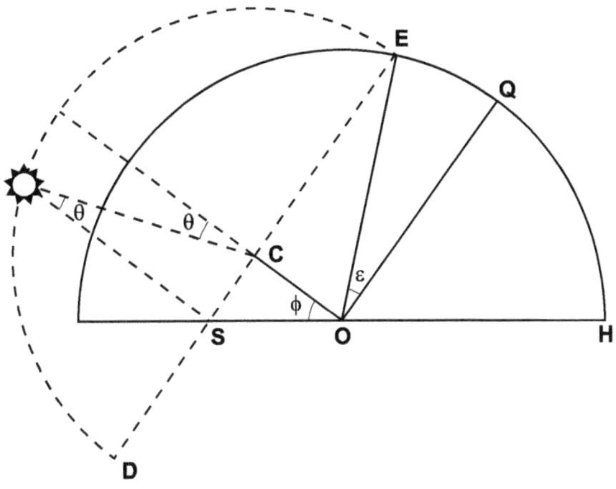

Figure 2.22
An analemma construction for computing local latitude from length of longest day

and so *SC/CO*, which is also equal to Crd (2ϕ)/ Crd (180°−2ϕ), is equal to

$$\frac{CE\,\mathrm{Crd}(2\theta)}{\mathrm{Crd}(2\varepsilon)}.$$

Since θ and ε are given, we may solve for ϕ.[82]

Another way that spherical astronomical results might have been obtained is through the use of stereographic projection of the sphere onto a plane. Best known today in the context of complex analysis, it was originally invented to transform the celestial sphere; the astrolabe, an ancient astronomical instrument which we shall see later, is a physical manifestation. In figure 2.23, the sphere is mapped to a flat surface by drawing lines through the south pole which intersect the horizontal plane through the equator. For instance, A on the sphere is sent to A' on the plane, while B is sent to B'. This process has the nice properties that circles on the sphere map to circles on the plane, and that angles are preserved (although, strangely, this latter fact was not known in any generality in antiquity[83]). Once the circles relevant to the problem have been transferred, plane trigonometry suffices to solve it.[84] Ptolemy, for example, solves the rising times problem in this way in his minor work the *Planisphaerium* (although it has been questioned whether

[82] The equivalent argument using modern functions is as follows: if *HQE* is a unit circle, then $CE = \cos\varepsilon$ and $CO = \sin\varepsilon$. Hence $SC = CE \sin\theta = \cos\varepsilon \sin\theta$, so that $\tan\phi = SC/CO = \cos\varepsilon \sin\theta / \sin\varepsilon$. The substitution $\sin\theta = -\cos(\theta + 90°) = -\cos(M/2)$ leads to the result.

[83] See the historical discussion of circle and angle preservation in stereographic projection in [Sidoli/Berggren 2008].

[84] [Donnay 1945], an intriguing little text on spherical trigonometry, bases its fundamental results on stereographic projection.

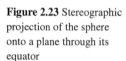

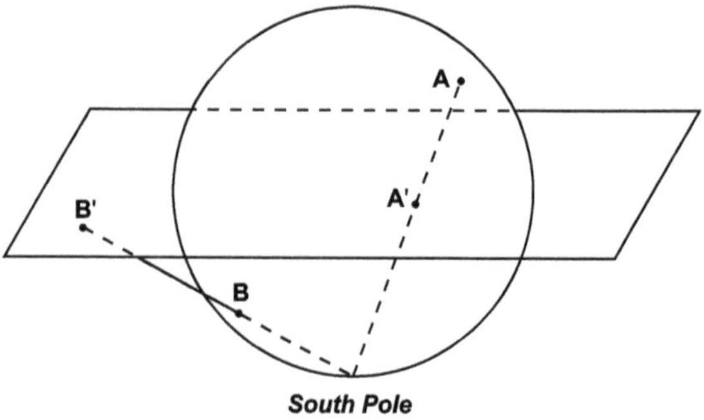

Figure 2.23 Stereographic projection of the sphere onto a plane through its equator

such computations were really Ptolemy's goal[85]). This technique goes back at least a couple of centuries before Ptolemy, and may originate with Hipparchus.[86]

Another bit of evidence for the origins of spherical trigonometry, a passage from Hipparchus's only surviving work *Commentary on Aratus*, records the point on the ecliptic that culminates when a certain star sets, given that star's equatorial coordinates. Recently, arguments have been raised that this calculation could not have been made using an analemma construction.[87] Since Hipparchus says that he used geometrical methods, spherical trigonometry is a strong candidate as an alternate solution.

The evidence remains circumstantial, but on the whole it seems likely that spherical trigonometry existed before Menelaus. A scholarly verdict on whether it goes back as far as Hipparchus may never arrive.

Claudius Ptolemy

Our fragmentary knowledge of the applied mathematical sciences before Claudius Ptolemy could hardly contrast more starkly than with the wealth of writings that we possess authored by Ptolemy himself. A number of his complete scientific works, some very extensive, have come down to us in a variety of textual traditions. They reveal a tremendous range of interests and an analytical depth perhaps never witnessed elsewhere before modern times. One wishes that his predecessors' works were better preserved, so that we

[85] [Berggren 1991a] argues that the word "projection" is not a faithful description of what is going on in the *Planisphaerium*; rather than mapping *points* on the sphere to the plane, Ptolemy is trying to represent *circles* on the plane. The calculation of rising times is not the point of the *Planisphaerium*, but is there simply to reassure readers of the faithfulness of the representation.

[86] An account of the solution to rising times may be found in [Neugebauer 1975, 860–865]; the possibility of Hipparchus's knowledge of stereographic projection is discussed in pp. 868–869.

[87] [Sidoli 2004].

could more clearly assess Ptolemy's achievements. What we have from Ptolemy must serve therefore as a report both of the giants upon whose shoulders Ptolemy stood, and of his own original contributions. As we have already seen, in many situations original authorship is difficult to decide.

We know from his reported astronomical observations that Ptolemy flourished around the middle of the second century AD, working in Alexandria, Egypt.[88] He is most widely known for his astronomy; his name will be forever associated with his geocentric model of the solar system, later opposed by Copernicus's heliocentric model. These attributions are not very apt, since a number of Ptolemy's predecessors had posed or developed geocentric models (including Eudoxus, Apollonius, and Hipparchus), and one astronomer had even suggested a heliocentric system (Aristarchus). But Ptolemy's astronomical system is by far the most sophisticated, successful, and truly scientific of all the extant ancient hypotheses. It is the only Greek scheme we have that gives a decent quantitative fit to the motions of the planets, predicts eclipses accurately, and handles a variety of other astronomical phenomena. Whether or not Ptolemy is responsible for its discovery, its blend of geometry and quantitative precision deserves a place at the apex of the ancient exact sciences.

Several of Ptolemy's other contributions were no less successful. His *Geography* established a standard for cartographic work through the medieval period and, eventually, in the West. Much the same can be said for his *Optics* (although its influence was indirect in the West). Most of Ptolemy's work on music theory, the *Harmonica*, also survives to today. However, his first commitment to astronomy is established by the existence of eight treatises, all of which survive at least in part.

Chief among these works is the *Almagest*.[89] Although not his first astronomical work,[90] the *Almagest* was the foundation for his most important later books. The *Handy Tables* extracted the mathematical tables and improved upon them; the *Planetary Hypotheses* relied on some *Almagest* results to compute the dimensions of the solar system; the *Tetrabiblos* turned its astronomical theory into astrological practice. Minor astronomical works

[88] See [Toomer 1975] for a technical overview of Ptolemy's scientific work.

[89] In the past century the *Almagest* has been translated twice into English, [Ptolemy (Taliaferro) 1952] and [Ptolemy (Toomer) 1984]; the latter is considered the standard today. Two excellent accounts of the contents of the *Almagest* are [O. Pedersen 1974b], whose notation we shall generally follow, and [Neugebauer 1975, vol. 1, 21–343].

[90] This honor appears to belong to the Canobic Inscription, a list of astronomical parameters originally engraved in stone at Canopus. Once thought to be spurious, the Canobic Inscription was recently found to represent an early stage in Ptolemy's astronomical work [Hamilton/Swerdlow/Toomer 1987].

included the *Analemma*[91] and the *Planisphaerium*, both topics of which we have already encountered.

Ptolemy named his masterwork *The Mathematical Collection*; it was later called "The Great Collection," in Arabic *Kitāb al-majistī*, from which the Latin title *Almagest* followed.[92] Its goal is almost startlingly modern: beginning with observations, generate a mathematical model for the motions of all celestial objects (the fixed stars, the planets, the Sun and the Moon), so that their positions and phenomena may be predicted. Indeed, the scientific appearance of the *Almagest* has often led to its being judged by modern standards. The slim observational data set from which Ptolemy develops his theory functions almost as an axiomatic base for the work, and the data generally provide far too perfect a match with the astronomical parameters Ptolemy is hoping to arrive at than one would reasonably expect. Thus every so often Ptolemy has been accused of scientific fraud, and defended on the basis that he likely selected nice observations from a larger set to fit his didactic purposes.[93] Our interest, thankfully, is not in the controversy but rather in the mathematics, so we shall move on.

Ptolemy's Chord Table

Ptolemy clearly needed a substantial trigonometric tool set to accomplish his goal, and he provides it in the first of the thirteen books in the *Almagest*. At its base is a table of the lengths of chords in a circle and a detailed set of instructions on how to construct one; since the instructions contain some of the earliest extant derivations of common trigonometric results, they have become favorite topics of popular books on ancient mathematics.[94] The *Almagest* uses the base circle radius $R = 60$ and gives chords to three sexagesimal places for arcs in $\frac{1}{2}^\circ$ increments from 0 to 180° (see figure 2.24, and extracts in figure 2.25). In terms of documentary evidence we are far ahead of where we were with Hipparchus—we have here the earliest extant trigonometric table—but we must be careful not to jump to conclusions. From Ptolemy's introduction:

[91] See [Luckey 1927].

[92] See [O. Pedersen 1974b, 15] or [Toomer 1975, 187].

[93] The latest episode began with the publication of *The Crime of Claudius Ptolemy*, [R. R. Newton 1977]. Various replies and arguments have appeared more or less continuously ever since; see "Ptolemy on Trial" [Swerdlow 1979a] for an early rebuttal. The journal *DIO* has continued the offensive, referring to Ptolemy's book as the *Almajest* (!). A recent attempt to encourage collegial debate on the issue was the back-to-back appearance of articles from both sides in the journal *Isis* [Thurston 2002] and [Gingerich 2002].

[94] The best general account of Ptolemy's methods for computing chords is [Aaboe 1964, 101–126]; see also [O. Pedersen 1974b, 56–65]. [Hultsch 1901/1902] argues that Ptolemy put two sources together in constructing his chord table.

LIBER I

Arcũ partes	m̃	Chordarum partes	m̃	2ª	trigesimarum m̃	2ª	3ª	Arcũ partes	m̃	Chordarum partes	m̃	2ª	trigesimarum m̃	2ª	3ª
0	30	0	31	25	1	2	50	23	0	23	55	27	1	1	33
1	0	1	2	50	1	2	50	23	30	24	26	13	1	1	30
1	30	1	34	15	1	2	50	24	0	24	56	58	1	1	26
2	0	2	5	40	1	2	50	24	30	25	27	41	1	1	22
2	30	2	37	4	1	2	48	25	0	25	58	22	1	1	19
3	0	3	8	28	1	2	48	25	30	26	29	1	1	1	15
3	30	3	39	52	1	2	48	26	0	26	59	8	1	1	11
4	0	4	11	16	1	2	47	26	30	27	30	14	1	1	8
4	30	4	42	40	1	2	47	27	0	28	0	40	1	1	4
5	0	5	14	4	1	2	46	27	30	28	31	20	1	1	0
5	30	5	45	27	1	2	45	28	0	29	1	50	1	0	56
6	0	6	16	49	1	2	44	28	30	29	32	18	1	0	52
6	30	6	48	11	1	2	43	29	0	30	2	44	1	0	48
7	0	7	19	33	1	2	42	29	30	30	33	8	1	0	44
7	30	7	50	54	1	2	41	30	0	31	3	30	1	0	40
8	0	8	22	15	1	2	40	30	30	31	13	50	1	0	35
8	30	8	53	35	1	2	39	31	0	32	4	8	1	0	31
9	0	9	24	54	1	2	38	31	30	32	34	22	1	0	27
9	30	9	56	13	1	2	37	32	0	33	4	33	1	0	22
10	0	10	27	32	1	2	35	32	30	33	34	46	1	0	17
10	30	10	58	49	1	2	33	33	0	34	4	55	1	0	12
11	0	11	30	5	1	2	32	33	30	34	35	1	1	0	8
11	30	12	1	21	1	2	30	34	0	35	5	5	1	0	3
12	0	12	32	36	1	2	28	34	30	35	35	6	0	59	57
12	30	13	3	50	1	2	27	35	0	36	5	5	0	59	52
13	0	13	35	4	1	2	25	35	30	36	35	1	0	59	48
13	30	14	6	16	1	2	23	36	0	37	4	55	0	59	43
14	0	14	37	27	1	2	21	36	30	37	34	47	0	59	38
14	30	15	8	38	1	2	19	37	0	38	4	36	0	59	32
15	0	15	39	47	1	2	17	37	30	38	34	22	0	59	27
15	30	16	10	56	1	2	15	38	0	39	4	5	0	59	22
16	0	16	42	3	1	2	13	38	30	39	33	46	0	59	16
16	30	17	13	9	1	2	10	39	0	40	3	25	0	59	11
17	0	17	44	14	1	2	7	39	30	40	33	0	0	59	5
17	30	18	15	17	1	2	5	40	0	41	2	33	0	59	0
18	0	18	46	19	1	2	2	40	30	41	32	[illegible]	0	58	54
18	30	19	17	21	1	2	0	41	0	42	1	30	0	58	48
19	0	19	48	21	1	1	57	41	30	42	30	54	0	58	42
19	30	20	19	19	1	1	54	42	0	43	0	15	0	58	36
20	0	20	50	16	1	1	51	42	30	43	29	33	0	58	31
20	30	21	21	12	1	1	48	43	0	43	48	49	0	58	25
21	0	21	52	6	1	1	45	43	30	44	28	1	0	58	18
21	30	22	22	58	1	1	42	44	0	44	57	10	0	58	12
22	0	22	53	49	1	1	39	44	30	45	26	16	0	58	6
22	30	23	24	39	1	1	36	45	0	45	55	19	0	58	0

Figure 2.24 The first page of the chord table in Ptolemy's *Almagest*, George of Trebizond's edition (1528) (courtesy of the Burndy Library)

> But first we shall show how one can undertake the calculation of [the chords'] amounts by a simple and rapid method, using as few theorems as possible, the same set for all. We do this so that we may not merely have the amounts of the chords tabulated unchecked, but may also readily undertake to verify them by computing them by a strict geometrical method.[95]

It seems to have escaped notice that Ptolemy does not claim either to have constructed the table in the manner outlined, or even to have authored the

[95] [Ptolemy (Toomer) 1984, 48].

Arcs	Chords	Sixtieths
½°	0;31,25	;1,2,50
1°	1;2,50	;1,2,50
1½°	1;34,15	;1,2,50
2°	2;5,40	;1,2,50
2½°	2;3,7,4	;1,2,48
3°	3;8,28	;1,2,48
⋮	⋮	⋮
12°	12;32,36	;1,2,28
⋮	⋮	⋮
24°	24;56,58	;1,1,26
⋮	⋮	⋮
36°	37;4,55	;0,59,43
⋮	⋮	⋮
48°	48;48,30	;0,57,21
⋮	⋮	⋮
60°	60;0,0	0;54,21
⋮	⋮	⋮
72°	70;32,3	;0,50,45
⋮	⋮	⋮
90°	84;51,10	;0,44,20
⋮	⋮	⋮
120°	103;55,23	;0,31,18
⋮	⋮	⋮
178½°	119;59,24	;0,0,41
179°	119;59,44	;0,0,25
179½°	119;59,56	;0,0,9
180°	120;0,0	;0,0,0

Figure 2.25 Extracts from Ptolemy's table of chords

table himself. As we shall see, there is cause to question at least the former.

Several chords are easily obtained using simple geometry (figure 2.26(a)–(c)); for instance $\text{Crd}\,60° = R = 60$, $\text{Crd}\,90° = \sqrt{R^2 + R^2} \approx 84;51,10$, and $\text{Crd}\,120° = 2\sqrt{R^2 - \left(\frac{R}{2}\right)^2} = \sqrt{3R^2} \approx 103;55,23$. A couple, Crd 36° and

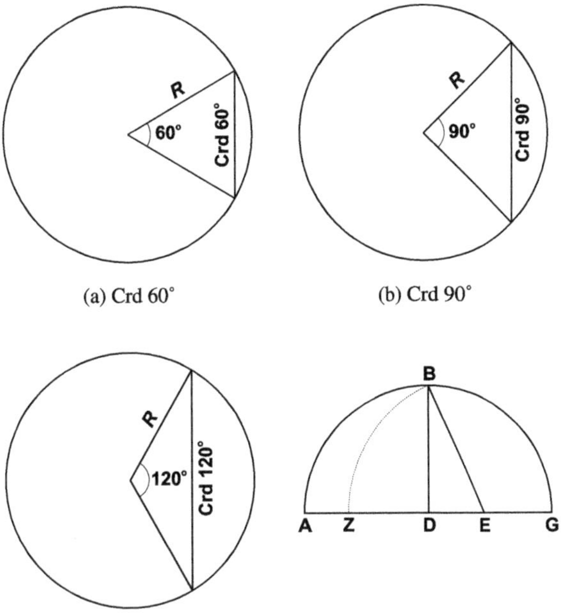

Figure 2.26
Calculations of simple chords

(a) Crd 60° (b) Crd 90° (c) Crd 120° (d) Crd 36° and Crd 72°

Crd 72°, are harder but accessible, and rely on two of the most beautiful consecutive theorems in Euclid's *Elements*[96]:

- **XIII.9:** The sides of a regular hexagon and a regular decagon inscribed in a circle are in "mean and extreme ratio" to each other (i.e., the ratio of the smaller to the larger is equal to the ratio of the larger to their sum—which turns out to be what is now called the "golden ratio" $\phi \approx 1.618:1$).
- **XIII.10:** The square of the side of a regular pentagon inscribed in a circle is equal to the sum of the squares of the sides of the hexagon and decagon—i.e., they form a right triangle.

In figure 2.26(d) let D be the center of the circle, let E bisect the radius DG, connect EB, and choose Z so that $EZ = EB$. Then $EZ^2 = ED^2 + GZ \cdot ZD$ (by *Elements* II.6[97]); but $EZ^2 = EB^2 = ED^2 + DB^2$, so $DB^2 = DG^2 = GZ \cdot ZD$. Since this latter equality implies that $DZ/DG = DG/ZG$, we know that DG and DZ form a mean and extreme ratio—but DG is the radius of the circle, hence the side of the hexagon. So, by XIII.9, DZ is the side of the decagon,

[96] These theorems may be found in [Euclid (Heath) 1925, vol. 3, 455–461].

[97] In modern terms: let $a = DE = EG$ and $b = ZD$; then this equation says simply that $(a + b)^2 = a^2 + (2a + b)b$.

and its length is Crd(360°/10) = Crd 36°. But then since DB is also a radius, ΔDBZ satisfies XIII.10, so BZ must be the side of the pentagon with length Crd 72°.

From here the desired results follow easily:

$$\text{Crd}\,36° = DZ = EZ - DE = EB - DE = \sqrt{R^2 + \left(\frac{R}{2}\right)^2} - \frac{R}{2} \approx 37;4,55,$$

and

$$\text{Crd}\,72° = BZ = \sqrt{DZ^2 + DB^2} = \sqrt{(\text{Crd}\,36°)^2 + R^2} \approx 70;32,3.$$

Ptolemy's Theorem and the Chord Subtraction/Addition Formulas

Next Ptolemy develops tools that allow him to generate more chords from existing values. The chord supplement formula (2.2) allows the quick determination of Crd(180° − θ), but on its own this does not get him much further. Much more powerful are the chord sum and difference formulas. To derive them, Ptolemy could have used Archimedes' Theorem of the Broken Chord (from the previous chapter), but instead he applies another theorem:

> **Ptolemy's Theorem:** [See figure 2.27.] In a quadrilateral inscribed in a circle, the product of the diagonals ($AC \cdot BD$) is equal to the sum of the products of the opposite sides ($AB \cdot CD + AD \cdot BC$).

The proof is elementary,[98] leading some to suspect that it is not original to Ptolemy.[99]

The chord difference formula is easily obtained from this by letting AD be the diameter of the circle. Let $\alpha = \overset{\frown}{AC}$ and $\beta = \overset{\frown}{AB}$. An application of Ptolemy's Theorem and a bit of manipulation gives

$$\text{Crd}(\alpha - \beta) = \frac{\text{Crd}\,\alpha \cdot \text{Crd}(180° - \beta) - \text{Crd}\,\beta \cdot \text{Crd}(180° - \alpha)}{2R}, \quad (2.17)$$

equivalent to the modern sine subtraction law

[98] Draw BE so that $\angle ABE = \angle DBC$. Then $\Delta ABD \sim \Delta EBC$ ($\angle ABD = \angle EBC$ by addition of $\angle EBD$, and $\angle BDA = \angle BCE$ since they both subtend chord AB—*Elements* III.20), so $BD \cdot CE = BC \cdot AD$. Also $\Delta ABE \sim \Delta BCD$ ($\angle ABE = \angle DBC$, and $\angle BAE = \angle BDC$ since they both subtend chord BC), so $BD \cdot AE = BA \cdot DC$. Therefore $AC \cdot BD = (AE + CE) \cdot BD = BA \cdot DC + BC \cdot AD$.

[99] See for instance [O. Pedersen 1974b, 57–58].

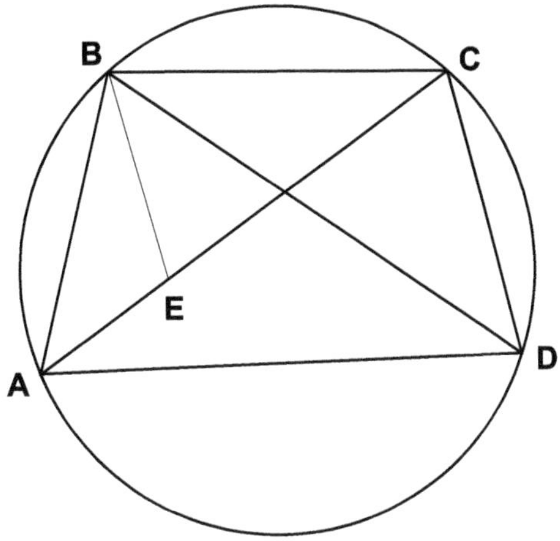

Figure 2.27
Ptolemy's Theorem

$$\sin(x-y) = \sin x \cos y - \sin y \cos x. \tag{2.18}$$

Using Ptolemy's Theorem to find the sum of two chords is not quite as straightforward. In figure 2.28 BZ is extended to E so that $\alpha = \widehat{AB} = \widehat{DE}$, and $\beta = \widehat{BC}$. Application of Ptolemy's Theorem to $BCDE$ gives

$$\text{Crd}(180° - (\alpha + \beta)) = \frac{\text{Crd}(180° - \alpha) \cdot \text{Crd}(180° - \beta) - \text{Crd}\,\alpha \cdot Crd\,\beta}{2R}, \tag{2.19}$$

from which Crd $(\alpha + \beta)$ may be found using the supplement formula (2.2). This is equivalent not to the sine addition law, but rather to the cosine law

$$\cos(x+y) = \cos x \cos y - \sin x \sin y.^{100} \tag{2.20}$$

The significance of this is that the supplement formula is not numerically stable when working from large arcs to small; in fact, it has been shown that the chord table is too accurate for (2.19) and (2.2) to have been used as prescribed in the *Almagest*.[101]

Finally, Ptolemy gives a procedure on how to find the chord of half a given arc, equivalent to the sine half-angle formula (2.4). Curiously, although he could have used Ptolemy's Theorem to derive it, he chooses instead to use a procedure similar to that which we saw in Archimedes.

[100] [Ptolemy (Toomer) 1984, 53]; [Aaboe 1964, 119–120].

[101] For arcs close to 180°, the chords are very close to each other and to 120^p, so small errors in rounding of these chords can result in large errors in the chords of the corresponding small arcs. See [Van Brummelen 1993, 67–68].

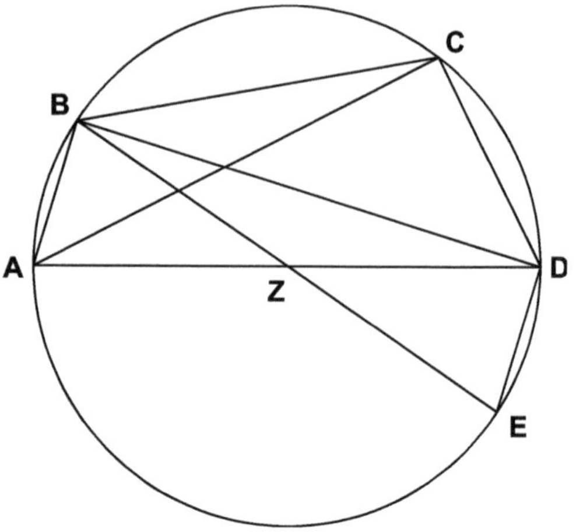

Figure 2.28
Derivation of the chord addition law

Perhaps it is a remnant of the tools available to build chord tables before Ptolemy.[102]

The Chord of 1°

The tools at hand allow us to compute Crd 12° (from Crd 72° and Crd 60°); from this the half-angle formula takes us down to Crd 6°, Crd 3°, $\mathrm{Crd}\,1\frac{1}{2}^\circ$ and $\mathrm{Crd}\,\frac{3}{4}^\circ$. But it is not possible to generate any chords of whole-numbered arcs other than multiples of 3°; to do so would be equivalent to trisecting the angle, as Ptolemy suggests. To get an estimate of Crd 1° Ptolemy proves a theorem similar to lemma (1.1), which Aristarchus and Archimedes had used to capture trigonometric quantities between bounds: if $\beta<\alpha<90^\circ$, then

$$\frac{\mathrm{Crd}\,\alpha}{\mathrm{Crd}\,\beta} < \frac{\alpha}{\beta}. \tag{2.21}$$

This is intuitively obvious; as the arc increases so does the chord, but as the arc increases the chord gets larger at a slower and slower rate.[103]

Substituting $\alpha = 1^\circ$, $\beta = \frac{3}{4}^\circ$ in (2.21), and then $\alpha = \frac{3}{2}^\circ$, $\beta = 1^\circ$, Ptolemy arrives at

$$\tfrac{2}{3}\mathrm{Crd}\,\tfrac{3}{2}^\circ < \mathrm{Crd}\,1^\circ < \tfrac{4}{3}\mathrm{Crd}\,\tfrac{3}{4}^\circ; \tag{2.22}$$

[102] This is suggested in [Toomer 1973a, 18–19].

[103] Hence (2.21) is equivalent to stating that the second derivative of the chord function is less than zero for $\theta < 90^\circ$.

both bounds turn out to be equal to 1;2,50.[104] Armed with this value, the chords of the multiples of 3°, and the sum, difference and half-angle identities, Ptolemy had the tools he needed to compute the remaining chords. Although the values in the table are too accurate to be certain precisely how it was done, it has been suggested that he began with a grid of entries for every 12°, then filled in the entries below each of these nodes using the chord subtraction law.[105]

The Interpolation Table

To be able to convert *all* arcs to chords, Ptolemy needed to use the table to find the chords of arcs that are *not* multiples of $\frac{1}{2}^\circ$. The purpose of the column labeled "Sixtieths" in figure 2.25 is to tabulate the function

$$s(\theta) = \frac{1}{30}[\mathrm{Crd}(\theta + \tfrac{1}{2}^\circ) - \mathrm{Crd}\,\theta], \tag{2.23}$$

the average amount by which the chord's value increases for every one-minute increment of arc. To find the chord of $n^\circ + x$, where n is a multiple of $\frac{1}{2}^\circ$ and x is measured in minutes, simply calculate

$$\mathrm{Crd}(n^\circ + x) \approx \mathrm{Crd}\,n^\circ + x \cdot s(\theta). \tag{2.24}$$

In other words, apply linear interpolation. Curiously, the entries in the table of $s(\theta)$ are actually more accurate than what can be obtained from the chords themselves. This observation suggests that more precision was used than Ptolemy shows us, at least in some stages of the calculations.[106]

Chords in Geography: Gnomon Shadow Length Tables

One of the earliest uses of chords in the *Almagest* is geographical: given the local terrestrial latitude ϕ, determine the length of the shadow cast by a ***gnomon*** at noon at the equinoxes and solstices. The gnomon, nothing more

[104] The upper bound should be 1;2,50,40, prompting an accusation of fudging ([Newton 1977, 27–28] and [Newton 1982, 197–203]). But it has been pointed out that if Ptolemy used a little more accuracy than he shows in the *Almagest*, the value he reports is correct to all three places [Glowatski/Göttsche 1976, 70–71].

[105] [Van Brummelen 1993, 46–73].

[106] Bafflingly, [Newton 1985, 16–18] asserts that since the chords and sixtieths do not match perfectly, the table $s(\theta)$ must represent "some other quantity," which Ptolemy attached mistakenly to the chord table. [Glowatski/Göttsche 1976] recompute the chord table according to the method spelled out in the *Almagest*, and demonstrate that more accuracy must have been used than is displayed, although the analysis in [Van Brummelen 1993, 46–73] suggests that the *Almagest* prescription may not have been carried out to the letter.

than a vertical stick in the ground, casts shadows of different lengths at different times of the day and year depending upon the Sun's altitude. Hence one should be able to determine ϕ using the length of a gnomon's shadow, as well as find the time of day—which is, after all, what a sundial does. In fact the gnomon is the most enduring of astronomical instruments; it predates the Greeks,[107] and can be found in several prominent European cathedrals.[108]

At noon on the day of the equinox, the arc from the Sun to the zenith is equal to ϕ (see figure 2.29).[109] On the summer solstice it is $\phi - \varepsilon$ (where ε is the obliquity of the ecliptic); at the winter solstice it is $\phi + \varepsilon$. Our goal is to find the lengths of the gnomon's shadow at the equinox (*GZ*) and at the solstices (*GN* and *GK*).

Text 2.3
Ptolemy, Finding Gnomon Shadow Lengths
(from the *Almagest* II.5)

Then, since $\widehat{GD}$, which is equal to the elevation of the north pole from the horizon, is 36° (where meridian *ABG* is 360°) at the latitude in question, and both $\widehat{\Theta D}$ and $\widehat{DM}$ are 23;51,20°, by subtraction $\widehat{G\Theta} = 12;8,40°$, and by addition $\widehat{GM} = 59;51,20°$.

Therefore the corresponding angles

$$\left.\begin{array}{l} \angle KEG = 12;8,40^\circ \\ \angle ZEG = 36^\circ \\ \angle NEG = 59;51,20^\circ \end{array}\right\} \text{where 4 right angles} = 360^\circ$$

and

$$\left.\begin{array}{l} \angle KEG = 24;17,20^{\circ\circ} \\ \angle ZEG = 72^{\circ\circ} \\ \angle NEG = 119;42,40^{\circ\circ} \end{array}\right\} \text{where 2 right angles} = 360^{\circ\circ}.$$

[107] [Kennedy 1969b, 335–337] gives examples of shadow length tables from four cultures, the oldest (Egyptian) dating back at least to 1300 BC. We shall not consider early examples of these tables to be trigonometric since they lack a measurable notion of angle or arc; nor do they exist beyond this particular astronomical situation.

[108] For an account of European cathedrals used as large gnomons, see [Heilbron 1999].

[109] The altitude of the North Pole, measured upward from the horizon toward the zenith, is always ϕ. At the equinox the Sun is on the equator, and at noon it reaches its maximum altitude. Therefore the Sun is 90° removed from the North Pole, which is ϕ above the horizon; hence $\widehat{A☼} = \phi$.

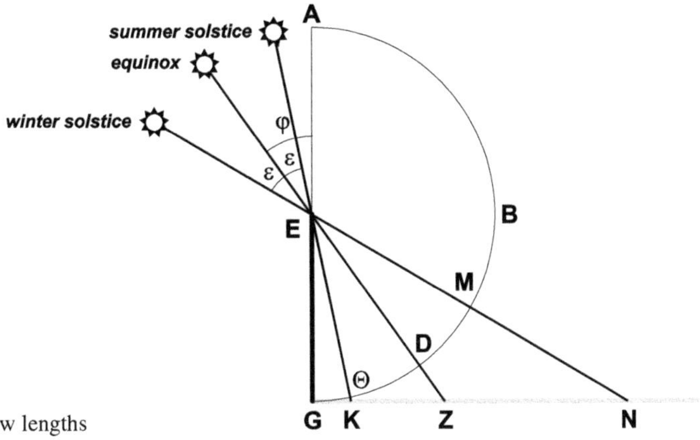

Figure 2.29
Gnomon shadow lengths

Therefore in the circles about right-angled triangles *KEG*, *ZEG*, *NEG*,

$$\overset{\frown}{GK} = 24;17,20^\circ$$

$$\text{and } \overset{\frown}{GE} = 155;42,40^\circ \text{ (supplement)},$$

$$\overset{\frown}{GZ} = 72^\circ$$

$$\text{and } \overset{\frown}{GE} = 108^\circ\text{, similarly [as supplement]},$$

$$\overset{\frown}{GN} = 119;42,40^\circ$$

$$\text{and } \overset{\frown}{GE} = 60;17,20^\circ \text{ (again as supplement).}$$

Therefore where $\text{Crd}\,\overset{\frown}{GK} = 25;14,43^p, \text{Crd}\,\overset{\frown}{GE} = 117;18,51^p$, and where $\text{Crd}\,\overset{\frown}{GZ} = 70;32,4^p, \text{Crd}\,\overset{\frown}{GE} = 97;4,56^p$, and where $\text{Crd}\,\overset{\frown}{GN} = 103;46,16^p, \text{Crd}\,\overset{\frown}{GE} = 60;15,42^p$.

Therefore, where the gnomon *GE* has 60^p, in the same units

the summer [solsticial] shadow, $GK \approx 12;55^p$,
the equinoctial shadow, $GZ \approx 43;36^p$,
and the winter [solsticial] shadow, $GN \approx 103;20^p$.[110]

Explanation: Ptolemy uses "demi-degrees" (indicated here by °°) alongside ordinary degrees throughout the *Almagest*, which allows him to use a clever

[110] Reprinted from [Ptolemy (Toomer) 1984, 80–81].

device to find ratios of sides in a triangle. To find the length of the equinoctial shadow we must first calculate GZ/GE. Imagine a circle circumscribed around ΔGZE and assign it a radius of 60 units. Then GZ is a chord in that circle corresponding to $\angle GEZ$ at the far side of the circle. By *Elements* III.20 we know that this chord corresponds to an angle at the center of the circle exactly twice as large as $\angle GEZ$—in other words, $\angle GEZ$ measured in demi-degrees! Once the two sides GZ and GE are found using the chord table Ptolemy finds the ratio between them, thereby eliminating the units local to the circumscribed circle. The calculation is completed by converting the units to a gnomon of length 60.

Ptolemy goes on to list shadow lengths for 39 climate zones, defined (as in Hipparchus) by a regular progression of lengths of longest daylight. For instance, for the sample results above he writes:

> The eleventh [climate zone] is the parallel with a longest day of 14½ equinoctial hours. This is 36° from the equator, and goes through Rhodes. In this region, for a gnomon of 60^p, the summer [solsticial] shadow [at noon] is $12\frac{11}{12}^p$, the equinoctial shadow $43\frac{3}{5}^p$, and the winter [solsticial] shadow $103\frac{1}{3}^p$.[111]

The equinoctial shadow length is just 60 tan ϕ so there is a sense in which these shadow lengths can be thought of as a sort of tangent table. In fact, when the tangent was eventually defined explicitly in medieval Islam it was called the "shadow." However, here the shadow lengths are simply another geographical table, and are never used with mathematical intent. Indeed, the representation of the numbers in common fractions would have made them difficult to use systematically in other contexts.

Spherical Astronomy in the *Almagest*

Spherical astronomy in the *Almagest* is confined mostly to Books I, II and VIII. Other books manage the motions of the Sun, Moon and planets by considering their ecliptic coordinates λ and β separately. As far as the rest of the *Almagest* is concerned, then, calculations on the sphere are needed primarily for conversion between coordinates—especially ecliptical (in which the planetary models function), equatorial (where the daily rotation occurs), and with respect to the local horizon (altitude and azimuth). Other uses of spherical astronomy included

[111] [Ptolemy (Toomer) 1984, 86].

rising times, whose astrological significance made them a favorite topic of study; they are given full treatment in Book II. Finally, simultaneous risings, culminations, and settings of fixed stars are handled in Book VIII.

All of these tasks are accomplished using only Menelaus's Theorem, (2.7) and (2.10) in figure 2.16. Ptolemy proves it in essentially the same way as Menelaus, but as we have noted, he does not attribute it to anyone. The theorem may be applied in a surprising number of variations; in fact, it is not hard to show that all the fundamental relations on which spherical trigonometry is now based fall quite easily from one or the other of the two Menelaus statements—even the spherical Law of Sines.[112]

We have already seen how Menelaus's Theorem may be used to convert coordinates from ecliptic to equatorial. Ptolemy's solution of the problem of rising times exemplifies nicely its flexibility when used appropriately. In figure 2.30, the local latitude $\phi = \overset{\frown}{ND}$ and position on the ecliptic $\lambda = \overset{\frown}{\Upsilon\star}$ are given; our goal is to find the oblique ascension $\theta = \overset{\frown}{\Upsilon E}$, the arc on the equator that rises as the ecliptic arc from ♈ to ☼ rises.

Both α and $\delta = \overset{\frown}{\star M}$ may be found using the change of coordinates procedures described earlier, but it is not immediately clear how to proceed from there until the meridian *NDG* is considered. Then several Menelaus configurations appear, among them *GME☼ND*. Applying Menelaus I (2.7), we obtain

$$\frac{\operatorname{Crd} 2\overset{\frown}{ND}}{\operatorname{Crd} 2\overset{\frown}{DG}} = \frac{\operatorname{Crd} 2\overset{\frown}{N\star}}{\operatorname{Crd} 2\overset{\frown}{\star M}} \cdot \frac{\operatorname{Crd} 2\overset{\frown}{ME}}{\operatorname{Crd} 2\overset{\frown}{EG}},$$

or

$$\frac{\operatorname{Crd} 2\phi}{\operatorname{Crd}(180^\circ - 2\phi)} = \frac{\operatorname{Crd}(180^\circ - 2\delta)}{\operatorname{Crd} 2\delta} \cdot \frac{\operatorname{Crd} 2\overset{\frown}{ME}}{\operatorname{Crd} 180^\circ}$$

(equivalent to $\sin \overset{\frown}{ME} = \tan\phi \tan\delta$), which gives the ***ascensional difference*** $\overset{\frown}{ME}$. Then the oblique ascension is simply $\theta = \alpha - \overset{\frown}{ME}$. (To find the rising time of an arc that does not start at ♈ but starts and ends at arbitrary points *A* and *B* on the ecliptic, simply find θ for the arcs ♈*B* and ♈*A* and subtract.)

[112] We do not deal with the specifics of this here, due to the danger of anachronism. The interested reader may find these derivations in [O. Pedersen 1974b, 75–78] and [Neugebauer 1975, 29–30].

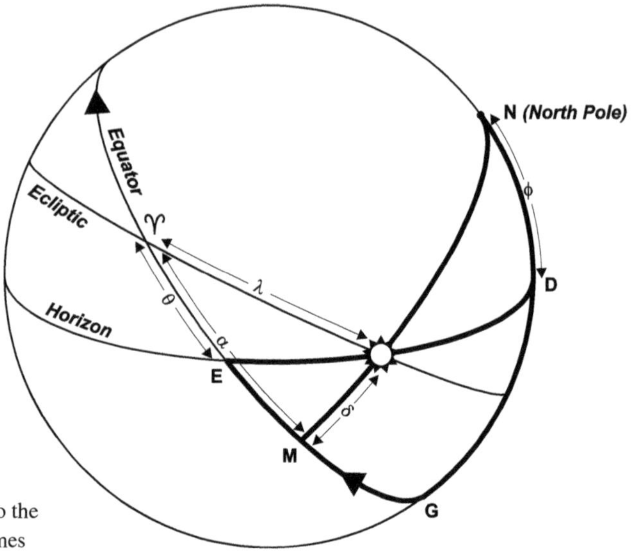

Figure 2.30
Ptolemy's solution to the problem of rising times

The addition of the meridian circle (or some other appropriately chosen great circle) to the diagram gives the Menelaus theorems almost as much power as we have in modern spherical trigonometry. They are used in this way throughout the *Almagest*; Menelaus proved his more advanced results this way in the *Spherics*; and as we shall see, the Menelaus theorems continued to be fundamental to Muslim astronomers until the early eleventh century.

Ptolemy on the Motion of the Sun

What we have seen so far of the *Almagest* is what Ptolemy calls "those aspects of heaven and earth which required, in outline, a preliminary mathematical discussion."[113] The bulk of the work from here is the production and use of mathematical models of the motions of the stars and planets. Once the geometric models have been postulated, two tasks must be performed: the determination of values for the numeric parameters (lengths of certain line segments and various angular velocities), and the construction of a scheme that allows one to predict a body's position for an arbitrary time *t*. For Ptolemy these are the primary functions of trigonometry.

[113] [Ptolemy (Toomer) 1984, 131].

The Sun must come first, since studies of the other planets depend on it in various ways (for instance, the crucial observations of the Moon are made at eclipses). We have seen already how Ptolemy found the eccentricity of the Sun's orbit given the lengths of the seasons; in figure 2.4 $e = EZ = 2;29\frac{1}{2}$, which he rounded to 2;30. Equally important is the orientation of the eccentric circle, which is found by determining the location of the Sun's apogee (its furthest point from the Earth). To do this calculation Ptolemy employs a familiar technique. In figure 2.4 the ecliptic, if drawn, would be a large circle centered at *E*, enclosing the entire diagram. Imagine a circle circumscribed around ΔEZX with hypotenuse 120^p. Since we found earlier that $ZX = 1;2$, in the new units we have $ZX = 120 \cdot (1;2 / 2;29\frac{1}{2}) = 49;46^p$. Using the chord table backwards, we get $\widehat{ZX} \approx 49°$. So the arc on the ecliptic upwards from the summer solstice to the Sun's apogee is half that, or 24;30°. Hence the Sun's apogee measured from the vernal equinox *N* is $\lambda_A = 65;30°$.

Of more mathematical interest are the tools that Ptolemy introduces to provide an answer to the question of prediction: given arbitrary *t*, where is the Sun? Clearly it is on the ecliptic, so its latitude $\beta = 0$; we must only find its longitude $\lambda(t)$. If need be, we may then apply our coordinate conversion techniques to locate the Sun with respect to the equator or the horizon.

Our first step is to separate the Sun's longitude into a mean position and a correction factor which Ptolemy calls the "prosthaphairesis," or as it became known in the medieval period, the "equation." In figure 2.31 *EZH* is the Sun's orbit circle centered at *C*; and circle *AG*, centered at the Earth *D*, is concentric with the ecliptic. The apogee *E* has longitude $\lambda_A = 65;30°$, as we calculated previously. The Sun travels at constant speed on *EZH*, so the ***mean anomaly*** a_m increases linearly. Thus $\widehat{EZ}$ is a linear function of time:

$$\widehat{EZ} = a_m(t) = a_m(t_0) + v \cdot \Delta t, \qquad (2.25)$$

where t_0 is the ***epoch*** (a fixed instant for which the Sun's mean anomaly $a_m(t_0)$ is already known),[114] Δt is the time difference from the epoch to *t*, and *v* is the Sun's mean velocity.

However, the Sun's position seen from the Earth is its ***true anomaly*** $a = \angle ADZ$. We can get from a_m to *a* by finding the ***solar equation***

[114] For Ptolemy this was the first day of the Egyptian calendar of the first year in the reign of Nabonassar in 747 BC. He chooses such an early date because he says he has records that go back that far.

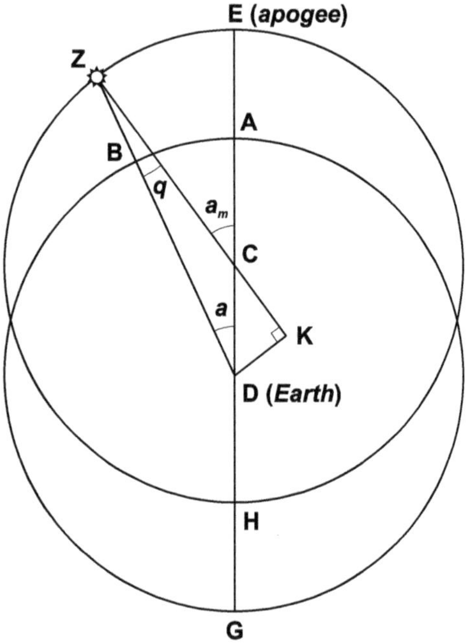

Figure 2.31
The solar equation q

$q(a_m) = \angle CZD$. Once we have q then $a(t) = a_m(t) - q$,[115] and the Sun's position on the ecliptic may be found simply by adding the longitude of the apogee E:

$$\lambda(t) = \lambda_A + a(t) = \lambda_A + a_m(t) - q. \tag{2.26}$$

Text 2.4
Ptolemy, Determining the Solar Equation
(from the *Almagest* III.5)

[See figure 2.31.] First, let $\overset{\frown}{EZ}$ be given, e.g. as 30°. Produce ZC and drop the perpendicular to it from D, DK.

Then, since $\overset{\frown}{EZ}$ is, by hypothesis, 30°,

$$\angle ECZ = \angle DCK = \begin{cases} 30^\circ \text{ where 4 right angles} = 360^\circ \\ 60^{\circ\circ} \text{ where 2 right angles} = 360^{\circ\circ}. \end{cases}$$

Therefore, in the circle about right-angled triangle DCK, $\overset{\frown}{DK} = 60^\circ$ and $\overset{\frown}{KC} = 120^\circ$ (supplement).

[115] This is true because both a_m and $a + q$ are supplementary to $\angle ZCH$.

Therefore the corresponding chords $\left.\begin{array}{r} DK = 60^p \\ \text{and } KC = 103;55^p \end{array}\right\}$ where hypotenuse $DC = 120^p$.

Therefore, where $DC = 2;30^p$ and radius $ZC = 60^p$, $DK = 1;15^p$ and $CK = 2;10^p$.

Therefore, by addition [of CK to radius ZC], $KCZ = 62;10^p$.

Now since $DK^2 + KCZ^2 = ZD^2$, the hypotenuse $ZD \approx 62;11^p$. Therefore, where ZD $= 120^p$, $DK = 2;25^p$, and, in the circle about right-angled triangle ZDK, $\widehat{DK} = 2;18°$.

$$\therefore \angle DZK = \begin{cases} 2;18^{\circ\circ} \text{ where 2 right angles} = 360^{\circ\circ} \\ 1;9° \text{ where 4 right angles} =. 360° \end{cases}$$

That will be the amount of the equation of anomaly at this position. And $\angle ECZ$ was taken as 30°. Therefore, by subtraction, $\angle ADB$ (which equals $\widehat{AB}$ of the ecliptic) equals 28;51°.[116]

Explanation: Some of the procedures in this passage (such as the circumscribed circle and demi-degrees) were explained in Text 2.3.

$DC = e$, so $DK = e \sin a_m$ and $CK = e \cos a_m$. Then $KCZ = R + e \cos a_m$ (where $R = 60$ is the radius of the orbit circle), so by Pythagoras, $ZD = \sqrt{(e \sin a_m)^2 + (R + e \cos a_m)^2}$. Hence

$$\sin q = \frac{e \sin a_m}{\sqrt{(e \sin a_m)^2 + (R + e \cos a_m)^2}}, \tag{2.27}$$

and finally, $a = a_m - q = 28;51°$.

Ptolemy could have used instead DK and ZK to get the much simpler formula

$$\tan q = \frac{e \sin a_m}{R + e \cos a_m} \tag{2.28}$$

(as was done in later Islamic astronomy), but since the chord table did not permit him to evaluate arc tangents, he had to use (2.27).

To complete the project of finding the Sun's longitude λ for any time t, Ptolemy supplies a table of values for the solar equation q as a function of a_m

[116] From [Ptolemy (Toomer) 1984, 158–159].

Common Numbers (a_m)		Equation(q)
6	354	0;14
12	348	0;28
18	342	0;42
24	336	0;56
30	330	1;9
⋮	⋮	⋮
60	300	2;1
⋮	⋮	⋮
84	276	2;21
90	270	2;21
93	267	2;23
96	264	2;23
99	261	2;22
⋮	⋮	⋮
177	183	0;8
180	180	0;0

Figure 2.32
Extracts from the solar equation table

(figure 2.32). Linear interpolation allows the reader to calculate q for whatever value of a_m comes along.

It is not certain whether this method of computing $\lambda(t)$ was new with Ptolemy, although there are no other obvious candidates; only a few tables for solar motion are known before him, and they are structured differently. We have an uncertain reference to a table of solar longitudes composed by Hipparchus, and this table probably had a different configuration.[117] A recently published papyrus fragment from the ancient garbage piles at Oxyrhynchus, Egypt contains a pre-Ptolemaic table for finding solar longitudes, again designed differently, but computed trigonometrically.[118]

The Motions of the Planets

A full treatment of the motions of the Moon and the planets in the *Almagest* is beyond our scope, but a glance at the generic planetary model—the mathematical heart of the *Almagest*—will be worthwhile. The simple epicyclic

[117] See the discussions in [Jones 1991a] and [Jones 1991b, 449].

[118] See [Jones 1997, 217–224] for an analysis of the fragment. It is edited in [Jones 1999, vol. 2, 94–99] and commented upon in [Jones 1999, vol. 1, 150–151].

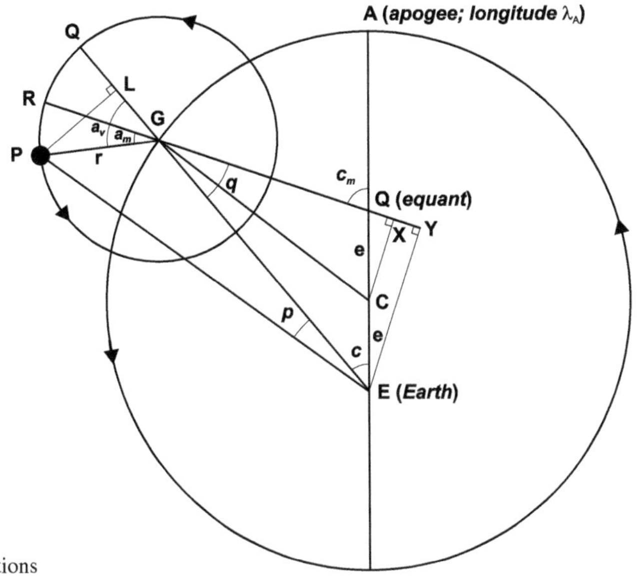

Figure 2.33
The planetary equations

model of figure 2.3 reproduces the planets' retrograde motions (figure 2.2) qualitatively. However, while it can be made to fit either the retrograde motions' *positions* or their *magnitudes*, it cannot fit both at the same time.[119] So Ptolemy modifies the model, by moving the Earth away from the center of the deferent C to E (figure 2.33), and by adding a new point, the ***equant*** Q, the same distance in the other direction from C.

The equant serves as the location for the clockwork of the mechanism; the angle at Q, the ***mean centrum*** c_m, increases at a constant rate. The planet is set in motion on the epicycle itself by having a_m increase linearly.[120] To calculate the planet's position Ptolemy must find two equations rather than one:

- the ***equation of center*** q (analogous to the solar equation), which when subtracted from c_m gives the ***true centrum*** c, the position of the center of the epicycle G; and
- the ***equation of anomaly*** p, which accounts for the position of the planet on the epicycle.

[119] For discussions of the motivation for the introduction of the equant point, see [Evans 1984], [Swerdlow 2004], [Jones 2004], and [Duke 2005c].

[120] For reasons beyond our scope here, he actually measures the planet's position on the epicycle using $a_v = a_m - q$. See [O. Pedersen 1974b, 283–285], [Neugebauer 1975, 183–184], and [Evans 1998, 376–377].

Once these are found, the planet's longitude is

$$\lambda(t) = \lambda_A + c(t) + p = \lambda_A + c_m(t) - q + p. \tag{2.29}$$

What follows is a modern account of Ptolemy's determination of the two equations. We first find $\rho = EG$, the distance from the Earth to the epicycle's center. Drop perpendiculars from E and C to GQ, extended if necessary. Then $CX = e \sin c_m$ and $QX = e \cos c_m$, and by Pythagoras

$$GX = \sqrt{GC^2 - CX^2} = \sqrt{R^2 - (e \sin c_m)^2}.$$

Then

$$GY = GX + XY = GX + QX = \sqrt{R^2 - (e \sin c_m)^2} + e \cos c_m,$$

and finally by Pythagoras applied to ΔGEY we have

$$\begin{aligned}\rho = EG &= \sqrt{GY^2 + EY^2} \\ &= \sqrt{\left(\sqrt{R^2 - (e \sin c_m)^2} + e \cos c_m\right)^2 + (2e \sin c_m)^2}.\end{aligned} \tag{2.30}$$

From here the equation of center is easy:

$$q = \angle YGE = \sin^{-1} \frac{EY}{EG} = \sin^{-1} \frac{2e \sin c_m}{\rho}. \tag{2.31}$$

As for the equation of anomaly p, we drop a perpendicular from P onto EG; then $PL = r \sin a_v$ and $GL = r \cos a_v$, where r is the radius of the epicycle. We could determine p using our knowledge of PL, $EL = EG + GL = \rho + r \cos a_v$, and an arc tangent, but again the table of chords forces us around the long way. First we find EP using Pythagoras; then

$$p = \sin^{-1} \frac{PL}{EP} = \sin^{-1} \frac{r \sin a_v}{\sqrt{(\rho + r \cos a_v)^2 + (r \sin a_v)^2}}. \tag{2.32}$$

Tabulating Astronomical Functions and the Science of Logistics

The completion of Ptolemy's planetary project requires that he provide his readers an easy means to evaluate p, and this task is harder than it was for q. By (2.32), p is obviously a function of the planet's position a_v on the epicycle. But it is also a function of the epicycle's position c_m on the deferent, since ρ varies with respect to c_m. It is one thing to tabulate a column of values of a single-argument function, but quite another to tabulate a two-dimensional array.

Ptolemy solves this problem by exploiting the fact that a_v has a much stronger effect on p than does the other argument c_m. In the plot of $p(a_v, c_m)$ in figure 2.34, the goal is to determine the height of a randomly chosen point X. Ptolemy tabulates only the four single-argument functions drawn in bold, each of which is easily done since the corresponding values of c_m are chosen strategically to simplify the calculation. He then assumes that the increase in the function from A to B to C proceeds in the same way (although with different magnitudes) as the increase in the tabulated values from D to E to F. Thus, where pairs of letters stand for differences in heights in the diagram, $AX/AB = DY/DE$. Since all points but X in this equation have been tabulated, we can solve for the height of X. This technique, now called ***Ptolemaic interpolation***,[121] was to be used throughout the medieval period and was applied to several astronomical functions.

One might wonder what all this talk of interpolation has to do with trigonometry, and indeed whether our use of functional notation unduly imposes a modern notion onto an ancient procedure. The former question is answered easily enough: what made ancient trigonometry distinct from its precursors was its ability to transform arcs to lengths (and vice versa) *systematically*. Without appropriate interpolation procedures one could not claim to do this with a table alone. So, interpolation completed the ancient trigonometric project.

Interpolation and the accompanying notion of function belonged to the somewhat murky discipline known as logistics, one of the applied branches of arithmetic. Defined vaguely as the art of practical calculation, logistics included Ptolemaic interpolation among its more advanced topics.[122] Although no treatises on logistics survive, it is referenced by several late Greek authors as one of four mathematical topics beyond the traditional quadrivium (arithmetic, geometry, music, astronomy), along with geodesy, optics, and mechanics. Our tables and interpolation methods assume familiarity with the idea of a continuous function, although perhaps "continuous" is not the right word here. Ptolemy's functions represent geometric

[121] More thorough expositions of Ptolemaic interpolation may be found in [O. Pedersen 1974a, 41–46] and [Van Brummelen 1994].

[122] For a modern account of logistics and the notion of function in Greek mathematics see [O. Pedersen 1974a]; a detailed exposition of logistical calculation may be found in [Vogel 1936], and a study of its development in Greece and relation to early Western algebra in [Klein 1934a and 1934b].

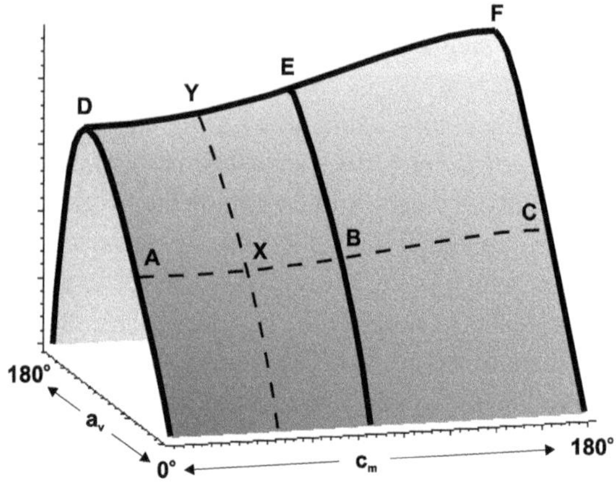

Figure 2.34
Ptolemaic interpolation

relationships, whether they happen to be continuous or not (although, of course, they generally are). It is an irony that the need to cover every geometric situation that might arise in astronomy forced ancient authors to introduce interpolation, hence the computational science of logistics—a mathematical subject removed as far as possible from the geometry that motivated it.

Trigonometry in Ptolemy's Other Works

One might get the impression from what has gone before in this chapter that ancient trigonometry is essentially what is found in the *Almagest*; and if one keeps in mind that the *Almagest* is a culmination of a period of development, this assertion is not far from the truth. There are, however, scraps of trigonometry in Ptolemy's other works. Oddly enough Ptolemy did not discover the sine law of refraction in his *Optics*; instead he postulated a scheme of second differences, reminiscent of Babylonian astronomical calculation, to model the phenomenon.[123] The *Geography*, in which Ptolemy sets himself the challenge of mapping a substantial portion of the Earth's sphere onto a flat surface, does appeal to trigonometry. He solves this problem in two different ways; the second requires an argument which will sound familiar.

The projection, illustrated in figure 2.35, works as follows. We are assumed to be watching the Earth from a point far away, directly above the parallel of the town of Soēnē in Lower Egypt on the Tropic of Cancer; thus its

[123] See [Smith 1982] for an analysis of the reasons for Ptolemy's failure, as well as a summary of the substantial literature on this topic.

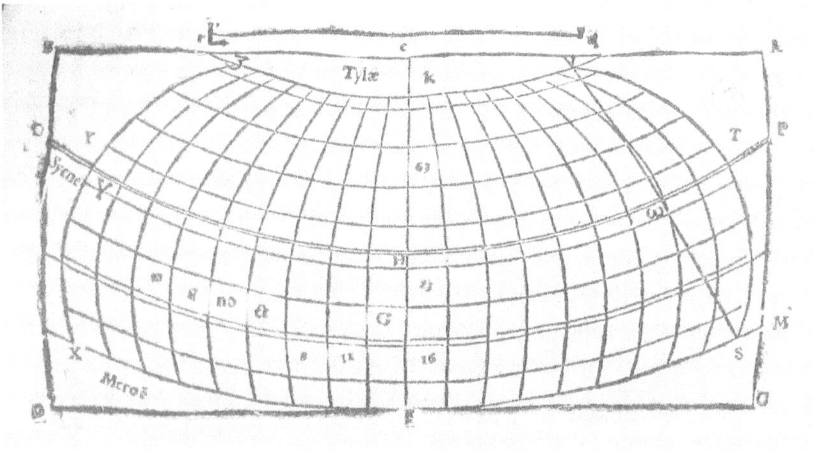

Figure 2.35 Ptolemy's second projection of the Earth's surface, from Johannes Ruremundanus's 1540 edition of Book VIII of the *Geography* (courtesy of the Burndy Library)

latitude is $\varepsilon \approx 23\frac{5}{6}°$. Where the radius $BE = 90$ units, the point nearest to us on the equator will appear to be $EZ = 23\frac{5}{6}$ units below E (figure 2.36). In order to draw a circular arc representing the equator (which must pass through B, Z and D), Ptolemy must find the center H of this circle. Once it is found, the other latitude circles will also be centered at H.

Text 2.5
Ptolemy, Constructing Latitude Arcs on a Map
(*Geography* I.24)

Now let *AEZG* and *BED* be imagined as straight lines representing arcs, such that *BE* has a ratio to *EZ* of $90{:}23\frac{5}{6}$. And let *GA* be produced, and let the center about which the circular segment *BZD* is to be described be at *H*, and let it be required to find the ratio of *HZ* to *EB*.

Let straight line *ZB* be drawn, and bisected at Θ, and let ΘH (which is of course perpendicular to *BZ*) be drawn. Then since *EZ* was assumed to be $23\frac{5}{6}$ of such [units] as straight line *BE* is 90, the hypotenuse *BZ* will be $93\frac{1}{10}$ of the same [units]. And angle *BZE* will be $150\frac{1}{3}$ of such [units; i.e., half-degrees] as two right angles are 360, and the remaining angle ΘHZ will be $29\frac{2}{3}$ of the same [half-degrees]. Consequently the ratio of *HZ* to $Z\Theta$ is $181\frac{5}{6}{:}46\frac{11}{20}$. But of such [units] as ΘZ is $46\frac{11}{20}$, straight line *BE* is 90; so that also of such [units]

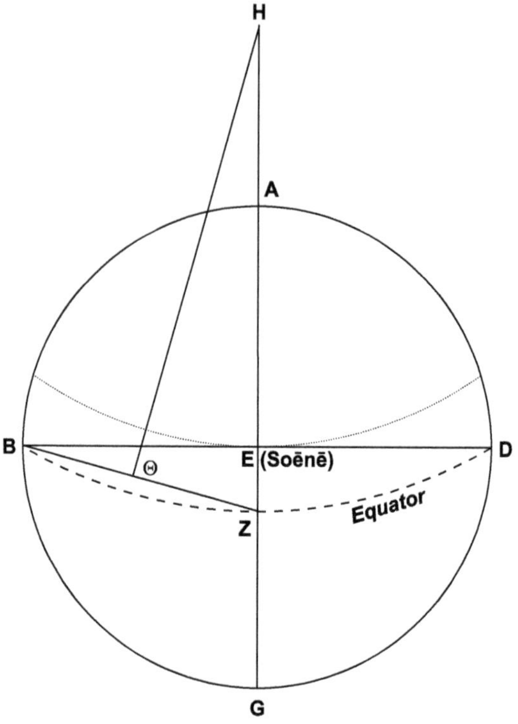

Figure 2.36
Generating Ptolemy's projection trigonometrically. The goal is to determine the equator circle; the latitude circle through Soēnē is also drawn in this image.

as straight line *BE* is 90 (and *ZE* is $23\frac{5}{6}$ of the same), we will have straight line *HZ* too as $181\frac{5}{6}$. And [we will thus obtain] point *H*, about which all the parallels in the plane map are to be described.[124]

Explanation: To understand Ptolemy's use of half-degrees, see Text 2.3.

First Ptolemy wants $\angle BZE$, which we can find immediately since $\angle EBZ = \tan^{-1}(EZ/EB)$. However, once again the chord table does not allow the equivalent of an arc tangent, so Ptolemy first computes $BZ = \sqrt{BE^2 + EZ^2} = \sqrt{90^2 + (23\frac{5}{6})^2} \approx 93\frac{1}{10}$, and then applies an arc sine: $\angle BZE = \sin^{-1}(BE / BZ) \approx 75\frac{1}{6}^\circ$ (equal to $150\frac{1}{3}^{\circ\circ}$). Then $\angle \Theta HZ = \angle EBZ \approx 14\frac{5}{6}^\circ$[125] (equal to $29\frac{2}{3}^{\circ\circ}$).

The rest of the argument proceeds without trigonometry. Since $\Delta H\Theta Z, \sim \Delta BZE,$

[124] Reprinted from [Ptolemy (Berggren/Jones) 2000, 90–91].

[125] This is true because $\Delta BZE \sim \Delta H\Theta Z$—both triangles are right, and they share another angle.

$$\frac{HZ}{Z\Theta} = \frac{BZ}{ZE} = \frac{93\frac{1}{10}}{23\frac{5}{6}}.$$

But $Z\Theta = \frac{1}{2}BZ = 46\frac{11}{20}$, so $HZ = 181\frac{5}{6}$. So H is found, and the latitude arcs may be drawn.

After Ptolemy

There is very little to say about Greek trigonometry, or indeed astronomy, after Ptolemy. One late fourth-century author, Theon of Alexandria, composed commentaries on both the *Almagest* and the *Handy Tables*, although for the most part they are disappointing to the historian—long-winded, and with little information beyond the original texts themselves. One theorem in his commentary to the *Almagest* is worth a note: the successive differences of the chords of any increasing arithmetic sequence of arcs ($<180°$) always decrease.[126] This result is equivalent to the fact that the second derivative of the chord function is negative—as is the lemma in (1.1), which Ptolemy had already used to get his estimate Crd $(1°) \approx 1;2,50$. As we shall see in the chapter on Islam, the use of chord *differences* in Theon's inequality rather than Ptolemy's chord *magnitudes* has the potential to improve the bounds around Crd $(1°)$ considerably, but Theon simply uses his result to verify Ptolemy's estimate.[127]

As far as is known, the trigonometrically-based astronomical methods found in the *Almagest* were (perhaps not surprisingly) not an immediate success with the astrologers, who continued to use arithmetical schemes as late as the fourth century AD.[128] Perhaps the more sophisticated astronomy was just too difficult to approach, or the extra precision did the astrologers little practical good. In any case, it is clear that the first real successors to Greek trigonometry were Indian, and those to pick up the torch of the *Almagest* were Islamic.

[126] See the account of this theorem in [von Braunmühl 1900, 27–28].
[127] [Rome 1931/1936/1943, vol. 2, 495].
[128] [Jones 1996, 144].

3 India

Transmission from Babylon and Greece

The roots of Indian astronomy go far back into recorded history. They are also extremely controversial, inciting both national pride and strong skepticism. Evidence of origins in many subjects is already unclear, and our information concerning transmission to India of mathematical and astronomical knowledge from other cultures is particularly spotty.[1] Ancient Vedic instructions giving the times to perform various rituals have led to a tremendous variety of interpretations concerning the extent of the astronomical knowledge encoded within them, as well as a wide range of datings. About all that we can say with some certainty is that interest in astronomy was present at least by the first millennium BC and possibly earlier, dealing especially with calendar periods. It became considerably more sophisticated and precise with the arrival of data on astronomical period relations from Babylon, probably in the fifth century BC. The arithmetic character of Babylonian astronomy (calculations of parameters, and representing continuously changing quantities with zigzag functions; see figure 1.3) was received well; as we shall see, Indian mathematicians were as skilled at computation as their Greek counterparts had been at geometric proof.

Nevertheless, Greek astronomy began to make its presence felt in India as early as the third and fourth centuries AD, partly through various trade routes. Works such as the *Romakasiddhānta* and the *Pauliśasiddhānta*[2] (lost today but known partially to us through later works such as Varāhamihira's sixth-century AD *Pañcasiddhāntikā*[3]) combined Babylonian arithmetical schemes with Greek procedures that had been based on geometry. For instance, the determination of solar eclipses in the *Romakasiddhānta* contains computations of solar and lunar positions using tables of equations of center, presumably based on Greek sources.

A remarkable fact of this transmission is that the astronomical theory received in India was not Ptolemy's, but something cruder. Most obvious is

[1] [Pingree 1978] is a starting point for an account of Indian mathematical astronomy that argues in favor of transmissions from Babylon and Greece. [Kak 2000] is a survey from the other side of the argument, taking the view that Indian astronomy is old, and more home-grown. See also [Ōhashi 2000] in the same volume. We tend to side with transmission theories, but some of the evidence is subject to different interpretations, and often the transmission is impossible to prove.

[2] The word *siddhānta* means a treatise.

[3] The *Pañcasiddhāntikā* is available in an edition, translation and commentary in [Varāhamihira (Neugebauer/Pingree) 1970/1971]; see also the analyses in [Shukla 1973] and [Shukla 1977]. [Pingree 1963, esp. 236–238] summarizes the works of this period.

the fact that the equant point in planetary models, prominent in the *Almagest*, does not arise explicitly (although see below). Hardly any documentation exists for the earliest arrival of Greek astronomical models in India, or for that matter what those Greek models would have looked like. So, it is very difficult to ascertain the extent to which what has come down to us represents transmitted knowledge, and what is original with Indian scientists. If it is all the former, we have a unique window on an otherwise-lost pre-Ptolemaic astronomy; if it is all the latter, we have the birth of a new science. The truth is probably a tangled mixture of both.[4]

The First Sine Tables

From the earliest extant works, especially Āryabhaṭa's *Āryabhaṭīya*[5] (ca. AD 500) and the *Pañcasiddhāntikā*, Indian astronomers set out to do almost everything that the Greeks had done—but with different emphases. The topics are familiar: determining solar, lunar and planetary positions; predicting eclipses; converting celestial coordinates; finding gnomon shadow lengths; and so on. From combinations of traditional chronologies and Babylonian goal-year texts, they derived planetary positions for intervals of up to billions of years since the epoch.[6] The calculations that led to these numbers may have spurred Indian interest in the mathematics of indeterminate equations. However, the main distinction between Greece and India is not in what they chose to study, but in what they chose to write. In the case of India, this is not much—many texts are simply prescriptions for calculations, given in verse for the purpose of memorization. The reasonings behind the laws often need to be inferred, and since the verses often can be interpreted in different ways, controversies inevitably result.

These calculations required a trigonometric base, and almost every Indian astronomical work had at least the rhetorical equivalent of a trigonometric table. The first surprise is that the chord function was no longer tabulated, replaced by the sine in even the most ancient of extant texts. Although no explicit comment on the substitution survives, it is not hard to see how it might have happened from some of our examples in the previous chapter. For instance, in the *Almagest* solar theory (Text 2.1), in order to find length *XE*

[4] On the Greek origins of the planetary models, see [Pingree 1971b] and [Pingree 1976]. On Āryabhaṭa in particular, [Chatterjee 1949] emphasizes his originality; [Pingree 1993] acknowledges the "Indianness" of his success in simplifying the mathematical computations while synthesizing existing theories.

[5] See [Āryabhaṭa (Shukla/Sarma) 1976].

[6] The ***epoch*** is a fixed starting point in time from which planetary positions at other times are calculated.

Ptolemy must first find $\widehat{\Theta NY}$, take the chord, and divide by two. Some early Indian astronomer, repeatedly doubling arcs and halving the resulting chords, must have realized that he could save time simply by tabulating this, and the *jyā-ardha* (or *ardha-jyā*, Sanskrit for "half-chord"), eventually shortened to *jyā*, was born.[7]

Like the chord, the *jyā* was defined as the length of a certain line segment in a circle. In figure 3.1, $\widehat{CD}$ is the argument θ and BD is the *jyā*. Thus the relation between the *jyā* and the modern sine is

$$jyā(\theta) = R\sin\theta, \tag{3.1}$$

where R is the radius of the base circle. Since this function will appear repeatedly from now on we shall represent it with the now-standard Sin θ (the capital letter signifying that the function is R times the modern one).

As early as the fifth century AD, other functions were defined and used. The *koṭijyā*, or cosine, is length AB in figure 3.1. The *utkrama-jyā* ("reversed sine"—also called the *śara*, or "arrow"), or versed sine, is length BC. This function, out of favor these days, is

$$\text{vers}\ \theta = 1 - \cos\theta. \tag{3.2}$$

As before, all functions are R times as large as their modern counterparts.

Different astronomers chose different values for the radius R (the *trijyā*[8] or *sinus totus*), although some values were more common than others.[9] One of the earliest choices may be found in the *Pañcasiddhāntikā* ("Five *Siddhāntas*"), a sixth-century work by Varāhamihira which summarizes several earlier *siddhāntas*, some of which are thought to be of Greek origin (for the trigonometry, especially the *Pauliśasiddhānta*). Although R is never defined explicitly there is no doubt from the mathematics in the text that it uses $R = 120$.[10]

Varāhamihira sets out in the standard Sanskrit verse a set of values for his Sine, tabulated in figure 3.2. The oddity of using $R = 120$ is explained by comparing Varāhamihira's table (figure 3.2) to Ptolemy's table of chords: if we divide the arguments by two and double the radius, we can transfer the chords directly to sines. Indeed, comparing with Ptolemy's values rounded to one sexagesimal place has led to the suggestion that Varāhamihira's values

[7] [Datta/Singh 1983, 40].

[8] *Trijyā* means "sine of three *signs*," i.e., three zodiacal signs of 30° each.

[9] [Gupta 1978] contains a staggering list of 25 different values of R and a discussion of how some of them transmitted to other cultures.

[10] [Varāhamihira (Neugebauer/Pingree) 1970/1971, part 1 pp. 52–57, part 2 pp. 36–37].

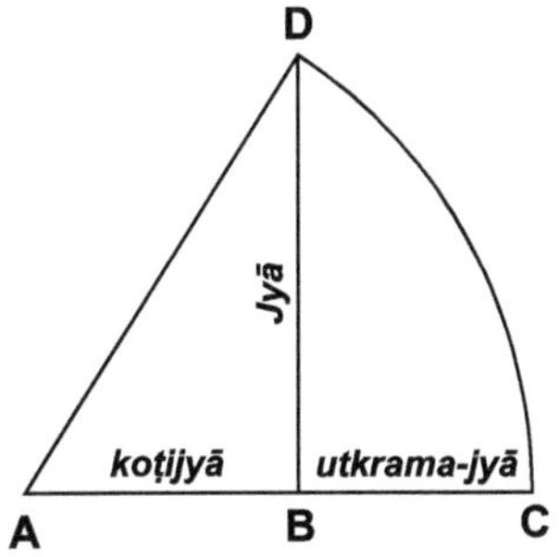

Figure 3.1
Indian trigonometric functions

in fact may have been taken from Ptolemy.[11] However, since the *Almagest*'s values are correct to the level of rounding of the *Pañcasiddhāntikā*, one cannot rule out the possibilities that the latter had an earlier source (Greek or Indian), or even that it is an original work.

Perhaps an even more peculiar feature of Varāhamihira's sine table, at least at first glance, is the step size of $3°45'$. This structure is standard in early Indian tables (for instance, see Āryabhaṭa's table also in figure 3.2), and we have seen it before. The reconstruction of Hipparchus's chord table in figure 2.6 uses a step size of $7\frac{1}{2}°$, which we have seen is easy to complete with a minimum of trigonometric tools. Converted to sines where the arcs are half as large, the Hipparchian process leads to a table with steps of $3\frac{3}{4}°$.

Many Indian sine tables also use $R = 3438$, which as we saw in the previous chapter results from measuring the circumference of the circle in $360 \cdot 60 = 21{,}600$ minutes. This value is found, for example, in the sine table in Āryabhaṭa's *Āryabhaṭīya* (ca. AD 499)[12] and in many important works that followed. As can be seen in figure 3.2 it holds the significant advantage that sines of small arcs are almost equal to the arcs themselves (since $3;45° = 225'$),[13] which is also a motive for the modern use of radian measure.

Whether or not the similarity between the early Indian sine tables and Hipparchus's reconstructed chord table proves transmission from Greece to

[11] [Yano 1977, 84–85]; see also [Bag 1990] for a similar comparison. Seventeen of the twenty-four entries in Varāhamihira's table match rounded chord values from the *Almagest* if we round up when the last place is 30. Also, seventeen entries in Varāhamihira's table are correct to both sexagesimal places.

[12] Two of the most widely available translations of the *Āryabhaṭīya* are in [Shukla 1976] and [Clark 1930], both of which contain extended commentaries. See also [Rodet 1879, 399, 410–413], which contains a translation into French of parts of the relevant passage and one of the many interpretations of Āryabhaṭa's difference methods for computing sines.

[13] [Woepcke 1854b] asserted that the step size of $225'$ was chosen deliberately as the largest convenient arc for which Sin $x = x$ when rounding to minutes of arc, but the possible link to Hipparchus and the natural division of the right angle into 24 segments are better explanations.

Figure 3.2
Varāhamihira and Āryabhaṭa's sine tables

θ (degrees)	Sin θ (Varāhamihira)	Sin θ (Āryabhaṭa)
3;45	7;51	225
7;30	15;40	449
11;15	23;25	671
15	31;4	890
18;45	38;34	1105
22;30	45;56	1315
26;15	53;5	1520
30	1,0;0	1719
33;45	1,6;40	1910
37;30	1,13;3	2093
41;15	1,19;7	2267
45	1,24;51	2431
48;45	1,30;13	2585
52;30	1,35;12	2728
56;15	1,39;46	2859
60	1,43;55	2978
63;45	1,47;37	3084
67;30	1,50;52	3177
71;15	1,53;37	3256
75	1,55;55	3321
78;45	1,57;42	3372
82;30	1,58;59	3409
86;15	1,59;44	3431
90	2,0;0	3438

India is a thorny question. The case for transmission suggested initially by the structure of the Indian tables recently has been bolstered long after its initial discrediting.[14] Given other evidences of astronomical transmission, the circumstantial evidence is now compelling. However, since the values in any reconstructed Hipparchan table do not come directly from any historical document, the entries in the tables themselves cannot be compared. Also, given that the circle divided into 360°, it seems plausible (perhaps almost

[14] See Chapter 2, note 30.

inevitable) that some Indian mathematician might have arrived independently at the same 24-part division of the 90° quadrant. So it is likely that some elements of Greek chord tables found their way to India (the 360° circle, the $3\frac{3}{4}^\circ$ step size, and $R=3438$), where they were converted to sines. As for the tables themselves, there is simply no way to decide whether the sine values were merely borrowed or computed independently in India.

Āryabhaṭa's Difference Method of Calculating Sines

In fact, a uniquely Indian approach to tabulating sines is found in some of the earliest works. Beginning with the assumption that the first entry in the table is $\sin(225') = 225$, various authors give schemes (in verse) prescribing the successive computation of the remaining 23 sine values, one at a time, using only those that have already been found. Rules of this type may be found in several texts, such as the *Paitāmahasiddhānta*[15] ("*Siddhānta* of *Pitāmaha*," ca. sixth century AD) and in the later *Sūryasiddhānta* ("*Siddhānta* of the Sun").[16]

A similar prescription in the *Āryabhaṭīya* has attracted considerably more attention, partly due to differing understandings of the method and its origin. We follow the most recent published interpretation, which comes with its own translation of the crucial passage. It is based on an explanation by the much later fifteenth-century Indian scientist Nīlakaṇṭha, who also does us the rare favor of accompanying the computation with a geometric derivation.[17]

[15] A short account of the sine difference method of the *Paitāmahasiddhānta* is in [Pingree 1978, 559], and a translation of a corrupt version of the text is in [Pingree 1967/68].

[16] The *Sūryasiddhānta* has an extremely complicated history of revisions and transmissions; the earliest version is at least as old as the fifth century AD. The introduction to [Shukla 1957] (a version based on Parameśvara's fifteenth-century commentary) goes into some detail on its history. Descriptions of the differences method for computing sines in the *Sūryasiddhānta* are in [Burgess 1860, 52–57] and [Chakravarty 2001, 3–6]. Apart from the *Sūryasiddhānta*, the method is described in a number of places; one accessible source is [Kennedy 1969b, 346–347].

[17] The interpretation we follow is by [Hayashi 1997], who also summarizes thirteen different published understandings of Āryabhaṭa's rather vague prescriptions. The mathematics that follows is based on Hayashi's paper. For some other takes see [Rodet 1879, 412–413], [Ayyangar 1923], [Naraharayya 1924, 107–108], [Gupta 1972a], and [Shukla/Sarma 1976, 51–54]. The latter also applies Nīlakaṇṭha, whose writing on the subject was published in [Āryabhaṭa (Śāstrī) 1931]. [Gupta 1971] describes a ninth-century improvement of Āryabhaṭa's tabular differences done by extending the numeric values to fractional parts, leading to $R = 3437;44,19$ instead of 3438.

Text 3.1
Āryabhaṭa, Computing Sines
(from the *Āryabhaṭīya*)

When the second half-[chord] partitioned is less than the first half-chord, which is [approximately equated to] the [corresponding] arc, by a certain amount, the remaining [sine-differences] are less [than the previous ones] each by that amount of that divided by the first half-chord.[18]

Explanation: Especially considering that the words in square brackets are modern insertions, it is no wonder that there are so many ways to read this passage!

In this discussion we divide into 24th parts of a right angle (hence $\mathrm{Sin}(1) = R\sin 3;45°$), and we write the first differences of sines as $\Delta(i) = \mathrm{Sin}(i) - \mathrm{Sin}(i-1)$. In figure 3.3 the 24 quadrant division points are the A_i's, while the midpoints are the B_i's. The sines that Āryabhaṭa seeks are the lengths P_iA_i; the distances $A_i A_{i+1} = B_i B_{i+1}$ are all equal to Crd(1).

Since the chord A_iA_{i+1} is perpendicular to OB_{i+1}, $\Delta A_i CA_{i+1} \sim \Delta\, OT_{i+1} B_{i+1}$; therefore

$$\frac{A_{i+1}C}{A_iA_{i+1}} = \frac{OT_{i+1}}{B_{i+1}O}, \text{ or } \frac{\Delta(i+1)}{\mathrm{Crd}(1)} = \frac{OT_{i+1}}{R}.$$

This gives us an equation for the first difference,

$$\Delta(i+1) = \frac{\mathrm{Crd}(1)}{R} \cdot OT_{i+1}. \tag{3.3}$$

If we apply (3.3) twice (with i and $i+1$ as arguments), we arrive at an equation for the second difference:

$$\Delta(i) - \Delta(i+1) = \frac{\mathrm{Crd}(1)}{R} \cdot (OT_i - OT_{i+1}) = \frac{\mathrm{Crd}(1)}{R} \cdot T_iT_{i+1}. \tag{3.4}$$

But what is T_iT_{i+1}? The similarity of ΔB_iDB_{i+1} and ΔA_iP_iO gives us the answer:

$$\frac{B_iD}{B_iB_{i+1}} = \frac{A_iP_i}{A_iO}, \text{ so } T_iT_{i+1} = B_iD = \frac{\mathrm{Crd}(1)}{R} \cdot \mathrm{Sin}(i). \tag{3.5}$$

Substituting (3.5) into (3.4) gives

[18] From [Hayashi 1997, 399].

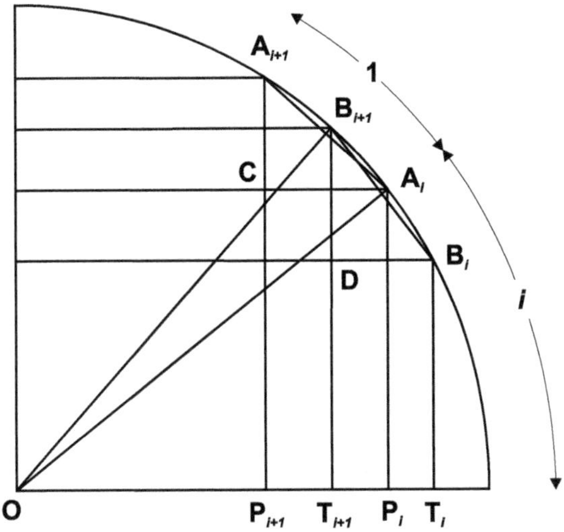

Figure 3.3
Diagram for the proof of Āryabhaṭa's procedure for sine differences

$$\frac{\Delta(i)-\Delta(i+1)}{\mathrm{Sin}(i)}=\left[\frac{\mathrm{Crd}(1)}{R}\right]^2; \tag{3.6}$$

thus the left side is independent of i.[19] Whether or not Āryabhaṭa knew this surprising equation is a matter for debate. Nīlakaṇṭha did; he noticed that applying it twice (for $i = 1$ and for arbitrary i) gives

$$\frac{\Delta(i)-\Delta(i+1)}{\mathrm{Sin}(i)}=\frac{\Delta(1)-\Delta(2)}{\mathrm{Sin}(1)}, \text{ or } \Delta(i)-\Delta(i+1) = [\Delta(1)-\Delta(2)]\cdot\frac{\mathrm{Sin}(i)}{\mathrm{Sin}(1)}. \tag{3.7}$$

From (3.7) all sines may be generated in sequence, since $\Delta(i+1)$ is the only quantity not known immediately from Sin(1), . . . , Sin(i); and $\mathrm{Sin}(i+1) = \mathrm{Sin}(i) + \Delta(i+1)$.

This is rather abstract; a numerical example will help to intuit the method in practice. Suppose we have found the first four sines in figure 3.2, and need to find the fifth. Then, according to (3.7), the quantity that we need to increase from 890 to the new sine differs from the previous sine-difference $(890 - 671 = 219)$ by

[19] Since the numerator of the left side of (3.6) is a second difference of sines, this difference equation is analogous to the fact that the second derivative of the sine function is the negative sine.

$[\Delta(1)-\Delta(2)]\cdot\frac{\text{Sin}(i)}{\text{Sin}(1)}=[225-224]\cdot\frac{\text{Sin}(4)}{\text{Sin}(1)}=\frac{890}{225}\approx 4.$ The new sine-difference is thus $219-4=215$, and so the fifth sine is $890+215=1105$. With some generosity in our interpretation, we can see that this pattern of calculation corresponds to the prescription in Āryabhaṭa's verse.

The ingenuity of this method is underlined by the fact that once the constants Sin(1) and $\Delta(1)-\Delta(2)$ are given, the results it generates are, at least in theory, not approximate but exact. As the sample calculation illustrates, it is also surprisingly efficient—although due to the way one value is calculated from another, sine values found from this method tend to decrease in accuracy as the computation continues.

It is quite possible that Āryabhaṭa and others discovered their difference formulas empirically, simply by observing patterns in the differences found in geometrically-generated sine values.[20] In fact, we must also admit the possibility that Āryabhaṭa simply borrowed the method from an even earlier source, Indian or Greek. We leave it to the reader to decide whether or not it is plausible that the above argument could have been found on its own at Āryabhaṭa's time. Difference schemes like this depend only on basic geometrical theorems and so would have been accessible to any culture familiar with Euclid.[21] But as far as we know they were not rediscovered until François Viète's work in the late sixteenth century—whereupon they transformed the construction of trigonometric tables.

Bhāskara I's Rational Approximation to the Sine

Even more audacious was Bhāskara I's method for computing the sine using no geometry whatever. Little is known about him, although the fact that he wrote a commentary on Āryabhaṭa's work[22] indicates that he worked in the same astronomical tradition. One of his best known works is the *Mahābhāskarīya* ("Great [Work of] Bhāskara," around AD 600), which describes a

[20] The idea that early Indian astronomers began working with previously existing sine tables and found difference patterns by experiment has been around for a long time; see for instance [Davis 1799, 245–249].

[21] [Playfair 1798, 97–100] discusses the relation between the scheme in the *Sūryasiddhānta* and Proposition 93 of Euclid's *Data*.

[22] On these commentaries see the series of four articles [Shukla 1971–1972], as well as [Keller 2006]. For a survey of Bhāskara I's works see [Pingree 1970–1994, vol. A4, 297–299]. Bhāskara's available treatises have been edited and translated in a series of three books published by Lucknow University, including the *Mahābhāskarīya* [Bhaskara I (Shukla) 1960].

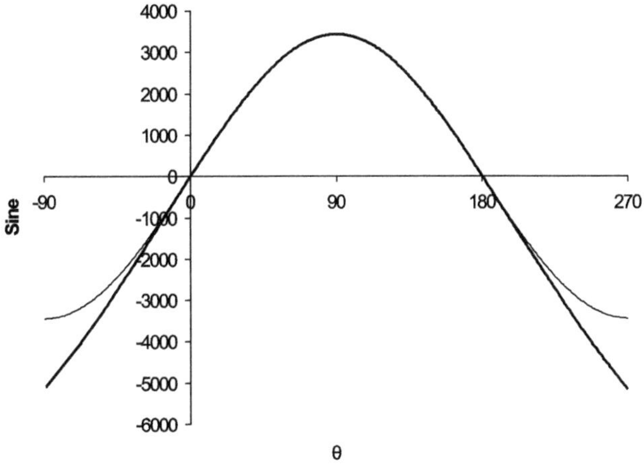

Figure 3.4
The Sine function (thin curve) and Bhāskara's rational approximation (thick curve)

number of innovations, not least of them a remarkable approximation to the sine—a ratio of two quadratic functions.

As usual Bhāskara states his rule for computing sines without any explanation; in modern notation, he asserts that

$$\operatorname{Sin}\theta = R \cdot \frac{4\theta(180-\theta)}{40{,}500-\theta(180-\theta)}. \tag{3.8}$$

This function is extremely close to the Sine over the range of values $0 \leq \theta \leq 180°$ (and better than the obvious parabolic approximation $(R/8100) \cdot \theta(180-\theta)$)[23]; the graphs are almost indistinguishable (see figure 3.4). The sine values are of course approximate, but they never err by more than six from the correct values (in units where $R = 3438$; compare with figure 3.2).

Bhāskara's approximation was picked up and occasionally modified by many subsequent Indian astronomers, as late as the sixteenth century;[24] however, they do not answer the question that presses the modern reader: how did Bhāskara arrive at it? A number of explanations have been proposed, including for instance assuming that the sine function is a ratio of two quadratics, and solving for these quadratics' coefficients by substituting known sine values.[25]

[23] This formula has roots at 0 and 180°; the $R/8100$ term scales it to reach a maximum value of R.

[24] [Gupta 1967] and [Gupta 1972b] provide quotations of versions of Bhāskara's formula from several works.

[25] [Gupta 1967] contains five explanations; [Gupta 1986] improves some of the arguments. See [Naimpally 1987] for a similar derivation. [Ayyangar 1950] argues that Bhāskara's formula is the best ratio of two quadratic expressions to represent the sine by forcing an arbitrary such ratio to satisfy certain properties of the sine function, including certain commonly known sine values.

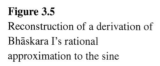

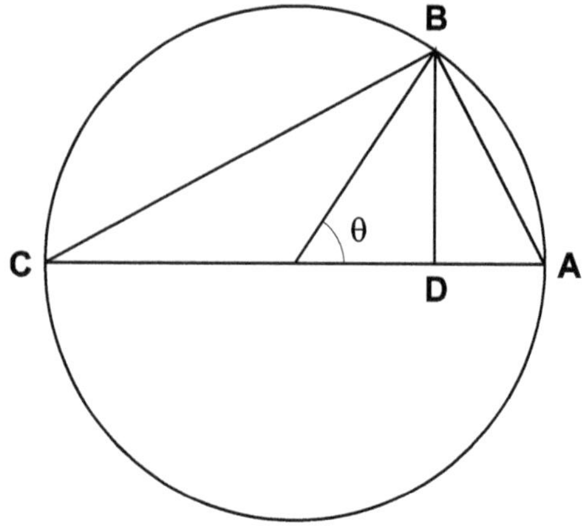

Figure 3.5
Reconstruction of a derivation of Bhāskara I's rational approximation to the sine

However, the most likely explanation (and the only one based on geometry) is as follows.[26]

In figure 3.5 the diameter AC is equal to $2R$. The area of ΔABC is equal to both $\frac{1}{2}AB \cdot BC$ and $\frac{1}{2}AC \cdot BD$, so $AB \cdot BC = AC \cdot BD$. But $BD = \text{Sin}\,\theta$, so

$$\text{Sin}\,\theta = BD = \frac{AB \cdot BC}{AC} < \frac{\overset{\frown}{AB} \cdot \overset{\frown}{BC}}{AC} = \frac{\overset{\frown}{AB} \cdot \overset{\frown}{BC}}{2R} = \frac{\theta(180^\circ - \theta)}{2R}. \qquad (3.9)$$

The replacement of sines with their corresponding arcs would have been familiar, since sines and arcs were measured in the same units. We convert the resulting inequality to an equality by considering the reciprocal of (3.9) and inserting two unknowns x and y:

$$\frac{1}{\text{Sin}\,\theta} = x\left(\frac{2R}{\theta(180^\circ - \theta)}\right) + y. \qquad (3.10)$$

Substituting the two most obvious sines, $\text{Sin}\,30^\circ = R/2$ and $\text{Sin}\,90^\circ = R$, into (3.10) results in a pair of linear equations for x and y. Their solution, and substitution back into (3.10), leads precisely to (3.8). We leave the details to the interested reader.

If this train of reasoning is indeed how Bhāskara came up with his approximation, it remains, in part, a puzzle. If he had used $\text{Sin}\,\theta$ in (3.10) rather

[26] This explanation was originally posed in [Inamdar 1950] and enhanced by Shukla in [Bhāskara I (Shukla) 1960, 208]; it is the first of [Gupta 1967]'s five approaches. See also [Hayashi 1990] for a speculation that Bhāskara's formula is ultimately related to Āryabhaṭa's trigonometry.

than its reciprocal, his approximation would have been only a quadratic function rather than a ratio of quadratics, and the result would have been inferior. The insertion of the parameters x and y does reflect Indian mathematical practice elsewhere. In any case, the early Indian reticence to provide motive leaves us, as often, with only this plausible reconstruction.

Improving Sine Tables

As clever as Āryabhaṭa's and Bhāskara's methods were, a sine table with entries only for every 3;45° of arc rounded only to the nearest integer (with $R = 3438$) leaves ample room for improvement. Ptolemy's method for constructing a chord table has significant advantages. One of these is the use of equivalents of sin 18° and sin 36°; this leads to the sines of all multiples of 3°, a grid of 30 entries over a span of 90° rather than the Indians' more cumbersome 24 entries. Furthermore, Ptolemy's approximation of an equivalent to sin 1° allowed him to find the sine of every degree.

The sine of 18° entered Indian mathematics in the twelfth century, in the *Jyotpatti* of Bhāskara II. One of the best and most popular scientists of medieval India,[27] Bhāskara II wrote a number of important works. If the number of editions and commentaries is any guide, his most widely read were the astronomical *Siddhāntaśiromaṇi* ("Crest-Jewel of *siddhāntas*") and the mathematical *Līlāvatī* (on arithmetic, algebra and geometry), traditionally considered to be part of the *Siddhāntaśiromaṇi*. The *Jyotpatti* ("Rationale of Sines") is a trigonometric appendix to the *Siddhāntaśiromaṇi*;[28] if one thinks of it as a separate work, then it is the first composition devoted solely to trigonometry in India. There is some reason to think this way: uncharacteristically for the *Siddhāntaśiromaṇi* it contains a number of results but no justifications. This has led to some speculation that the material contained within it is foreign,[29] perhaps transmitted from Islam.

In modern notation the *Jyotpatti* prescribes $\mathrm{Sin}\,18^\circ = \left(\sqrt{5R^2} - R\right)\Big/4$, equivalent to $\sin 18^\circ = (\sqrt{5} - 1)/4$, and $\sin 36^\circ = \sqrt{(5 - \sqrt{5})/8}$;[30] both values are correct. Bhāskara II gives no clue how he arrived at them, and this omission led to a number of derivations and explanations (not all correct) in subsequent commentaries. Some of these inferences are geometric; others

[27] The list of manuscripts of Bhāskara II's works in Pingree's *Census* [Pingree 1970–1994, vol. A4, 299–326] extends to twenty-seven pages!

[28] The *Siddhāntaśiromaṇi*, including the *Jyotpatti*, is available in English translation in [Wilkinson 1861]. See [Pingree 1970–1994, vol. A4, 318–319] for its extensive publication history.

[29] M. M. Sudhakara Dvivedi, quoted in [Gupta 1980, 45].

[30] [Yano 1997, 990].

are algebraic; two are similar to derivations in Euclid and Ptolemy.[31] Perhaps one or more of these methods reflect Bhāskara II's thoughts; once again, we simply do not know.

From here Bhāskara II needs only two steps to build a sine table comparable in quality to Ptolemy's chord table. The first step, an approximation to Sin 1°, is a much easier chore than Ptolemy's: if Sin 225′≈225, then even more accurately Sin 60′≈60. (Indeed, this approximation is actually a little better than Ptolemy's Crd 1°≈1;2,50.) The second, a method of finding the sines of sums and differences of angles, allows Bhāskara II to complete a sine table for every degree from 1° to 90°. His rule, given in the *Jyotpatti* in verse, is as follows:

> The Sines of the two given arcs are crossly multiplied by (their) Cosines and (the products are) divided by the radius. Their (that is, of the quotients obtained) sum is the Sine of the sum of the arcs; their difference is the Sine of the difference of the arcs.[32]

This is

$$\mathrm{Sin}(\alpha \pm \beta) = \frac{\mathrm{Sin}\,\alpha \cdot \mathrm{Cos}\,\beta}{R} \pm \frac{\mathrm{Cos}\,\alpha \cdot \mathrm{Sin}\,\beta}{R}, \tag{3.11}$$

which is in turn equivalent to the modern formula. This first Indian appearance of the sine sum and difference laws is yet again unaccompanied by justification. The analogous cosine laws are not in the *Jyotpatti*; according to the seventeenth-century *Marīci* commentary by Munīśvara, the reason for this exclusion is that once Sin $(\alpha \pm \beta)$ are known, Cos $(\alpha \pm \beta)$ are more easily found simply using the Pythagorean Theorem.[33]

The later commentaries include a variety of derivations of the sine and cosine addition laws, some of which are geometrical; one is based on Ptolemy's Theorem. However, we prefer to show a striking illustration of the number-theoretic aspect of Indian mathematics (in the *Marīci* commentary on *Jyotpatti*). It relies on a lemma from indeterminate analysis, a subject of great fascination in India.[34]

Lemma: If x_1, y_1 is a solution to the equation $Nx^2 + a = y^2$, and x_2, y_2 is a solution of $Nx^2 + b = y^2$, then $x_1y_2 + x_2y_1$, $y_1y_2 + Nx_1x_2$ is a solution of $Nx^2 + ab = y^2$.

[31] See [Gupta 1976b] for a translation of the relevant section of the *Jyotpatti*, as well as summaries of the various derivations of sin 18° in the commentaries.

[32] Translated in [Gupta 1974a, 165] from [Bhāskara II (Sastri) 1929, 283].

[33] [Gupta 1974a, 165]; the *Marīci* commentary is summarized in [Gupta 1980].

[34] The following derivation is based on [Gupta 1974a, 167–169]. See also [Datta/Singh 1935/1938, vol. 2, 141–181] for the lemma, related results, and their appearances in the works of other writers.

The relevance of this lemma to our situation becomes clear if one sets $N = -1$ and $a = b = R^2$, which turns the indeterminate equations into the Pythagorean Theorem. In this case both $x_1 = \text{Sin}\,\alpha$, $y_1 = \text{Cos}\,\alpha$ and $x_2 = \text{Sin}\,\beta$, $y_2 = \text{Cos}\,\beta$ satisfy the conditions of the lemma. Substitution gives us solutions

$$x = \text{Sin}\,\alpha \cdot \text{Cos}\,\beta + \text{Cos}\,\alpha \cdot \text{Sin}\,\beta \tag{3.12}$$

and

$$y = \text{Cos}\,\alpha \cdot \text{Cos}\,\beta - \text{Sin}\,\alpha \cdot \text{Sin}\,\beta \tag{3.13}$$

to the equation $x^2 + y^2 = R^4$. Dividing the latter through by R^2 results in the Pythagorean Theorem once again—$(x/R)^2 + (y/R)^2 = R^2$. The leap to the conclusion that x/R and y/R represent $\text{Sin}(\alpha + \beta)$ and $\text{Cos}(\alpha + \beta)$ respectively seems to have been taken with confidence.

Other Trigonometric Identities

The sum and difference formulas are vital to building trigonometric tables finer than the traditional 24 entries per 90°. But once one has the formulas, they also can be used to generate many other identities. In particular, formulas for Sin 2θ, Cos 2θ, Sin 3θ, Cos 3θ, and higher multiples may be generated simply by writing $n\theta = \theta + \theta + \cdots + \theta$ and applying the sum formulas repeatedly. This was done by the later astronomer Kamalākara in his *Siddhānta-Tattva-Viveka* (1658) up to the sine and cosine of 5θ; he quotes Bhāskara II (who clearly knew that this could be done) for the addition and subtraction laws.[35] Kamalākara's sine triple-angle formula, for instance, was

$$\text{Sin}\,3\theta = \text{Sin}\,\theta\left(3 - \frac{(\text{Sin}\,\theta)^2}{(\text{Sin}\,30°)^2}\right), \tag{3.14}$$

equivalent to the modern formula

$$\sin 3\theta = 3\sin\theta - 4\sin^3\theta; \tag{3.15}$$

while his cosine quintuple angle formula was

$$\text{Cos}\,5\theta = \frac{5(\text{Sin}\,\theta)^4\text{Cos}\,\theta - 10(\text{Sin}\,\theta)^2(\text{Cos}\,\theta)^3 + (\text{Cos}\,\theta)^4(\text{Cos}\,\theta)}{R^4}, \tag{3.16}$$

equivalent to

[35] A full account of Kamalākara's work on multiple angle formulas may be found in [Gupta 1974c].

$$\cos 5\theta = 5\sin^4\theta\cos\theta - 10\sin^2\theta\cos^3\theta + \cos^5\theta. \tag{3.17}$$

The identity (3.14)/(3.15) has special significance, since it may be used to get an accurate estimate of sin 1° from sin 3°—provided that one is able to solve cubic equations. As we shall see, this method was first applied by the Persian astronomer Jamshīd al-Kāshī in the early fifteenth century to generate an extremely accurate value of sin 1° (using an iterative scheme to solve the cubic), leading eventually to the impressive sine table in Ulugh Beg's *Sulṭānī Zīj*. Kamalākara, also interested in trisecting arcs, chose another path. Rather than solving the cubic, Kamalākara took (3.15) and replaced $\sin^3\theta$ with $[(\sin 3\theta)/3]^3$. Solving for $\sin\theta$ and setting $\alpha = \theta/3$, this substitution leads to an approximate formula, but an extremely serviceable one for small angles:

$$\sin\frac{\alpha}{3} \approx \frac{1}{3}\sin\alpha - \frac{4}{81}\sin^3\alpha.^{36} \tag{3.18}$$

Now that Kamalākara had an improved formula for $\sin\alpha/3$, he could apply it to (3.15) rather than the cruder $[(\sin 3\theta)/3]^3$, ending up with an even better formula for $\sin\alpha/3$. Repeating iteratively, he could derive formulas as accurate as he wished.[37]

Of course some of the "submultiple" angle formulas were known from the beginning; as we have seen, an equivalent to the sine half-angle formula goes back to ancient Greece. India's introduction of new trigonometric functions had already produced alternate formulations of basic identities like this, easier to express and use. Already in the sixth-century *Pañcasiddhāntikā* of Varāhamihira, we find a half-angle formula in the directions for building a sine table, immediately after he gives values for Sin 30° and Sin 60°:

Text 3.2
Varāhamihira, a Half-angle Formula
(from the *Pañcasiddhāntikā*)

When the remaining [Sines] are desired, the radius is diminished by the Sine of the remainder of the subtraction of twice the arc from a quadrant; the square of half of that [remainder] is to be added to the square of half [the Sine] of double [the arc].

The square root of that is the desired Sine.[38]

[36] [Datta/Singh 1983, 63–64]; Kamalākara applied a similar technique to arrive at an approximate one-fifth angle formula. Other late Indian mathematicians such as Munīśvara (1627) (see [Majumdar 1978]) and Ghulam Husain Jaunpuri (1790) (see [Rizvi 1984]) also recognized the utility of the triple angle formula for computing sines.

[37] For text, translation, and analysis of Kamalākara's scheme, see [Gupta 1974d].

[38] From [Varāhamihira (Neugebauer/Pingree) 1970/1971, part 1, 53].

Explanation: The goal of this passage is to determine the first sine in the table—of 3;45°—which requires three successive halving operations from 30°. The clause up to the semicolon tells us to compute $R - \text{Sin}(90° - \theta)$, i.e., the versed sine of θ. The rest of the sentence squares $\text{Sin}(\theta)/2$ and applies the Pythagorean Theorem. So Varāhamihira's formula is

$$\text{Sin}^2\left(\frac{\theta}{2}\right) = \left(\frac{\text{Vers}\,\theta}{2}\right)^2 + \left(\frac{\text{Sin}\theta}{2}\right)^2. \tag{3.19}$$

This is easy to verify. In figure 3.6 $BD = \text{Sin}\ \theta$ and $BC = \text{Vers}\ \theta$, so the two terms inside the squares on the right side of (3.19) are equal to EC and EF respectively. Thus the Pythagorean Theorem is being applied to ΔCEF. Since F is the midpoint of CD, joining AF makes it obvious that $CF = \text{Sin}\ (\theta/2)$.[39]

(3.19) is equivalent to the well-known modern trigonometric identity

$$2\sin^2(\theta/2) = 1 - \cos\theta. \tag{3.20}$$

The critical role played by half-angle formulas in the computation of sine tables should be clear by now, so it is no surprise that variants are found in many Indian astronomical texts.[40]

We have seen before that it is often possible to "find" trigonometric identities seemingly applied within geometric procedures where they are not expected.[41] One Indian example is in Brahmagupta's *Khaṇḍakhādyaka* ("Eating Candy," AD 665), in which we find an apparent use of the Law of Sines within its planetary theory.

Text 3.3
Brahmagupta, the Law of Sines in Planetary Theory?
(from the *Khaṇḍakhādyaka*)

Multiply the 'sine' of the [*Śīghra*] anomaly by the 'sine' of the maximum *Śīghra* equation and divide by the 'sine' of the corresponding *Śīghra* equation; the result is the '*Śīghra* hypotenuse'. When the [*Śīghra*] anomaly is half a circle

[39] Although my explanation is mathematically equivalent to [Varāhamihira (Neugebauer/Pingree) 1970/1971, part 2, 36–37] and [Datta/Singh 1983, 50–51], it may represent the original author's thought process more accurately. Neugebauer/Pingree and Datta/Singh derive (3.19) but do not explain the way the formula is presented, and they appeal unnecessarily to the chord function.

[40] [Datta/Singh 1983, 50–56] describes a number of them.

[41] For instance the appearance of the Law of Cosines in *Elements* II.13 and the formula $\tan a = \sin b \tan A$ within the proof of a theorem in Theodosius's *Spherics*.

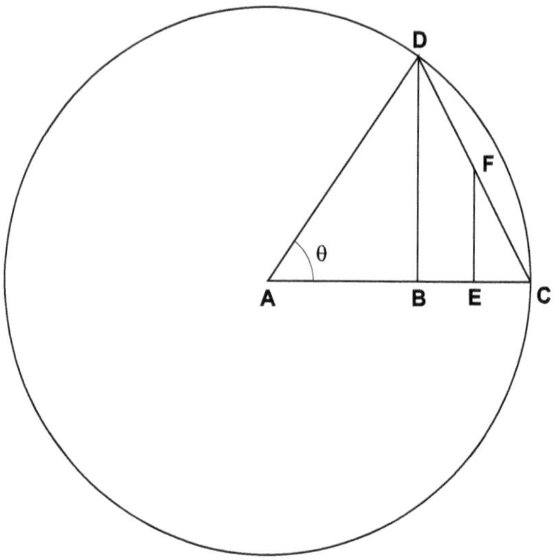

Figure 3.6
A derivation of Varāhamihira's Sine half-angle formula

> this *Śīghra* hypotenuse is equal to the radius diminished by the 'sine' of the maximum equations; when the anomaly is equal to the whole circle, the same is equal to the radius increased by the same 'sine' of the maximum equation.[42]

Explanation: (See figure 3.7.) We shall deal with Indian planetary theory later in this chapter. For now we need to know only that the *śīghra* is one of two epicycles in the standard model, $a = \angle PMK$ is its anomaly (the angle that locates the planet on the epicycle), and $\sigma = \angle PEM$ is the planet's displacement caused by the *śīghra* epicycle.

When the *śīghra* equation is at its maximum $\sigma = \sigma_{max}$, $\angle EPM$ is a right angle. If *EM* is set to be equal to the radius *R*, then *PM*, the radius of the epicycle, is the Sine of σ_{max}. So, our text prescribes

$$\frac{\text{Sin}\, a \cdot PM}{\text{Sin}\, \sigma} = EP, \tag{3.21}$$

the distance from the Earth to the planet. Since Sin a = Sin(180° - a), this is equivalent to applying the Law of Sines to ΔEPM.

Whether or not this apparent use of the Law of Sines allows us to claim independent knowledge of it for Brahmagupta is a delicate matter, which we shall discuss further in the section in Chapter 5 on the Venetian *marteloio* system of navigation.

[42] [Brahmagupta (Sengupta) 1934, 115]. A couple of punctuation marks have been corrected.

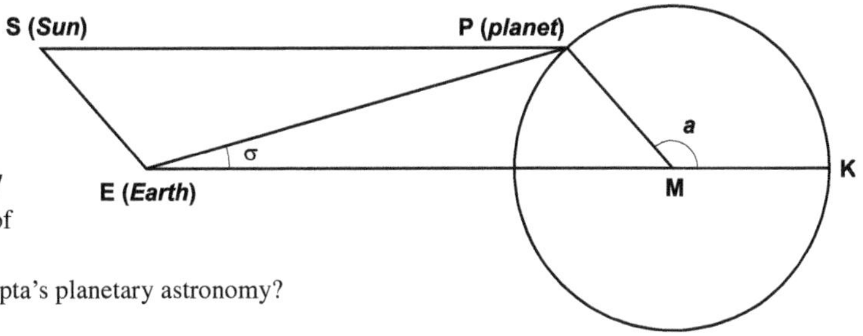

Figure 3.7 The Law of Sines in Brahmagupta's planetary astronomy?

Brahmagupta's Second-order Interpolation Scheme for Approximating Sines

Brahmagupta's accomplishments did not end with the Law of Sines. One of his more remarkable mathematical achievements was a method that began with a sine table with a mere six entries, and computed the sine of any given arc in a single calculation, with surprising accuracy.[43] His approach boils down to a second-order finite difference interpolation scheme, mathematically equivalent to the modern second-order Newton-Stirling formula; it may be the earliest use of finite difference interpolation. Second order schemes are often explained as a fitting of parabolic arcs between the given data points. We shall resist that temptation, and instead follow Bhāskara II's explanation of Brahmagupta's thinking. This couplet describing the procedure comes from Brahmagupta's *Dhyānagraha*, an early work; it would be repeated in his later and more well-known *Khaṇḍakhādyaka*.

Text 3.4
Brahmagupta, Interpolating Sines
(from the *Dhyānagraha*)

Multiply half the difference of the tabular differences crossed over and to be crossed over by the residual arc and divide by 900′. By the result [so obtained] increase or decrease half the sum of the same (two) differences, according as this [semi-sum] is less or greater than the difference to be crossed over. We get the true functional difference to be crossed over.[44]

[43] Brahmagupta's interpolation method is examined in [Sengupta 1931a] and [Gupta 1969]. See also [Gupta 1979b] for a seventeenth-century modification by Munīśvara that applied iteration and often led to improved results.
[44] From [Gupta 1969, 88].

θ	Sin θ
0°	0
15°	39
30°	75
45°	106
60°	130
75°	145
90°	150

Figure 3.8
Brahmagupta's sine table

Explanation: In the *Khaṇḍakhādyaka* Brahmagupta works with the very small sine table of figure 3.8, which uses the somewhat common $R = 150$.[45] Suppose we wish to find Sin 39°. Bhāskara II argues that we need a difference between sine values separated by $15°(=900')$ that corresponds to $\theta = 39°$; the difference of 31 between Sin 30° = 75 and Sin 45° = 106 is a quantity more appropriate to the sine's rate of change in the middle of the interval ($\theta = 37\frac{1}{2}°$). What is required is a rate of change appropriate to $\theta = 39°$.

As Bhāskara II justifies it, we consider the average of the difference between the given sines in the interval containing 39° (from 30° to 45°) and the one before it (from 15° to 30°) as the sine difference appropriate to $\theta = 30°$, i.e., $(36 + 31)/2$. We then take the difference in the interval from 30° to 45° (i.e., 31) as the difference appropriate to $\theta = 45°$.[46] To find the difference appropriate to $\theta = 39°$ we use linear interpolation. This calculation is represented in our text as follows:

$$-540\left(\frac{36-31}{2}\right)/900 + \left(\frac{36+31}{2}\right) = 32 \tag{3.22}$$

(where 540, the "residual arc," is the excess of 39° over 30° expressed in minutes). The appropriate difference for $\theta = 39°$ is thus 32, and so the desired sine of 39° is

$$\text{Sin}\,39° = \text{Sin}\,30° + \frac{540}{900}(32) = 94.2. \tag{3.23}$$

[45] Oddly, in Brahmagupta's other well-known astronomical work, the *Brāhmasphuṭasiddhānta*, written quite early in life, he computes a sine table for the common increment of 3;45°, but using the peculiar and as yet unexplained value $R = 3270$.

[46] The first difference calculation (for $\theta = 30°$) is reasonable, but the second difference calculation (for $\theta = 45°$) is not, and it is puzzling why he did not use the same approach for both differences. The seventeenth-century commentator Kamalākara took Bhāskara II to task for this error [Gupta 1969, 89], although Bhāskara II himself seems to have been aware of room for improvement [Datta/Singh 1983, 98].

In terms of the modern sine, this is 39° = 0.628 as opposed to the correct value of 0.6293 (linear interpolation gives 0.624), good enough for astronomical purposes. Given that Brahmagupta begins with sine values for arcs all of 15° apart, this accuracy is impressive.

The phrasing of Brahmagupta's couplet suggests that he meant it to apply beyond the sine, and to decreasing functions as well as increasing. In fact, in the *Dhyānagraha* he uses his method with both the sine and the versed sine, and he and later commentators use it with both the *śīghra* and *manda* corrections of planetary theory (on which we will see more later). In the case of the *śīghra* Brahmagupta takes his work even further, applying the interpolation to a table with entries separated by unequal argument intervals.

Taylor Series for Trigonometric Functions in Mādhava's Kerala School

Perhaps the most dramatic episode in Indian trigonometry happened late in the medieval period, and produced the most accurate trigonometric table in medieval India. The story begins in the late fourteenth century with the mathematician Mādhava, from the region of Kerala on the southwest tip of India. Very little is known about Mādhava and not one of his works on the topics we shall discuss survives.[47] Mādhava would have taught in an *illam*, a family compound. Although he wrote some of his mathematics, the small group of students who gathered around him memorized many of his findings in verse.[48] Followers (second, third, and even fourth generation) wrote the verses down and provided commentaries; since none of the original verses provided any hint of derivations, some of these later authors filled in explanations.

These verses contain some of the most sophisticated and unique mathematics of the medieval period. Parameśvara, who wrote a number of treatises in the early fifteenth century, is one of the best known writers.[49] Other major figures include Nīlakaṇṭha, who has come up before in this chapter—his *Tantrasaṅgraha*[50] ("Epitome of Astronomical Treatises") deals with some of the Mādhavan mathematics we shall see shortly; he also wrote the

[47] On Mādhava and his works see [Pingree 1970–1994, vol. A4, 414–415].

[48] A nice description of life and work in a typical *illam* may be found in [Pingree 1985, 28–29].

[49] See [Pingree 1970–1994, vol. A4, 187–192] for information on Parameśvara's works. [Gupta 1974b] gives a translation and explanation of his description of a third-order series approximation to the sine which differs from the Taylor series.

[50] The *Tantrasaṅgraha* is available in a Sanskrit edition in [Nīlakaṇṭha (Sarma) 1977] and has been translated into English in a series of supplements to the *Indian Journal of History of Science*; see [Nīlakaṇṭha (Sarma/Narasimhan) 1998].

Āryabhaṭīyabhāṣa, a commentary on the works of Āryabhaṭa—and Jyeṣṭhadeva, whose sixteenth-century *Yuktibhāṣā*[51] ("Vernacular [Exposition] of Rationales") is related to the *Tantrasaṅgraha*. Finally, Śaṅkara, a temple-sweeper who became the last of the gurus we know to have descended from Mādhava, wrote two works containing extensive commentary on Mādhava's series, the *Yuktidīpikā* ("Lamp of Rationales," a Sanskrit redaction of the *Yuktibhāṣā*),[52] and the *Kriyākramakarī* ("Performing the Steps for Calculation," a commentary on Bhāskara II's *Līlāvatī*).[53] Like most modern commentators,[54] we shall concentrate on the mathematics of the sine and cosine power series.[55] Nevertheless it is important to remember the ultimate purpose of these deliberations: to provide a sound trigonometric basis for mathematical astronomy.

We follow Jyeṣṭhadeva's *Yuktibhāṣā*, the earliest work that contains a demonstration of the sine and cosine series.[56] Modern trigonometric functions

[51] For a summary of the *Yuktibhāṣā* see [Sarma/Hariharan 1991].

[52] See [Nīlakaṇṭha (Sarma) 1977] for an edition of the *Yuktidīpikā*.

[53] The text of the *Kriyākramakarī* related to the π series has recently been made available, with further explanation, in [Plofker 2007, 481–493].

[54] Mādhava's power series were mostly ignored until the 1940s, but have been the focus of overwhelming interest since then. The Gregory and π series were first brought to the attention of the West early in the nineteenth century by C. M. Whish [Whish 1835], but modern studies began in earnest with a series of papers by C. T. Rajagopal and his colleagues. An initial paper on the Gregory series [Marar/Rajagopal 1944] was followed by a sequence of three papers in *Scripta Mathematica* expressing the argument for the Gregory series in modern terms ([Rajagopal 1949], [Rajagopal/Aiyar 1951], and [Rajagopal/Aiyar 1952]). This and a similar article dealing with the sine and cosine series [Rajagopal/Venkataraman 1949], based on the *Yuktibhāṣā*, are useful but tend to introduce unnecessary notions of limits and the integral calculus. Two later papers, [Rajagopal/Rangachari 1978] and [Rajagopal/Rangachari 1986], deal with the enunciations of the trigonometric series in the *Tantrasaṅgraha*, and [Gupta 1976a] describes the relevant passage in Nīlakaṇṭha's commentary to the *Āryabhaṭīya*. Since this series of papers a number of authors have dealt with various aspects of the series, particularly the question of how much calculus is contained within them; we shall address this briefly later. A couple of anonymous Kerala texts on the subject have also received modern treatment: the *Mahājyānayanaprakāra*, edited and translated in [Gold/Pingree 1991], and the *Karaṇapaddhati* (see the description of the power series passage in [Bag 1966]). For those who can read Japanese, consult especially Chapters 3 and 4 of [Hayashi/Kusuba/Yano 1997]. What little biographical information that is known for certain about the Kerala scientists may be found in [Pingree 1970–1994] or in [Raja 1995].

[55] In fact, there are two distinct power series for the sine in the Kerala school; the other one begins to vary from the Taylor series in the third term, described in [Gupta 1974b]. See [Plofker 2001] for a possible geometric explanation for the difference, and [Plofker 2005] for a detailed comparison of the two series.

[56] The *Yuktibhāṣā*, written in Malayāḷam, is not available in English. This exposition is based partly on [Rajagopal/Venkataraman 1949] and especially on [Bag 1979, 276–282], which holds the advantage of avoiding the notation of the integral calculus. Bag had already described the enunciations of the series in the *Tantrasaṅgraha* and hinted at the *Yuktibhāṣā* derivation in [Bag 1976]. We have tried to follow the text as closely as we could; streamlined modern accounts are available in [Katz 1995] and [Bressoud 2002]. The *Yuktidīpikā* is an elaboration of the *Yuktibhāṣā* and contains a related argument, translated and analyzed in [Plofker 2005].

and radian measure are used, since the plethora of extra constants and technical annoyances that arise when using the original Indian functions—although they lead us to appreciate Mādhava's accomplishment all the more—tend to obscure the main ideas. Finally, we use the symbol ≈ only when an approximation is actually being made at that step; this convention mirrors how the material appears in the texts.

We have already seen Indian work on first and second differences of sines; the fact that first differences of sines are proportional to cosines, and that second differences of sines are proportional to the sines themselves, would have been known at least by observation. Our demonstration of the Taylor series for sine and cosine begins with a derivation of this fact, using the unit circle in figure 3.9. We assume that arc $\Delta\theta$ is sufficiently small that Crd $\Delta\,\theta \approx \Delta\,\theta$. In ΔXYZ, YZ is the difference $\Delta\ \sin\theta$ between and $\angle AOZ$ and sin $\angle AOX$; likewise, $XY = \Delta\ \cos\,\theta = \cos\,\angle AOX - \cos\,\angle AOZ$. Now $\Delta OAB \sim \Delta XYZ$, since all corresponding sides are perpendicular to each other. Hence

$$\frac{\Delta\theta}{1} \approx \frac{\Delta\sin\theta}{\cos\theta} = \frac{\Delta\cos\theta}{\sin\theta},$$

which gives us the following expressions for the first differences:[57]

$$\begin{aligned}\Delta\sin\theta &= \cos\theta\cdot\Delta\theta,\\ \Delta\cos\theta &= \sin\theta\cdot\Delta\theta.\end{aligned} \tag{3.24}$$

Statements like these had been around at least since Bhāskara II. They can be used to obtain an expression for the second difference of the sine:

$$\begin{aligned}\Delta^2\sin\theta &= \Delta\sin\theta - \Delta\sin(\theta+\Delta\theta)\\ &= \cos\theta\cdot\Delta\theta - \cos(\theta+\Delta\theta)\Delta\theta\\ &= \Delta\cos\theta\cdot\Delta\theta\\ &= \sin\theta(\Delta\theta)^2.\end{aligned} \tag{3.25}$$

The real work starts in the unit circle of figure 3.10, where the arc θ to be analyzed is divided into a large number n of equal small arcs $\Delta\theta$ (thus, $\theta = n\cdot\Delta\theta$). Vertical lines represent the sines of successive multiples of $\Delta\theta$;

[57] The statement as shown here leaves ambiguity concerning precisely where the cosine and sine are to be evaluated. When used later in this argument it is chosen correctly for the geometry of the situation, in the middle of the interval $\Delta\theta$, but this is not always done. In any case, since $\Delta\theta$ is small and quasi-infinitesimal arguments are already at work, it does not matter that θ be chosen precisely, as may be seen by the approximation applied in (3.31). See [Gupta 1979a] for a clever use of this result, from the Mādhava school, to interpolate backwards in a Sine table from a Sine value to its corresponding arc.

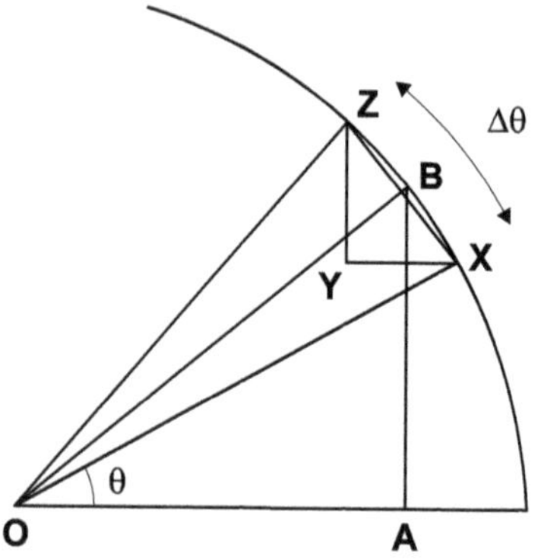

Figure 3.9
First differences of sines and cosines

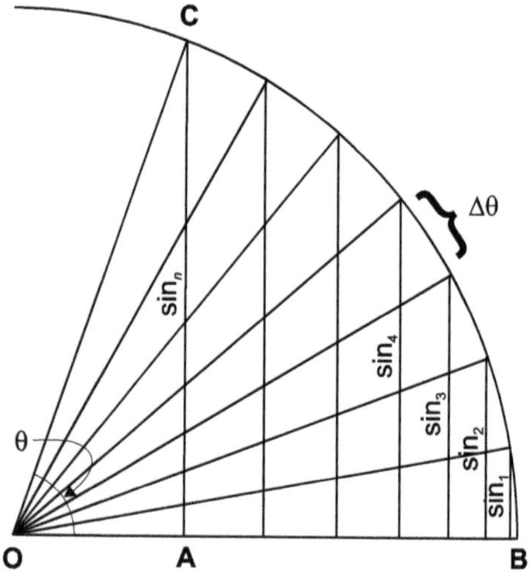

Figure 3.10
Derivation of Mādhava's Taylor series for sine and cosine

to save space we shall write $\sin_j = \sin(j\cdot\Delta\theta)$. Writing the first backward differences of these sines as $\Delta\sin_j = \sin_j - \sin_{j-1}$, notice that $\Delta\sin_1 = \sin_1$. Also the sought sine is $\sin_n = \sin\theta$, and finally, since n is large, $\theta = n\cdot\Delta\theta \approx n\cdot\sin_1$.

We are ready to take the first step in the iteration that will lead to our series: using the first-order sine approximation $\sin\alpha \approx \alpha$ to generate a second-order cosine approximation. Notice that the versed sine of θ, AB in figure 3.10, is equal to the sum of the cosine differences $\Delta\cos_j$. So, by, (3.24),

$$\begin{aligned}\operatorname{vers}\theta = 1-\cos\theta &= \sin_1\cdot\Delta\theta + \sin_2\cdot\Delta\theta + \cdots + \sin_n\cdot\Delta\theta \\ &= [\sin_1 + \sin_2 + \cdots \sin_n]\Delta\theta.\end{aligned} \tag{3.26}$$

Taking $\sin\alpha \approx \alpha$ for all sines in this expression, we get

$$\begin{aligned}\operatorname{vers}\theta = 1-\cos\theta &= \left[\frac{\theta}{n}+\frac{2\theta}{n}+\cdots+\frac{n\theta}{n}\right]\Delta\theta \\ &= \left[1+2+\cdots+n\right]\frac{\theta}{n}\Delta\theta = \left[1+2+\cdots+n\right]\left(\Delta\theta\right)^2 \\ &\approx \frac{n^2}{2}\left(\Delta\theta\right)^2.\end{aligned} \tag{3.27}$$

The approximation of the sum of positive integers from 1 to n in the last step was well known, as was the correct value $n(n + 1)/2$. Indeed, correct formulas for the sums of m-th powers of the first n positive integers had been described, for instance, by Nārāyaṇa Paṇḍita in the mid-fourteenth century.[58] In our argument all that will be needed are the approximations

$$1^m + 2^m + \cdots + n^m \approx \frac{n^{m+1}}{(m+1)}. \tag{3.28}$$

Equation (3.27) is actually most of the way to a second-order approximation of the cosine; since $\theta = n \cdot \Delta\theta$, a little rearrangement produces the familiar

$$\cos\theta = 1 - \frac{\theta^2}{2}. \tag{3.29}$$

Our next step is to convert the second-order cosine into a third-order sine. Let $\cos_{\frac{1}{2}}$ and $\operatorname{vers}_{\frac{1}{2}}$ be the cosine and versed sine of half of $\Delta\theta$, $\cos_{1\frac{1}{2}}$ and $\operatorname{vers}_{1\frac{1}{2}}$ the cosine and versed sine of $1\frac{1}{2}\cdot\Delta\theta$, and so on. Then by (3.24)[59]

$$\begin{aligned}\Delta\sin_1 &= \cos_{\frac{1}{2}}\cdot\Delta\theta &&= \left(1-\operatorname{vers}_{\frac{1}{2}}\right)\Delta\theta, \\ \Delta\sin_2 &= \cos_{1\frac{1}{2}}\cdot\Delta\theta &&= \left(1-\operatorname{vers}_{1\frac{1}{2}}\right)\Delta\theta, \\ &\vdots && \vdots \\ \Delta\sin_n &= \cos_{n-\frac{1}{2}}\cdot\Delta\theta &&= \left(1-\operatorname{vers}_{n-\frac{1}{2}}\right)\Delta\theta,\end{aligned} \tag{3.30}$$

[58] See [Saraswathi 1963] for a discussion of the development of the study of these arithmetic series in India, as well as others (including Mādhava's Taylor series).

[59] Here (3.24) is applied correctly, with the arguments of the cosines chosen in the middle of each interval. In any case, in the following step (3.31) an approximation is made that renders the choice of argument moot.

Subtracting each of these equations from the first, we obtain

$$\begin{aligned}
\Delta\sin_1 - \Delta\sin_n &= \Delta\theta(\text{vers}_{n-\frac{1}{2}} - \text{vers}_{\frac{1}{2}}) \approx \Delta\theta\cdot\text{vers}_n = \Delta\theta\cdot\frac{n^2}{2}(\Delta\theta)^2,\\
\Delta\sin_1 - \Delta\sin_{n-1} &\approx \Delta\theta\cdot\text{vers}_{n-1} = \Delta\theta\cdot\frac{(n-1)^2}{2}(\Delta\theta)^2,\\
&\vdots\\
\Delta\sin_1 - \Delta\sin_2 &\approx \Delta\theta\cdot\text{vers}_2 = \Delta\theta\cdot\frac{(2)^2}{2}(\Delta\theta)^2,
\end{aligned} \tag{3.31}$$

since $\text{vers}_{\frac{1}{2}} \approx 0$ and $\text{vers}_{n-\frac{1}{2}} \approx \text{vers}_n$, and applying (3.27) to the equalities at the end of each line. Adding the left sides of the equations in (3.31) results in

$$\begin{aligned}
&(n-1)\Delta\sin_1 - (\Delta\sin_2 + \Delta\sin_3 + \cdots + \Delta\sin_n)\\
&= n\cdot\Delta\sin_1 - (\Delta\sin_1 + \Delta\sin_2 + \Delta\sin_3 + \cdots + \Delta\sin_n)\\
&= \theta - \sin\theta.
\end{aligned} \tag{3.32}$$

So, if we add both sides of (3.31), we get

$$\begin{aligned}
\theta - \sin\theta &= \left[n^2 + (n-1)^2 + \cdots 2^2\right]\frac{(\Delta\theta)^3}{2}\\
&\approx \left(\frac{n^3}{3}\right)\cdot\frac{(\Delta\theta)^3}{2} = \frac{n^3}{3!}(\Delta\theta)^3.
\end{aligned} \tag{3.33}$$

We have now arrived at the third-order sine approximation: since $\theta = n\cdot\Delta\theta$, (3.33) becomes

$$\sin\theta = \theta - \frac{\theta^3}{3!}.$$

At this point in the iteration the process can be automated somewhat. We derived our previous cosine approximation by applying $\sin\alpha \approx \alpha$ to (3.26) to obtain (3.27); instead, we now apply our improved estimate $\sin\alpha \approx \alpha - \alpha^3/3!$. This adjustment gives us

$$\begin{aligned}
1 - \cos\theta &= \left[\left\{\frac{\theta}{n} - \frac{(\theta/n)^3}{3!}\right\} + \left\{\frac{2\theta}{n} - \frac{(2\theta/n)^3}{3!}\right\} + \cdots\right]\Delta\theta\\
&= \left[\frac{\theta}{n} + \frac{2\theta}{n} + \cdots\right]\Delta\theta - \left[\frac{(\theta/n)^3}{3!} + \frac{(2\theta/n)^3}{3!} + \cdots\right]\Delta\theta
\end{aligned} \tag{3.34}$$

$$= [1+2+\cdots]\frac{\theta}{n}\cdot\Delta\theta - [1^3+2^3+\cdots+n^3]\frac{(\theta/n)^3}{3!}\cdot\Delta\theta$$
$$\approx \left(\frac{n^2}{2}\right)(\Delta\theta)^2 - \left(\frac{n^4}{4}\right)\frac{(\Delta\theta)^4}{3!}.$$

Substituting (as usual) $\theta = n\cdot\Delta\,\theta$, we arrive at

$$\cos\theta = 1 - \frac{\theta^2}{2!} + \frac{\theta^4}{4!}.$$

Similarly for the fifth-order sine: we derived our previous sine approximation by applying our previous cosine approximation term (3.27) to (3.31), resulting in (3.33). If we use our updated cosine term (3.34) instead, we get

$$\begin{aligned}\theta - \sin\theta &= \left[\left\{\frac{n^2}{2!}(\Delta\theta)^2 - \frac{n^4}{4!}(\Delta\theta)^4\right\}\right.\\ &\quad \left. + \left\{\frac{(n-1)^2}{2!}(\Delta\theta)^2 - \frac{(n-1)^4}{4!}(\Delta\theta)^4\right\} + \cdots\right]\Delta\theta\\ &= \left[n^2+(n-1)^2+\cdots+1\right]\frac{(\Delta\theta)^3}{2!} - \left[n^4+(n-1)^4+\cdots+1\right]\frac{(\Delta\theta)^5}{4!}\\ &\approx \frac{n^3}{3}\cdot\frac{(\Delta\theta)^3}{2!} - \frac{n^5}{5}\cdot\frac{(\Delta\theta)^5}{4!} = \frac{n^3}{3!}\cdot(\Delta\theta)^3 - \frac{n^5}{5!}\cdot(\Delta\theta)^5.\end{aligned} \tag{3.35}$$

Since $\theta = n\cdot\Delta\theta$,

$$\sin\theta = \theta - \frac{\theta^3}{3!} + \frac{\theta^5}{5!}.$$

From this point, applying improved sine/cosine approximations to the cosine/sine derivation adds an extra term each time. Of course, the pattern in the coefficients is already clear at this stage.

It will come as no surprise that these arguments are the basis of several claims that the calculus was known in medieval India.[60] The flavor of some of

[60] See for instance [Sengupta 1932]; Bag's *Mathematics in Ancient and Medieval India* [Bag 1979] goes so far as to entitle one chapter "Infinitesimal Calculus." Bag suggests that calculus-based methods were transmitted to the West from Kerala [Bag 1979, 285]). [Katz 1995] discusses several examples of calculus-like methods in Islam and India (including the sine and cosine series), coming to much the same conclusion that we do. [Bressoud 2002] ("Was calculus invented in India?") describes the mathematical arguments leading to the Taylor series, and eventually concludes that since they apparently did not transmit to the West, the answer to the question in his title is "no."

the algebra is similar to those of Riemann sums arguments in integral calculus, and the connection between (3.24) and the differentiation rules for the sine and cosine is obvious. However, these claims cannot really be taken seriously; what made the calculus significant to Western science was its ability to handle derivatives and integrals of many functions systematically. Mādhava's and others' derivations work only on particular trigonometric functions, and cannot be associated with calculus any more than some of Archimedes' uses of the method of exhaustion to "integrate" circles and parabolic segments. This observation does not diminish Mādhava's accomplishment in any way (or Archimedes's for that matter), which ranks prominently among the most ingenious and powerful mathematics of the medieval period.

With his series Mādhava would have been able to compute sines relatively easily, and more importantly, to arbitrary accuracy. His sine table (see figure 3.11[61]) was by far the most accurate of its time in India; although the step size of the arguments in the table is still only 3;45° each entry is accurate to the equivalent of about seven decimal places, making it comparable to the best tables available in the Arabic world at the time.[62]

To achieve a table of this level of accuracy would require a better value for π than is implied by the traditional Indian base circle radius $R = 21{,}600/2\pi \approx 3438$, and we can see by the last row of figure 3.11 that Mādhava seems to have realized this fact. Indeed one of his other infinite series, which gives the circumference C of a circle in terms of its diameter d, is

$$C = \frac{4d}{1} - \frac{4d}{3} + \frac{4d}{5} - \cdots.^{63} \qquad (3.36)$$

This series would have allowed him to calculate π to arbitrary accuracy, at least in theory. Anyone who has worked with (3.36) is aware of its painfully slow convergence; however, Mādhava improved it by inventing several ingenious correction terms that greatly sped up the process.[64]

[61] Reprinted from [Yano 1997, 989] with corrections.

[62] The sine tables in Jamshīd al-Kāshī's *Khāqānī Zīj* (ca. AD 1410) were about as accurate; those in Ulugh Beg's *Sultānī Zīj* (AD 1440), on which al-Kāshī probably worked, gained about two decimal places. These tables, however, contained many more entries than Mādhava's.

[63] This is a special case of the Gregory series arctan $x = x - (x^3/3) + (x^5/5) - \ldots$, also known to Mādhava. See several references to papers on this topic in preceding footnotes; for a modern account of the origin of the series, see [Roy 1990]. Of course tangents did not exist in Indian trigonometry; the series was expressed with appropriate uses of sines and cosines.

[64] These correction terms were certain rational functions of n, where n is the number of series terms used. Several theories have been proposed regarding the origins of the corrections; see [Juschkewitsch 1964, 171–172] and especially [Hayashi/Kusuba/Yano 1990].

θ (degrees)	Sin θ
3;45	224;50,22
7;30	448;42,58
11;15	670;40,16
15	889;45,15 [−1]
18;45	1105;1,39
22;30	1315;34,7
26;15	1520;28,35
30	1718;52,24
33;45	1909;54,35
37;30	2092;46,3
41;15	2266;39,50
45	2430;51,15 [+1]
48;45	2584;38,6 [+1]
52;30	2727;20,52
56;15	2858;22,55
60	2977;10,34 [+1]
63;45	3083;13,17
67;30	3176;3,50
71;15	3255;18,22 [+1]
75	3320;36,30
78;45	3371;41,29
82;30	3408;20,11
86;15	3430;23,11 [+1]
90	3437;44,48

Figure 3.11
Mādhava's sine table

Applying Sines and Cosines to Planetary Equations

The Indian ability to make creative and innovative uses of received knowledge from other cultures is nowhere better exemplified than in their approach to the motions of the planets.[65] We find several different models, all involving combinations of eccentric circles and epicycles. Other than the

[65] Indian models for the motions of the planets are discussed in many sources, including [Chatterjee 1949, 59–70], the appendix (pp. 174–188) to [Neugebauer 1956], [Prakash 1968, 147–152], [Sen 1971, 111–116], [Sen 1974], [Pingree 1978, 557–59], and [Sen 1987]. Āryabhaṭa's contributions are in [Āryabhaṭa (Shukla/Sarma) 1976, xxxi–xxxiii and 104–112].

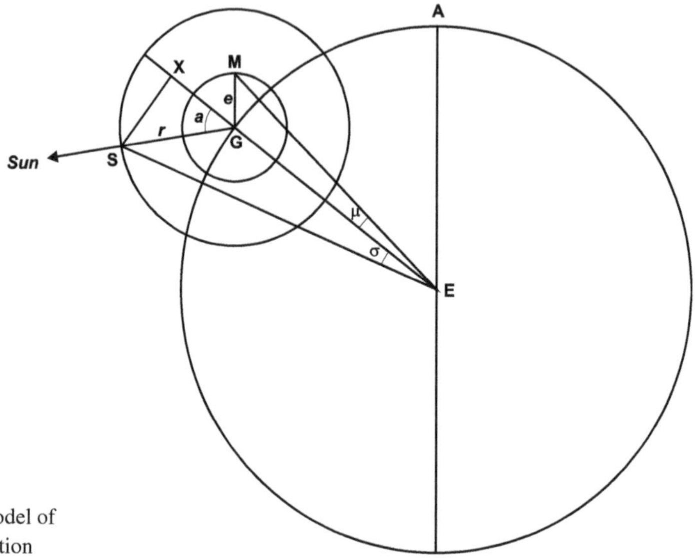

Figure 3.12
An Indian model of planetary motion

equant point, these Indian constructions contain elements familiar to Greek astronomy, and it is no stretch to see in them the traces of predecessors of Ptolemy. The earliest texts work as follows (see figure 3.12): *AG* is the deferent circle, centered at the Earth *E*. Centered at *G* are not one, but two epicycles. The *manda* or "slow" epicycle, with radius *e*, has a radius *GM* that always points parallel to *EA*, the apogee. *M*, the *mandocca*, determines the *manda* correction $\mu = \angle MEG$. The *śīghra* or "fast" epicycle, with radius *r*, has a radius *GS* that (for the superior planets) always points in the direction of the Sun.[66] *S*, the *śīghrocca*, determines the *śīghra* correction $\sigma = \angle GES$.

Clearly the *manda* correction plays the same role as the Greek equation of center, effectively moving the Earth away from the center of the deferent; and the *śīghra* correction does what the Greek epicycle does. But the location of the planet itself is obscure. The texts give a pattern of calculation that involves taking into account half the *śīghra* correction, then half the *manda* correction, then the *śīghra* again, and finally the *manda* again, recomputing the mean positions that are the arguments for the corrections at each stage. The reader may well be puzzled by this iterated calculation, which seems at

[66] For the inferior planets *EG* points in the direction of the mean Sun, and *GS* forms with *EG* an angle equal to the argument of anomaly *a*.

first to bear little relation to figure 3.12. The factors of $\frac{1}{2}$ in the first two steps suggest a procedure that Ptolemy had performed when he introduced the equant point (a bisection of the eccentricity), and this correlation may be no coincidence.[67]

Variations of the Indian model even included pulsating epicycles (i.e., with varying radii), also possibly related to the equant. So it is clear that the Aristotelean requirement of uniform circular motion was not operating in any real sense; the theoretical basis so important to the Hellenistic astronomers (and to us) was to the Indians less significant than computational results. What might appear to us a jumbled mix of incompatible approaches nevertheless served its purpose, to give accurate numerical solutions to astronomical problems.

Calculation of the *śīghra* correction from its argument *a* proceeded as follows.[68] Form the *bhuja* Sin *a* ("angle/argument/side of right triangle"), and the *koṭi* Cos *a* ("upright/complement/other side of right triangle"); these correspond to the *bhujaphala SX* and the *koṭiphala GX* respectively (*phala* = "result"), although not in the correct units. The standard unit is 1/360th of the deferent; the circumference of the *śīghra* epicycle for Saturn, for instance, is given in the *Pañcasiddhāntikā* as $c = 40$ of these units. So

$$\begin{aligned} bhujaphala &= SX = \frac{c}{360} \cdot \text{Sin}\, a \quad \text{and} \\ kotiphala &= GX = \frac{c}{360} \cdot \text{Cos}\, a. \end{aligned} \tag{3.37}$$

Then the *śīghra* hypotenuse (of ΔESX) is

$$\begin{aligned} ES &= \sqrt{EX^2 + SX^2} + \sqrt{(EG + GX)^2 + SX^2} \\ &= \sqrt{(R + koṭiphala)^2 + bhujaphala^2}, \end{aligned} \tag{3.38}$$

and from here the *śīghra* correction may be found simply:

[67] That these calculations reflect a buried use of the equant point was first proposed in [van der Waerden 1961] (and his case analyzed in [Thurston 1992]); it was taken considerably further and more convincingly in [Duke 2005b]. If one accepts that Indian planetary astronomy came from Greece, then this is a strong case for the existence of the equant in Greece before Ptolemy.

[68] The following is based on [Varāhamihira (Neugebauer/Pingree) 1970/1971, part II, 102–104]. The *manda* correction may be found in a similar manner, but some texts, including the *Pañcasiddhāntikā* but not those emerging from Āryabhaṭa's school [Sen 1987, 120], used instead a rather crude approximation.

$$\sin\sigma = \frac{SX}{ES} = \frac{bhujaphala}{\acute{s}\bar{\imath}ghra\ \text{hypotenuse}}.^{69} \qquad (3.39)$$

Spherical Astronomy

Planetary positions were hardly the only use for trigonometry in India, although astronomy remained the central focus. A healthy tradition of gnomonics continued, presumably because of its immediate uses in timekeeping.[70] The mathematics required is usually relatively simple, involving plane trigonometry, similar triangles, and the Pythagorean Theorem. Much more depth is called for when the field of play transforms from a plane to a sphere.

We have seen that trigonometry's astronomical roots led its practitioners to work with arcs and distances on the surface of a sphere (usually, the celestial one). In the Greek science of spherics, some treatises separated the mathematical theory from its astronomical context—so transparently that the connection was as obvious as it was unstated. Nevertheless this may have led Menelaus or his predecessors to study spherical triangles independently, a step that would never be taken in India. Since the applications were strictly astronomical in any case, Indian astronomers might have considered the distinction to be moot as long as their methods were as effective as those developed by their Greek counterparts.

Unfortunately, as we have seen throughout this chapter, Indian scientists wrote much more on their results than on their methods. Several efforts have been made to reconstruct the mathematics of Indian spherical astronomy, but the reader can imagine that such rehabilitations often say more about the author's point of view than about the original sources.[71] We can say that there is no evidence of Indian knowledge of Menelaus's theorem or any other theorems that operate solely on the surface of the sphere. But such theorems are not necessary to solve the astronomical problems that faced Indian astronomers. Our first example illustrates how results may be obtained just as well from working with similar triangles within the sphere.

[69] As we have seen in Greek planetary theory, the lack of a tangent function caused extra complications. Here, with a tangent, the *śīghra* hypotenuse might have been bypassed altogether.

[70] The interested reader will find a thorough description of gnomons and their applications in Bhāskara I's commentary to the *Āryabhaṭīya*, in [Keller 2006].

[71] The two major papers on Indian spherical astronomy, [Kaye 1919] and [Sengupta 1931b], take sharply opposing points of view—Kaye in favor of Greek origins, Sengupta in favor of Indian independence. [Zeuthen 1900, 27]'s assertion on the side of Greek origins was particularly stark: "La seule chose que nous rencontrions chez les Hindous sans l'avoir rencontrée dans les ouvrages grecs qui nous sont parvenus, c'est l'usage de tables de sinus au lieu des tables de cordes."

Text 3.5
Varāhamihira, Finding the Right Ascension of a Point on the Ecliptic
(from the *Pañcasiddhāntikā*)

Multiply the diameter by the square-root of the difference between the squares of the Sines of the declinations [of the Sun] and [of the Sines of the ends of the several] signs and divide [the products] by the [respective] diameters of the day [-circles]. The arc from this . . . [equals the] . . . rising of [each] of the signs.[72]

Explanation: (See figure 3.13.) We wish to determine the right ascension α of a point P on the ecliptic, given its longitude λ and declination δ. Varāhamihira's instructions, translated into modern notation, are:

$$\text{Sin}\,\alpha = \frac{2R}{d}\sqrt{\text{Sin}^2\lambda - \text{Sin}^2\delta}\ , \qquad (3.40)$$

where R is the radius of the sphere, and d is the diameter of the "day-circle" (the circle through P parallel to the equator; thus $d\,/\,2 = OB$). There are several ways to get to this formula, and once again the text is of no help in distinguishing between the possibilities. The most natural method is the following.[73]

Drop perpendicular PB onto OD, and drop perpendiculars BA and CD onto $O♈$. Then $PA = \text{Sin}\ \lambda$ and $PB = \text{Sin}\ \delta$;[74] hence $AB = \sqrt{\text{Sin}^2\ \lambda - \text{Sin}^2\ \delta}$. Now $CD = \text{Sin}\ \alpha$.[75] So, since $\Delta OAB \sim \Delta OCD$,

$$\frac{CD}{OD} = \frac{AB}{OB},$$

and hence

$$\frac{\text{Sin}\,\alpha}{R} = \sqrt{\frac{\text{Sin}^2\lambda - \text{Sin}^2\delta}{d/2}}, \qquad (3.41)$$

from which (3.40) follows.

[72] From [Varāhamihira (Neugebauer/Pingree) 1970/1971, part 1, 61].

[73] Our explanation differs from [Varāhamihira (Neugebauer/Pingree) 1970/1971, part 1, 41–42]; but it is more direct, does not rely on other spherical trigonometric relations, and coincides with the appearance of the formula in the text.

[74] For PA, consider right triangle OPA where $\angle POA = \lambda$; for PB, consider right triangle POB where $\angle POB = \delta$.

[75] Since, in right triangle OCD, $\angle COD = \alpha$.

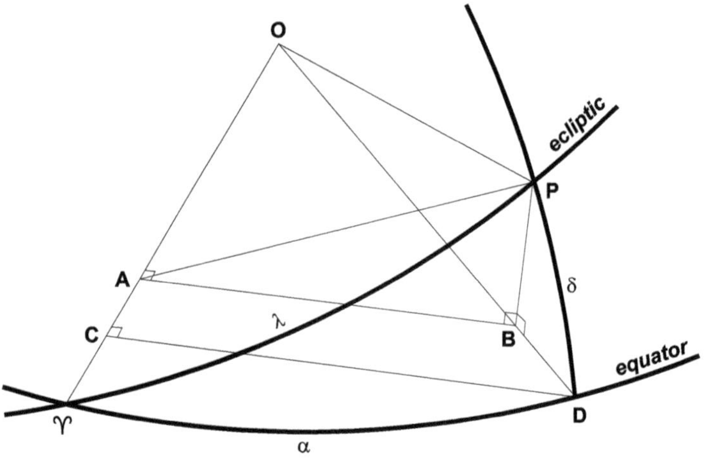

Figure 3.13 A derivation of Varāhamihira's formula for the right ascension

The tools required to effect this solution are very simple; indeed, they echo this passage by Nīlakaṇṭha:

> The whole of the planetary-mathematics is pervaded by two theorems, namely the so-called Pythagorean Theorem and the Rule of Three (the proportionality of sides in similar triangles).[76]

In fact, the commonly-stated opinion that Indian spherical astronomy uses the Greek method of the analemma (discussed in the previous chapter) has no direct basis in evidence. Many derivations even in modern spherical trigonometry are based on elementary geometry within the sphere, and are no less powerful for it. This leaves us with two possibilities: either Indian methods were in fact discovered independently, or they are witnesses to an earlier Greek theory without the analemma.[77] The truth of the matter may lie somewhere in between.

As we have seen, Indian solutions to astronomical problems often involve approximations of various sorts, and spherical problems are no exception. Most obvious is an assumption that the spherical triangle is actually plane—although it is often difficult to be sure about this, given the

[76] Quotation taken from [Gupta 1974e, 98].

[77] In this point, [Sengupta 1931] may well be correct in his case against [Kaye 1919] on the absence of the analemma in India—and the study [Shukla 1968] agrees. The general claim that Indian methods are independent of Greek methods simply cannot be evaluated fully without further evidence.

ambiguities of the Sanskrit texts and the intervention of commentators. Consider this example from the *Āryabhaṭīya*: in figure 3.14, the *nonagesimal N* is the highest point on the ecliptic, 90° removed from the rising point *R*.[78] Given arcs *n* and *y* (which is equal to $\angle MZN$, the ***hour angle*** of *N*—so named because it measures the number of hours[79] until noon, when the Sun reaches the meridian) find *m*, the zenith distance of *N*. Āryabhaṭa's first step is to find *z*:

$$\operatorname{Sin} z = \frac{\operatorname{Sin} n \cdot \operatorname{Sin} y}{R} \tag{3.42}$$

(a fifth-century use of the Rule of Four Quantities!). To find *m*, Āryabhaṭa computes

$$\operatorname{Sin} m \approx \sqrt{\operatorname{Sin}^2 n - \operatorname{Sin}^2 z}. \tag{3.43}$$

This statement seems to assume that the *Sines* of the sides of ΔMNZ form a plane right-angled triangle, which turns out to be no better than assuming that ΔMNZ itself is a plane right-angled triangle.

This is how Āryabhaṭa's method was interpreted by Brahmagupta (who criticized him harshly for the approximation), Bhāskara II, and recent commentators.[80] Evidence recently published from the Kerala astronomer Nīlakaṇṭha's commentary on the *Āryabhaṭīya*, however, suggests another way to understand it. Nīlakaṇṭha takes the right side of (3.43) to be equal not to the *dṛkkṣepjyā* (Sin *m*), but only an intermediate stage, the *svadṛkkṣepa*. It turns out to be a relatively simple matter—an application of similar triangles in the diagram—to go from there to the correct formula:

$$\operatorname{Sin} m = \mathit{svad\d{r}kk\d{s}epa} \cdot \frac{R}{\operatorname{Cos} z}. \tag{3.44}$$

Whether or not Nīlakaṇṭha was simply being kind to Āryabhaṭa will never be known for certain, but the latter's correct solutions to similar problems (indeed, similar to Varāhamihira's right ascension problem above) gives some reason for us to be charitable also.[81]

At the same time several hundred miles to the west, Muslim astronomers were developing spherical trigonometry to a high level of sophistication, basing their approach more closely on their inheritance from

[78] The nonagesimal point appears as early as Ptolemy's *Almagest*, and is used extensively in Indian astronomy. It is still important in astrology today.

[79] Converting from degrees to hours by the factor $24^h / 360°$.

[80] See for instance [Sengupta 1931, 17–19] and [Shukla/Sarma 1976, 142].

[81] This interpretation was brought to light in [Gupta 1987].

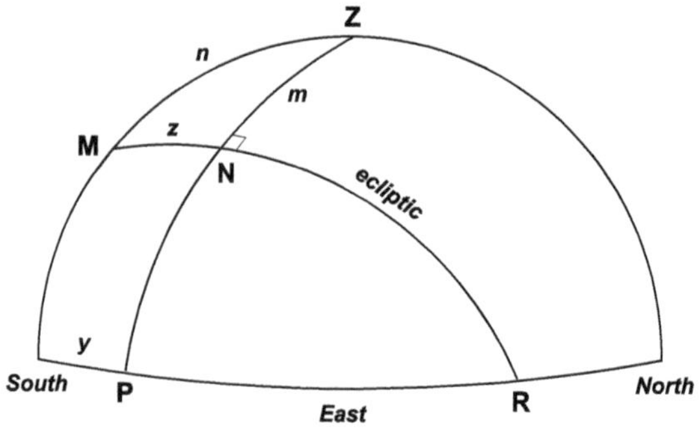

Figure 3.14
The nonagesimal point N

Greece (especially Ptolemy). Starting in the late fourteenth century, some of these methods found their way into India. However, they do not appear to have been well understood initially, and various errors and misconceptions crept into the Indian works. Without a firm theoretical basis, mistakes could not be removed easily.[82] Nevertheless there do seem to have been continuing efforts to "Sanskritize" Islamic spherical trigonometry, especially coordinate conversions, from this time onwards. Even so, the subject was not treated independently of astronomy, and the approximative techniques typical of Indian astronomy remained in use.[83]

Finally, a remarkable synthesis of Indian spherical astronomical knowledge occurs in a passage of Nīlakaṇṭha's *Tantrasaṅgraha* (AD 1500).[84] In figure 3.15 the triangle between points Z (the zenith), N (the celestial north pole), and ☼ (the Sun) is called the *astronomical triangle*, a concept that survives to today. Its significance arises from the fact that each of its sides and two of its angles are important astronomical quantities. The sides are $90° - \phi$ where ϕ is the observer's terrestrial latitude; $90° - \delta$ where δ is the Sun's declination; and $90° - a$ where a is the Sun's altitude above the horizon. As for angles, $\angle Z$ is the Sun's azimuth and $\angle N$ is the Sun's hour angle.

So five quantities in the astronomical triangle are significant; and, like plane triangles, spherical triangles usually require knowledge of three elements to be determined completely. Hence there are ten situations, dealing with each of the ten possible ways to specify three of the five elements as

[82] For an analysis of the first of these works (including explanations of some of the mistakes), the *Yantrarāja* by Mahendra Sūri, see [Plofker 2000].

[83] [Plofker 2002a]; see also [Ohashi 1997].

[84] For the Sanskrit see [Nīlakaṇṭha (Sarma) 1977, lxxiv–lxxv, 219–237]; an English translation and analysis appear in [Nīlakaṇṭha (Sarma/Narasimhan) 1998, S73–S82]; and the first modern study is [Gupta 1974e].

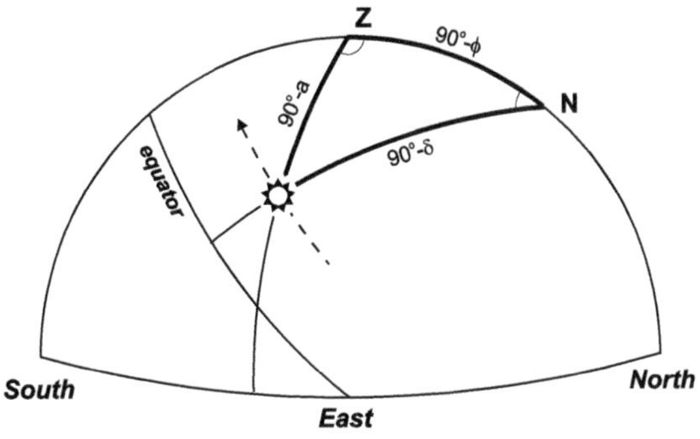

Figure 3.15
The astronomical triangle

given. Nīlakaṇṭha deals with each of these cases in turn, providing a rule to solve the remaining two elements. For instance, given δ, ϕ, and $\angle Z$, find a: we first find the "divisor"

$$d = \sqrt{\text{Sin}^2\,\phi + (\text{Cos}\,Z \cdot \text{Cos}\,\phi\,/\,R)^2}\,, \tag{3.45}$$

and then solve for a in

$$\text{Sin}\,a = \frac{R}{d^2}\left(\text{Sin}\,\delta \cdot \text{Sin}\,\phi \pm \sqrt{d^2 - \text{Sin}^2\delta} \cdot \sqrt{d^2 - \text{Sin}^2\phi}\right). \tag{3.46}$$

(This formula can be derived today using the spherical Law of Cosines and solving the quadratic equation in sin a that results when cos a is replaced with $1 - \sin^2 a$.)[85]

Neither the methods used (we are once again in the dark, and some of these methods seem to go back to much earlier times), nor the fact that Nīlakaṇṭha's solutions are correct, are particularly striking. Rather, what is remarkable is the handling of all ten cases of the astronomical triangle in one place; this is as close as one can get to a complete solution of the general spherical triangle. There is no particular reason why Nīlakaṇṭha's methods cannot be applied to any other triangle. So, although there is no attempt to generalize beyond the astronomical context, the spherical triangle is handled as systematically here as it is in modern textbooks.

Using Iterative Schemes to Solve Astronomical Problems

Science often thrives on inverse problems, that is, situations where the path to a solution is the reverse direction from the original development of the

[85] [Gupta 1974e, 88–89].

subject. For instance, in Greek and Indian astronomy a planet's longitude λ is usually found first by determining the mean longitude λ_m, a linear function of time t, and then altering it by one (or more) equations or correction factors, say, $q(\lambda_m)$. Then

$$\lambda = \lambda_m + q(\lambda_m). \tag{3.47}$$

But if one wishes to determine when a given body will be in a certain location (for the Sun and Moon, this is useful for eclipse calculations), (3.47) must be inverted: λ_m must be found from λ. Then we can backtrack to t easily enough, using its linear relation with λ_m.

So, the problem lies in solving (3.47) for λ_m.[86] One seventh-century example of this is in the *Brāhmasphuṭasiddhānta*: Brahmagupta wishes to find the Sun's mean longitude λ_m given its true longitude λ. The problem he faces is: in

$$\lambda = \lambda_m - \text{arcSin}\left(\frac{r \cdot \text{Sin}(\lambda_m - M)}{\sqrt{\left(\frac{r}{R} \cdot \text{Sin}(\lambda_m - M)\right)^2 + \left(R + \frac{r}{R} \cdot \text{Cos}(\lambda_m - M)\right)^2}}\right) \tag{3.48}$$

(where r, M, and R are known constants), solve for λ_m.[87]

Difficulties of this sort do arise occasionally in Ptolemaic astronomy, but in Greece there is a distinct preference for problems with direct solutions. In India there is hardly any distinction between precise and approximate solutions, so it is not surprising that approximations became popular. We might divide Indian approaches into two classes, *fixed-point* and *two-point* iterations, both of which go back to the earliest texts of Indian mathematical astronomy from the fifth century AD.

One example of fixed-point iteration may be found within a method of calculating Sines by Parameśvara, in his fourteenth-century *Siddhāntadīpikā* (a supercommentary on a commentary on Bhāskara I's *Mahābhāskarīya*).

[86] Readers familiar with Kepler's equation will recognize a kindred spirit here. Kepler's equation, which arises often in astronomy, is $y = x - a \sin x$; the goal is to solve for x given y.

[87] [Plofker 2002b, 171].

Text 3.6
Parameśvara, Using Fixed-point Iteration to Compute Sines
(from the *Siddhāntadīpikā*)

The Sine is to be derived from the arc by a method called "aviśeṣa."

One should halve the given arc, and the half of that, and so on in this way. [Successive] halving from the half of that [arc] is required until the half is less than 1/300 of the quadrant [$=18'$]. Then the Chord of the halved arc is equal to that [arc].

Because of the smallness of the Versine, the Sine too is nearly equal to that [arc]. The Cosine is [calculated] by means of the Sine, and the Versine from the Cosine. Repeatedly, the square root of the square of the Chord diminished by the square of the Versine is the corrected Sine, and from that, the Cosine [is calculated], and thus from the Cosine, the Versine.

The square root of the difference of the squares of the Chord and the Versine is the corrected Sine, and having done this repeatedly, one may make that indistinguishable [from the previous value, by iterative correction].[88]

Explanation: The term *aviśeṣa* means "without difference," implying that the iteration is to be carried out until successive iterates no longer differ. Parameśvara intends his process to be applied only to small arcs. For larger arcs he instructs the reader to halve the arc successively until it is less than $18'$. After our passage he specifies a successive doubling process, returning to the original arc.

Once $\theta < 18'$, we apply the usual useful fact that for small θ, $\theta \approx \text{Sin}\,\theta \approx \text{Crd}\,\theta$. This gives us our initial estimates Sin_0 for $\text{Sin}\,\theta$ and Crd_0 for $\text{Crd}\,\theta$. To derive an improved estimate Sin_1 we step through some familiar identities, beginning and ending with the Sine:

$$\text{Cos}_0 = \sqrt{R^2 - \text{Sin}_0^{\;2}}, \tag{3.49}$$

$$\text{Vers}_0 = R - \text{Cos}_0, \tag{3.50}$$

and

$$\text{Sin}_1 = \sqrt{\text{Crd}_0^{\;2} - \text{Vers}_0^{\;2}}. \tag{3.51}$$

If the original Sin_0 were equal to $\text{Sin}\,\theta$, then Sin_1 would simply be equal to Sin_0; but Sin_0 is only an approximation, so we hope that Sin_1 is a closer

[88] Quoted from [Plofker 1996, 248].

one. We take it through formulas (3.49)–(3.51) again, updating all the subscripts (except for Crd_0, which remains constant), obtaining Sin_2. Repeat as necessary until eventually $\text{Sin}_n = \text{Sin}_{n+1}$ to the level of accuracy that we are shooting for.

The modern (or even ancient Greek) reader might be puzzled at the assumption that Sin_1 is closer to $\text{Sin}\,\theta$ than is Sin_0, and might wish for a demonstration. If we combine (3.49)–(3.51) into a single formula, we get

$$\text{Sin}_{n+1} = \sqrt{\text{Crd}_0^{\,2} - \left(R - \sqrt{R^2 - \text{Sin}_n^{\,2}}\right)^2}. \qquad (3.52)$$

Since we are hoping that $\text{Sin}_{n+1} = \text{Sin}_n$, (3.52) may be thought of as an equation of the form

$$x = f(x), \qquad (3.53)$$

where we intend to solve for x. Considering both sides of this equation as functions of x (see figure 3.16), we are searching for the intersection of the line $y = x$ with the curve $y = f(x)$. Substituting our initial guess x_0 into $f(x)$ gives x_1; on the graph the height $f(x_0)$ may be transferred to a length x_1 by reflecting it about the line $y = x$. Repeated substitution in the example of figure 3.16 causes the iterates to spiral around the intersection point.

A moment's reflection (and perhaps a couple of sketches) should convince the reader that this method will converge if the slope of $f(x)$ near the intersection point is between -1 and 1. In practice convergence was probably verified simply by trial and error. There are some later texts (starting with Bhāskara II) that show evidence of concern over questions of convergence: usually alterations made to existing methods that improve their chances of success.[89] But the authors do not discuss convergence or divergence directly, and they never leave the astronomical context for more general mathematical discourse.

Fixed-point iteration is surprisingly flexible; with a little manipulation it can be made to solve many equations. In the case of Brahmagupta's solar motion (3.48) for instance, recalling that λ is given and λ_m is to be found, a little rearrangement gives

[89] [Plofker 2004, 565–569].

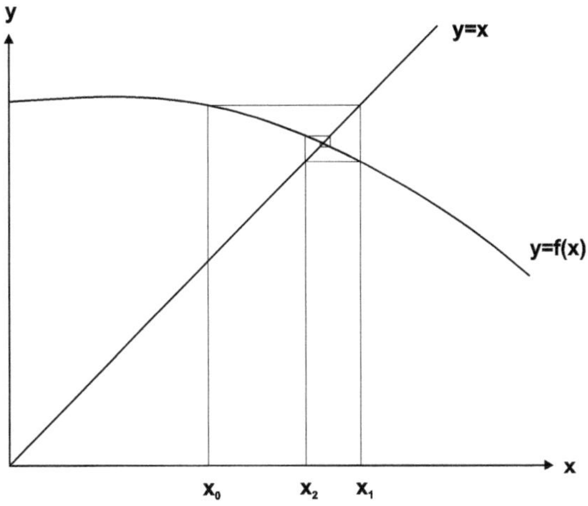

Figure 3.16
Fixed-point iteration

$$\lambda_m = \lambda + \arcsin\left(\frac{r \cdot \mathrm{Sin}(\lambda_m - M)}{\sqrt{\frac{r}{R} \cdot \mathrm{Sin}(\lambda_m - M)^2 + \left(R + \frac{r}{R} \cdot \mathrm{Cos}(\lambda_m - M)\right)^2}}\right).$$

This is now in the form $\lambda_m = f(\lambda_m)$, and fixed-point iteration may be applied.[90]

The two-point techniques used in India, especially *regula falsi*[91] but also (much later) the *secant method*, begin with a pair of initial estimates (or "seeds") rather than a single one. The next iterate is generated by applying linear interpolation between the two seeds. The choice of the next pair of estimates depends on the method. Again, although *regula falsi* was applied in different astronomical contexts, it was never discussed as an independent topic.

Conclusion

It is clear by now that the debate point behind many conversations about Indian astronomy is the degree to which it depends on its Greek antecedent, and astronomical mathematics is no exception to this. There seems little

[90] [Plofker 2002b, 172].
[91] Not to be confused with the "rule of false position" used in early mathematical history to solve linear equations, although it is related.

doubt that the spark for trigonometry in India came through the importing of some pre-Ptolemaic version of Greek mathematical astronomy. The earliest trigonometric tables share some features with their Greek forebears, although the precise nature of this connection may never be known. However, the Indian computational style clearly took the mathematical side of the subject in new directions—most obviously through the use of the sine, cosine and versed sine, but also through new methods for computing trigonometric quantities and various interpolation and iterative schemes. Methods used in Indian spherical astronomy may reflect Greek antecedents, but well-known Greek tools such as the analemma and Menelaus's Theorem did not leave direct traces in Indian texts so far as we know. Clearly, Indian trigonometry was full of its own creative ideas; and both in content and method, it became something quite distinct from its Greek heritage.

4 Islam

Foreign Junkets: The Arrival of Astronomy from India

The birth of Islam in the early seventh century and the spectacular, unprecedented growth of its culture over the next decades led in almost no time to the unification of a vast collection of smaller societies. What had been home to a number of different traditions soon became a single entity stretching from the borders of India westward through the middle East, across northern Africa, and eventually to northern Spain and the borders of France. By adopting both a common religion and Arabic as the common language, the Islamic world became a permanent part of the global cultural landscape. It endures and thrives still today even though its political unity began to dissolve within a century of its birth.

It did not take very long for traces of Indian learning, including astronomy, to find their way to the new society;[1] one of the more significant sources was the early eighth-century *Zīj al-Arkand*, based on Brahmagupta's *Khaṇḍakhādyaka*. In the early 770s a political delegation from Sind (modern Pakistan) to caliph al-Manṣūr in Baghdad took along an Indian astronomer; perhaps even more important than the diplomatic results, this trip led to the *Zīj al-Sindhind*,[2] a translation of a Sanskrit astronomical text influenced by both the *Paitāmahasiddhānta* and Brahmagupta's *Brāhmasphuṭasiddhānta*.[3] More than most early works, the *Zīj al-Sindhind* brought Indian computational and astronomical techniques to early Islam.

Some of these methods were developed in the most important astronomical treatise to emerge from the Indian tradition in Islam: Abū Jaʿfar Muḥammad ibn Mūsā al-Khwārizmī's early ninth-century *Zīj al-Sindhind*, based on the earlier work of the same name. Al-Khwārizmī's books on arithmetic (from which the word *algorithm*, based on his name, is derived) and algebra (derived from the word *al-jabr*, the name for a common algebraic operation) enjoyed considerable influence, particularly through Muslim Spain and into the West. Much the same may be said of his *Zīj*, which would be a major

[1] See [Goldstein 1986] for a more detailed summary of the arrival of astronomy in Islam, including both the Sasanian sources and the translation period in the ninth century. See also [Lorch 2001a], which contains a substantial bibliography on issues of transmission throughout the medieval period.

[2] Al-Fazārī, who aided in this translation, played a large role in the transmission of Indian methods to Islam. For instance, the Indian parameters for the base circle radius $R = 3438$, 3270, and 150 are all found in his works [Pingree 1971a, 555]. [Pingree 1970b] gathers together many of the sources and references on al-Fazārī's life and work.

[3] [Pingree 1971a, 555–556].

source for the Spanish *Toledan Tables* and existed in various forms for over a millennium. The *Zīj*'s history is correspondingly complicated; the original Arabic version no longer exists, and segments have been altered or replaced over the centuries.[4]

Al-Khwārizmī's *Zīj* uses sines and versed sines (*sahm*, in Latin *sagitta*) to solve astronomical problems, and implicitly applies tangents and cotangents to solve altitude problems using gnomons and shadows. The sine table in the existing version of this work gives entries to three sexagesimal places and uses $R = 60$, but the commentators al-Bīrūnī and al-Muthannā mention the presence of a sine table using the Indian parameter $R = 150$. It has been argued recently that al-Khwārizmī in fact included two sine tables in his *Zīj*, the existing $R = 60$ table for intervals of $1°$ as well as a small table for pedagogical purposes including only the *kardaja* increments of $15°$ each, something like Brahmagupta's figure 3.8.[5] If this claim is true, it is easy to see why the smaller table disappeared fairly quickly in favor of the more extensive one.

We have seen that the Indian astronomical works influencing early Islam in turn owe some debt to their Greek predecessors. In one of the history of science's more curious moments, a second path of transmission to the Islamic world opened up for Greek science in the early ninth century with the establishment of 'Abbāsid caliph al-Maʿmūn's "House of Wisdom" in Baghdad.[6] Possibly in connection with this institute, many of the best ancient scientific works were translated from Greek, Sanskrit, and Syriac into Arabic. The cream of Greek mathematics, including works by Euclid, Archimedes and Apollonius, soon found its way to Islam. In mathematical astronomy this collection included many books that we have seen so far: Euclid's *Phenomena*, Aristarchus's *On the Sizes and Distances of the Sun and Moon*, Hypsi-

[4] [Al-Khwārizmī (Suter) 1914] is an edition of a twelfth-century Latin translation of the *Zīj*; [al-Khwārizmī (Neugebauer) 1962] is a translation and commentary based on it. See [Toomer 1973b] for a survey of al-Khwārizmī's scientific achievements. [Van Dalen 1996] also sorts out the origin of the materials in the extant twelfth-century Latin translation of the Cordoba recension of the *Zīj* by al-Majrīṭī (ca. 1000).

[5] [McCarthy/Byrne 2003]. [Hogendijk 1991] reconstructs a sine table with $R = 150$ from a manuscript of a treatise on the astrolabe by al-Khwārizmī, and demonstrates that it is derived by quadratic interpolation from a table computed for every $15°$. The existing $R = 60$ table appears to come from the Ptolemaic tradition (this was first asserted in [Bjørnbo 1909, 11]), but whether or not it is due to al-Khwārizmī himself is uncertain: several authors assume the latter, but [McCarthy/Byrne 2003] argue the former. See also [van Dalen 1996] for an analysis of al-Khwārizmī's table of the equation of time as well as a valuable summary of previous research on all of al-Khwārizmī's tables, including the trigonometric ones.

[6] See [Berggren 2002] for a survey of the translation movements that brought Greek geometry to Islam, including mathematical astronomy. On the translation movement in general see [Gutas 1998], especially pp. 53–60 for an argument that the House of Wisdom did not play as dominating a role as has been claimed.

cles's *On the Rising of the Twelve Signs of the Zodiac*, and both Theodosius's and Menelaus's *Spherics*.[7] But by far the most important astronomical work to enter the Islamic world was Ptolemy's *Almagest*. Through three translations beginning in the eighth century, the *Almagest* was in the strange position of competing with the theories of its predecessors, in Indian garb. Before long, it won most of these intellectual battles.

Basic Plane Trigonometry

The construction of the basic trigonometric function was one struggle that the *Almagest* partly lost to the Indian works, and partly won. The natural advantage of the sine in astronomical computation was quickly recognized, and it became the standard trigonometric function. However, the radius of the base circle and the methods used to compute sine tables derived from the *Almagest*.

Since trigonometry served astronomy almost exclusively in early Islam, the basic identities of plane trigonometry were generally found in the preliminary chapters of astronomical works, especially the *zīj*. This genre, which we have mentioned several times already, was one of the most important forms of astronomical exposition in medieval Islam.[8] Over 225 of these handbooks were compiled, usually containing a wide variety of astronomical subjects and extensive numerical tables, mostly of trigonometrically defined functions. *Zīj*es included the usual topics of planetary motions, eclipses, and so forth; but they also dealt to some extent with astronomical functions related to Muslim religious practice, timekeeping, and astrological computations. Within planetary theory and spherical astronomy, *zījes* after the Greek influx were inspired by a mix of the styles of Ptolemy's *Handy Tables* and the *Almagest*. Although the tables were normally the centerpiece of a *zīj*, the presentation of topics took its lead from the more theoretical *Almagest*.

Zījes often commence their trigonometric discussions with the standard identities, particularly the sine sum and difference formulas, the sine half-arc formula,[9] and others. Some of the early *zījes*, reflecting their Indian roots, include formulas relating the sine to the surprisingly useful versed sine and the chord. (Indeed the Arabic word for sine, *al-jaib*, came

[7] On the history of these Greek works in Arabic translation see [Matvievskaya 1981].

[8] The study of *zījes* was given a proper foundation with the publication of the classic [Kennedy 1956a], which surveyed the contents of twelve *zījes* and described 125 of them. It has been supplemented recently by the extensive article [King/Samsó 2001]. Benno van Dalen will shortly be completing a new and considerably expanded *zīj* survey.

[9] Even quarter-arc and eighth-arc formulas, in al-Bīrūnī's case. See [Schoy 1927, 10–12].

about by a transcription of the Indian *jyā* [chord] to the Arabic *jiba*, and then *jaib*.[10])

Perhaps the earliest astronomer to attack trigonometric identities systematically was Abū'l-Wafā' al-Būzjānī (AD 940–998), a mathematician who made substantial contributions to both geometry and arithmetic. His masterwork of astronomy achieved enough admiration to be called, itself, the *Almagest*.[11] Abū'l-Wafā' continued to use the chord and versed sine as well as the sine, and many of his theorems deal with their interconnections. He discusses many of the basic theorems in his *Almagest*, although not necessarily in modern form. For instance, his

$$\frac{\text{Crd}\,\alpha}{\text{Crd}\,\alpha/2} = \frac{\text{Crd}\,(180^\circ - \alpha/2)}{R}, \tag{4.1}$$

corresponds to our $\sin\alpha = 2\sin(\alpha/2)\cos(\alpha/2)$.[12]

Abū'l-Wafā''s *Almagest*, to which we shall return later, was not primarily a mathematical work: it still had as its central purpose the study of astronomy. It would not be long before a treatise entirely devoted to plane trigonometry was written: *Book on the Derivation of Chords in a Circle*, by Abū'l-Rayḥān al-Bīrūnī.[13] One of the greatest of Muslim scientists, al-Bīrūnī (973–1048) wrote a staggering number of works on astronomy, mathematics, geography, Indian literature, and many other topics. *On Chords*, a novel approach to deriving the trigonometric identities, may have been intended as the mathematical background to his classic astronomical work, the *Qānūn al-Masʿūdī*.[14]

The main idea of *On Chords* is to replace Ptolemy's Theorem, used in the *Almagest* to derive several identities, with the Theorem of the Broken Chord. In figure 4.1, $\overset{\frown}{AB} > \overset{\frown}{BC}$ and D is the midpoint of $\overset{\frown}{ABC}$; we drop perpendicular

[10] See [von Braunmühl 1900/1903, vol. 1, p. 49] and [Gutas et al. 1998]. The Arabic *jiba* and *jaib* are written the same way when the vowels are removed (as they often are in Arabic script); the word *jaib* means a fold, pocket, or gulf.

[11] For a discussion of Abū'l-Wafā''s *Almagest* and a translation of some fragments into French see [Carra de Vaux 1892].

[12] [Von Braunmühl 1900/1903, vol. 1, pp. 55–56]. Pages 54–59 remain even today a vital contribution to our understanding of the plane trigonometry in Abū'l-Wafā''s *Almagest*. But see also [Debarnot 1996, 522] for a transcription of all of Abū'l-Wafā''s identities into modern notation, and [al-Bīrūnī (Debarnot) 1985, 15–21] on his spherical trigonometry.

[13] *On Chords* was translated into German in [Suter 1910/1911] and into Russian in [al-Bīrūnī (Rosenfeld/Krasnova/Rožanskaja) 1963]; summaries are available in [Saidan 1979] and more concisely in [Saidan 2000, 169–173]. See also the partial translation into English in [Saud 1979].

[14] Although Arabic editions exist, this massive work has not yet been translated into a Western language. However, extensive and detailed tables of contents were published in [Kennedy 1971b], and in Russian in [Rosenfeld/Rožanskaja 1969].

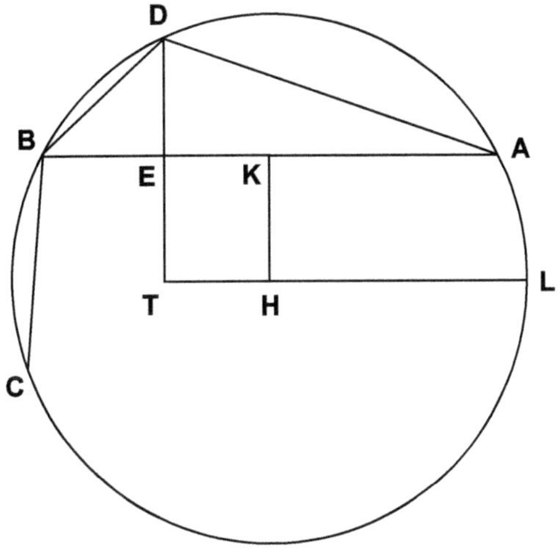

Figure 4.1
Al-Bīrūnī and the Theorem of the Broken Chord

DE onto *AB*. From this configuration al-Bīrūnī establishes the following propositions:

- $AE = EB + BC$ (the Theorem of the Broken Chord);
- $\text{Crd}\left(\widehat{AB}\right) \cdot \text{Crd}\left(\widehat{BC}\right) + \text{Crd}\left(\widehat{DB}\right)^2 = \text{Crd}\left(\widehat{AD}\right)^2$;
- If D is the midpoint of $\widehat{AB}$, then $\text{Crd}\,(\widehat{AB}) \cdot \text{Crd}(\widehat{BC}) + \text{Crd}(\widehat{CD})^2 = \text{Crd}(\widehat{DB})^2$;
- $DE \cdot EB = \text{area}\ (\Delta ADC) - \text{area}\ (\Delta ABC)$.

Al-Bīrūnī provides multiple proofs of each theorem, including a staggering twenty-five of the first—one of which he attributes to Archimedes in a work that is now lost. The reader may recall that this attribution has been called upon to assert that Archimedes himself had a form of trigonometry. However, the extraction of trigonometric identities from this theorem is not trivial, and al-Bīrūnī does not ascribe to Archimedes any trigonometric intent.

Equivalents to the standard identities are demonstrated, as well as a few less familiar ones. For instance, given Crd α and Crd β al-Bīrūnī shows how to find Crd $((\alpha + \beta)/2)$ as follows. In figure 4.1 let $\alpha = \widehat{AB}$, $\beta = \widehat{BC}$, and H be the center of the circle. Extend HK and HT perpendicular to AB and DE respectively. Then $HK = ET = \frac{1}{2}\text{Crd}(180° - \alpha)$,[15] which gives ET.

[15] This may be seen by extending *AH* to the far side of the circle, say to *X*, and joining *BX*. Then $\Delta AHK \sim \Delta AXB$, and since *AH* is a radius and *AX* a diameter, one triangle has half the dimensions of the other.

Also, from the Theorem of the Broken Chord $AE=(AB+BC)/2$,[16] so $KE=AE-AK=AE-AB/2=BC/2=\frac{1}{2}\text{Crd }\beta$ is also known. But if we have KE then we have TH, and hence LT. Now the product of LT and the rest of the diameter is equal to the square of DT,[17] so we know DT. But we know ET already, so by subtraction we have DE. Apply the Pythagorean Theorem to ΔADE, and we have AD, which is equal to Crd $((\alpha+\beta)/2)$.[18]

Building a Better Sine Table

The reason that the *zījes* studied identities, at least initially, is the same as Ptolemy's intent in the *Almagest*—to build a fundamental trigonometric table. But there is a problem that goes beyond the identities: as we have seen before, a mathematical problem arises if one divides the circle into 360 degrees. All geometrically derived sine values, such as Sin 30°, Sin 36°, and Sin 45°, are multiples of 3°. The sine sum and difference laws and the half-angle formula allow us to compute many other sines, but none for integer-valued arcs other than the multiples of 3°. This forces the aspiring table builder to use an approximation of some sort to find the crucial value of Sin 1°; once this step is taken, the identities allow the sines of all other integer-valued arcs to be found.

Now Ptolemy worked with chords, but it is trivial to convert his method for dealing with this problem (which we saw in chapter 2) to sines; the result can be written as

$$\frac{2}{3}\cdot\text{Sin}\frac{3}{2}^{\circ}<\text{Sin }1^{\circ}<\frac{4}{3}\cdot\text{Sin}\frac{3}{4}^{\circ}. \tag{4.2}$$

This inequality can be represented visually as in figure 4.2; on the graph of the sine function both the lower bound and upper bound are linearly interpolated between the origin and the known sine value, leading to the error bounds drawn in bold on the figure.[19]

In Ptolemy's case, the upper and lower bounds come to the same value to all three of the sexagesimal places that he uses. But if one wishes to go beyond that (and most Islamic table authors did), the method must be enhanced

[16] The Theorem gives $AE = EB + BC$; add AE to both sides of the equation and the result follows.
[17] By, for instance, *Elements* VI.13.
[18] From [Suter 1910/1911, 53–54].
[19] It is of course grossly anachronistic to draw these figures as functions in Cartesian coordinates (although see the section later in this chapter on astronomical instruments), but figure 4.2 and figure 4.5 will help the modern reader see why Islamic methods provided such sharp improvements on Ptolemy.

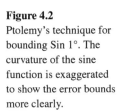

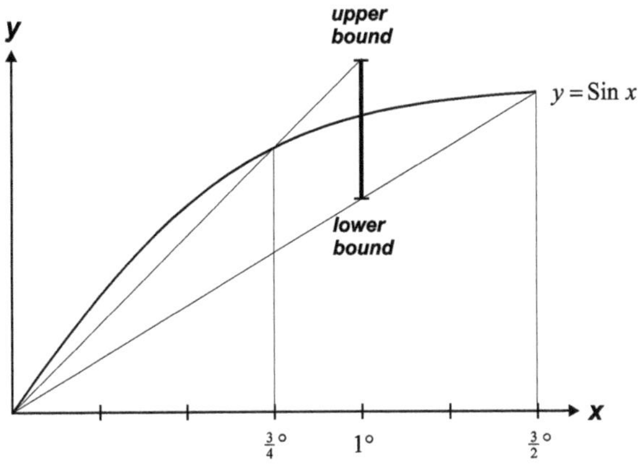

Figure 4.2
Ptolemy's technique for bounding Sin 1°. The curvature of the sine function is exaggerated to show the error bounds more clearly.

somehow. The earliest known improvement is by Ibn Yūnus, an Egyptian astronomer of the late tenth century, in his *Ḥākimī Zīj*. He begins simply enough by building the bounds from geometrically constructible sines of arcs closer to 1° than Ptolemy had used:[20]

$$1;2,49,40,4 \;=\tfrac{8}{9}\cdot \text{Sin}\tfrac{9}{8}^\circ < \text{Sin } 1^\circ < \tfrac{16}{15}\cdot \text{Sin}\tfrac{15}{16}^\circ = 1;2,49,45,10. \quad (4.3)$$

Since $\frac{15}{16}^\circ$ is half as distant from 1° as $\frac{9}{8}^\circ$ is, Ibn Yūnus takes as his estimate a sine value half as distant from the upper bound as from the lower:

$$\begin{aligned}\text{Sin } 1^\circ &= 1;2,49,40,4 + \tfrac{2}{3}(1;2,49,45,10 - 1;2,49,40,4)\\ &= 1;2,49,43,28,\end{aligned} \quad (4.4)$$

effectively interpolating linearly between the bounds. But Ibn Yūnus is not finished. He takes this value for Sin 1° as the basis for two computations of Sin 2°: one using the sine double-arc formula, the other using Sin 3° and the sine difference formula. The two values of Sin 2° that arise turn out to be upper and lower bounds for the correct value; hence, their difference (0;0,0,0,48) gives the sum of the two errors. Ibn Yūnus takes half of this number to be the error in his value for Sin 1° and subtracts it, giving

$$\text{Sin}1^\circ = 1;2,49,43,28 - \frac{1}{2}(0;0,0,0,48) = 1;2,49,43,4 \quad (4.5)$$

[20] Recall that sine of arcs of the form $(3m/2^n)^\circ$ are geometrically constructible.

θ	Sin θ	
0;10°	0;10,28,19	
0;20	0;20,56,38	
0;30	0;31,24,57	[+1]
⋮	⋮	
1;0	1;2,49,43	
	⋮	
2;0	2;5,38,19	[+2]
⋮	⋮	
10;0	10;25,8,1	[+1]
⋮	⋮	
40;0	38;34,2,8	[+1]
⋮	⋮	
89;30	59;59,51,47	
89;40	59;59,56,21	
89;50	59;59,59,5	
90	60;0,0,0	

Figure 4.3
An extract from the sine table in the *Ḥākimī Zīj*

(as opposed to the correct value 1;2,49,43,11).[21] Starting with this fundamental quantity, Ibn Yūnus goes on to compute an impressive table of sines for every 10′ of arc, extracted in figure 4.3.[22]

Around the same time in Baghdad, Abū'l-Wafā' came up with a way to approximate Sin 1° that was to be revisited every so often for four centuries. His method relies on the following lemma:

$$\text{Sin}\,(\alpha+\beta)-\text{Sin}\,\alpha<\text{Sin}\,\alpha-\text{Sin}\,(\alpha-\beta) \tag{4.6}$$

[21] Ibn Yūnus's work on this topic is available only in the unpublished doctoral dissertation [King 1972, 78–80], and is summarized in [Debarnot 1996, 525]. Although Ibn Yūnus's application is rather sloppy, the idea of using two calculations of Sin 2° to correct an estimate of Sin 1° is sound. If ε is the error in the original value for Sin 1°, then the double-arc formula generates an error of about 2ε in Sin 2°, whereas the difference formula generates an error of about $-\varepsilon$. Thus the difference in the two values of Sin 2° is about 3ε. So, had Ibn Yūnus replaced the 1/2 in (4.5) with 1/3, he would have had a very good estimate indeed for Sin 1° : 1;2,49,43,12. And now that a method to improve the estimate has been found, there is no reason not to repeat the process using the improved value of Sin 1° as a new starting point, thereby converging to the true value of Sin 1° iteratively (or at least, as close as the accuracy in the used value of Sin 3° allows). Of course this is all modern amusement; although it is in the spirit of Islamic work inspired by its Indian heritage (especially that of Ḥabash al-Ḥāsib) to do this sort of thing, there is no evidence to suggest that Ibn Yūnus or anyone else thought of this particular idea.

[22] Taken from [King 1972, 84]; entries recomputed anew.

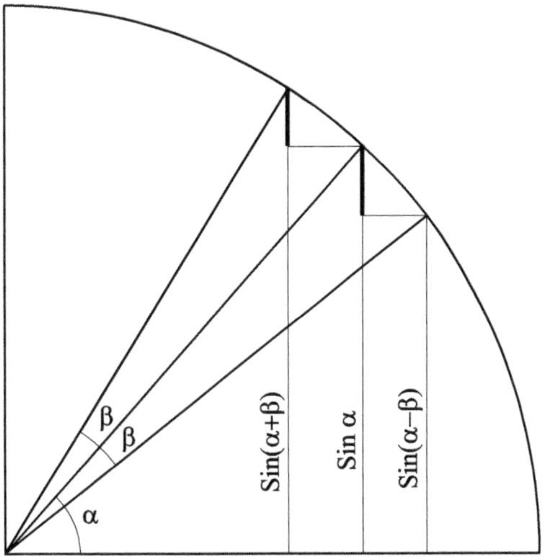

Figure 4.4
The lemma to Abū'l-Wafā''s calculation of $\mathrm{Sin}\,\frac{1}{2}^{\circ}$

(for $\beta < \alpha < \alpha + \beta < 90^{\circ}$). This statement is obvious once one examines the corresponding diagram: it means that successive differences of sines decrease as the arcs increase (the higher of the two bold line segments in figure 4.4 is smaller than the lower one)—i.e., that the graph of the sine function for $0^{\circ} < \theta < 90^{\circ}$ is concave down.

Abū'l-Wafā', actually searching for $\mathrm{Sin}\,\frac{1}{2}^{\circ}$, applies the lemma twice as follows. Two bracketing values, $\mathrm{Sin}\,\frac{15}{32}^{\circ}$ and $\mathrm{Sin}\,\frac{18}{32}^{\circ}$, are found by the usual geometric means. Since

$$\mathrm{Sin}\,\frac{16}{32}^{\circ} - \mathrm{Sin}\,\frac{15}{32}^{\circ} > \mathrm{Sin}\,\frac{17}{32}^{\circ} - \mathrm{Sin}\,\frac{16}{32}^{\circ} > \mathrm{Sin}\,\frac{18}{32}^{\circ} - \mathrm{Sin}\,\frac{17}{32}^{\circ},$$

it follows that $\mathrm{Sin}\,\frac{16}{32}^{\circ} - \mathrm{Sin}\,\frac{15}{32}^{\circ}$ (the largest of these three differences) is greater than $\frac{1}{3}$ of their sum, the computable $\mathrm{Sin}\,\frac{18}{32}^{\circ} - \mathrm{Sin}\,\frac{15}{32}^{\circ}$. Hence

$$\mathrm{Sin}\,\frac{1}{2}^{\circ} > \mathrm{Sin}\,\frac{15}{32}^{\circ} + \frac{1}{3}\left(\mathrm{Sin}\,\frac{18}{32}^{\circ} - \mathrm{Sin}\,\frac{15}{32}^{\circ}\right). \tag{4.7}$$

By a similar argument

$$\mathrm{Sin}\,\frac{1}{2}^{\circ} < \mathrm{Sin}\,\frac{15}{32}^{\circ} + \frac{1}{3}\left(\mathrm{Sin}\,\frac{15}{32}^{\circ} - \mathrm{Sin}\,\frac{12}{32}^{\circ}\right), \tag{4.8}$$

and we have bounded $\mathrm{Sin}\,\frac{1}{2}^{\circ}$.[23]

[23] This method was first discussed in the modern historical literature, and was analyzed minutely, in [Woepcke 1854a, 155–167] as it appeared in the work of the early sixteenth-century scientist Mirim Chelebi, grandson of Jamshīd al-Kāshī's successor at the observatory in Samarqand. Abū'l-Wafā''s work appears in [Woepcke 1860].

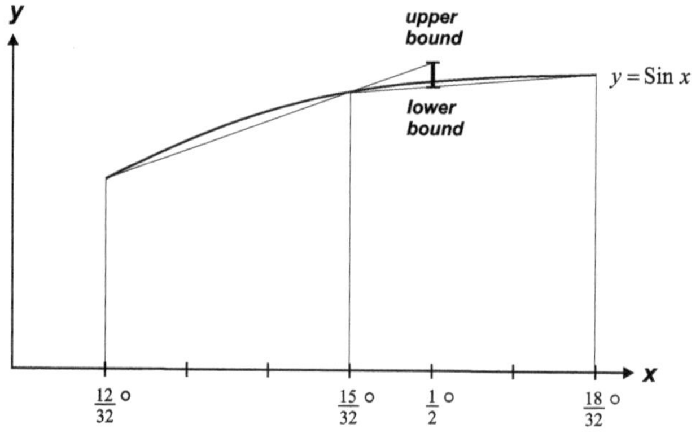

Figure 4.5
Abū'l-Wafā''s technique for bounding Sin $\frac{1}{2}^{\circ}$

Working with differences of sines, rather than the sines themselves, gave Abū'l-Wafā' a leg up on Ptolemy and Ibn Yūnus. His method is equivalent to approximating the sine curve by drawing a line between two known points on the curve, rather than drawing lines from the origin to the given points (see figure 4.5, and compare with figure 4.2). This leads to upper and lower bounds roughly six times tighter than Ptolemy's method is able to achieve with the same givens.[24] Abū'l-Wafā''s value $\text{Sin}\frac{1}{2}^{\circ} = 0;31,24,55,54,55$ is in error only in the fifth sexagesimal place,[25] an order of magnitude better than Ibn Yūnus's value of Sin 1°—and it could have served as a basis for his lost four-sexagesimal-place sine table. This method was used afterward by several mathematicians, as late as Jamshīd al-Kāshī in his early fifteenth-century *Khāqānī Zīj*—but as we shall see in a moment, al-Kāshī had considerably more to offer on this point.

One novel approach to the sine table problem, from around the same time as Ibn Yūnus and Abū'l-Wafā', attacks it with an iterative technique. A mathematical chapter of al-Bīrūnī's *Qānūn al-Masʿūdī* contains a discussion of the regular nonagon which turns out to be helpful in the production of a table of chords.[26] Of most interest to us is its use in the determination of Crd 1°.

[24] Using correct values of the sines of $\frac{12}{32}^{\circ}$, $\frac{15}{32}^{\circ}$, and $\frac{18}{32}^{\circ}$, the difference between Abū'l-Wafā''s upper and lower bounds is 0;0,0,0,5,40,38. Using the correct sines of $\frac{15}{32}$ and $\frac{18}{32}$, the difference between the upper and lower bounds using Ptolemy's method is 0;0,0,0,33,18,26. See [Debarnot 1996, 527, n. 50] for a similar calculation.

[25] The correct value is 0;31,24,55,54,0,13.

[26] The following discussion is based on [Schoy 1927, 18–30, esp. 21–22]; an account in English of the same material is in [Schoy 1926]. See also [Kazim 1951] for a survey of the trigonometry in the *Qānūn al-Masʿūdī*.

One side of a nonagon subtends an arc of 40°; its chord is just as unattainable as that of 1°. However, the chord of the nearby value 42° is available, and al-Bīrūnī computes it: 0;43,0,14,57,15. Apply the half-arc formula twice, and we have $\text{Crd}\,10\frac{1}{2}^\circ = 0;10,58,48,41,56$. Combine this value with Crd 30° using the chord summation formula, and we arrive at the chord of an arc even closer to 40°, namely Crd 40;30° = 0;41,32,2,34,6.

But there is no reason to stop here. Apply the half-arc formula twice and the summation formula again, and we arrive at Crd 40;7,30° = 0;41,9,15,26. Every time this process is repeated, the difference from 40° decreases by a factor of four. Apply it eight more times and we are nearly there:

$$\text{Crd}\,40;0,0,0,24,43,9,15,28,7,30^\circ = 0;41,2,32,42,29. \tag{4.9}$$

As an approximation to Crd 40° this value is quite good (the last place should be a 7). From here it is an easy matter to find Crd 1°; for instance, use the chord difference formula with 40° and 36° to get Crd 4°, and apply the half-arc formula twice.

There is a subtle but crucial change in al-Bīrūnī's trigonometry in this example: he uses $R = 1$, not 60. Indeed, the sine table in the *Qānūn al-Masʿūdī* uses a unit radius and is hence equivalent to a modern one. This possibility and its advantages had been known and implemented already in Abū'l-Wafā''s work[27] as well as in the *Table of Minutes* of al-Bīrūnī's teacher Abū Naṣr Manṣūr,[28] but it was seldom used afterward by Islamic astronomers. Since calculations were in base 60 in any case, the advantage in switching to $R = 1$ was minimal, equivalent to saving the occasional shift of a decimal point.

One curious episode in the history of Sin 1° is the novel approach taken by the twelfth-century mathematician al-Samaw'al ibn Yaḥyā al-Maghribī. Renowned for his contributions to algebra at a very young age, he also wrote a lengthy astronomical work late in life (*Exposure of the Errors of the Astronomers*) in which he takes to task a number of his predecessors and contemporaries for various sins.[29] Al-Samaw'al's take on the matter is as follows.

[27] See for instance [von Braunmühl 1900/1903, vol. 1, p. 58].

[28] The Table of Minutes applies the notion of auxiliary functions, which we shall discuss later. Abū Naṣr Manṣūr explicitly sets $R = 1$, so one of his functions is simply the modern sine. Indeed, the title "Table of Minutes" refers to the fact that his functions have their most significant sexagesimal place in the minutes position rather than the units position. See [Jensen 1972].

[29] For one example of al-Samaw'al's attitude, see [Berggren/Van Brummelen 2003] on his criticism of Abū Sahl al-Kūhī's work on finding the dip angle to the horizon when elevated above the sphere of the earth.

Text 4.1
Al-Samaw'al ibn Yaḥyā al-Maghribī,
Why the Circle Should Have 480 Degrees
(from *Exposure of the Errors of the Astronomers*)

They relied on the [bounds on the chord of 1°] from both sides, not for the inequality, but to determine the chord . . . between the two bounds. Nobody knows its value; for them it remains as this approximation and the truth is that there is no chord that they could rely upon.

But if they divided the circle not by their division, that is on 360, and they corrected it to 240 or to 480, they would be able to compute the chord of one [degree] . . . .

Because, if the circle is 480 [parts], then the chord of the [pentagon = 96 parts] and of the [hexagon = 80 parts] are known, and the chord of their difference is known, so the chord of 16 is known.

Then the chord of 8 [parts] is known, and the chord of 4 [parts] is known, and the chord of 2 [parts] is known and the chord of one [part] is known.

And if they did that, they would have been able to compute these chords according to this division and compute the chord for each [part].[30]

Explanation: Al-Samaw'al does not approve of the entire approach of trapping Crd 1° between bounds; for him this is an approximation and not appropriate for geometrical work. The only recourse is to redefine the number of degrees in a circle! By shifting from 360° to 480°, all chords and sines may be found geometrically without any need for bounds, since what was the geometrically accessible $\frac{3}{4}^{\circ}$ becomes 1° in al-Samaw'al's new units.

As far as we know al-Samaw'al's idea did not catch on, but he did construct a sine table based on this model, extracts of which are in figure 4.6.[31]

Our last venture into the problem of Sin 1° is also its ultimate solution. The Iranian astronomer Jamshīd al-Kāshī (?–1429) was without a doubt the greatest computational scientist of his time; his achievements are still being discovered today. His *Calculator's Key*, on arithmetic and algebra, contains many gems of elementary, yet sophisticated mathematics, including a method for computing the fifth root of an arbitrary number.[32] His astronomical work the *Khāqānī*

[30] Translation by Amani Ansari and Glen Van Brummelen.

[31] Al-Samaw'al's efforts on chords and the sine table are being prepared for publication in [Van Brummelen/Kerai, to appear].

[32] On the *Calculator's Key* see [Luckey 1951]; [Berggren 1986, 53–63] is a detailed account of the extraction of a 5th root.

Figure 4.6
An extract from al-Samaw'al's table of sines for a circle of 480°

θ	Sin θ	
1°	0;47,8	[+1]
2°	1;34,15	[+1]
3°	2;21,20	
⋮	⋮	
40°	30;0,0	
⋮	⋮	
60°	42;25,35	
⋮	⋮	
80°	51;57,41	
⋮	⋮	
118°	59;58,33	[−13]
119°	59;59,35	[−6]
120°	60;0,0	

Zīj[33] is full of mathematical treasures, a couple of which we shall revisit later in this chapter. He designed and built a number of ingenious astronomical instruments. Closer to our topic, he was the first to compute π beyond the equivalent of six decimal places, reaching a full sixteen.[34] Late in his relatively short life he became a leading member of Ulugh Beg's scientific court in Samarqand, and had a hand in the construction of one of the most accurate and heavily used of all Islamic astronomical handbooks, the *Sulṭānī Zīj*.

Al-Kāshī's original treatise on Sin 1° is lost, but it seems to have provoked a flood of commentaries and variants after his death, so the method can be reconstructed with some accuracy.[35] The first of its two central ideas is to recognize that Sin 1° is a root of a relatively simple cubic equation. One

[33] Several aspects of the *Khāqānī Zīj* have been studied; see for instance [Kennedy 1985] and [Van Brummelen 2006]. [Kennedy 1998] is a detailed description of the contents of this remarkable *zīj*.

[34] On this calculation (based on something similar by Archimedes) see [Luckey 1950]; it has been described a number of times since then, including a popular account in [Van Brummelen 1998].

[35] The five known historical treatises on al-Kāshī's method are summarized in [Rosenfeld/Hogendijk 2002–2003, 27]. Among the publications on this topic are the following: [Sédillot 1847/1853, vol. 2, 77–83] discusses Chelebi's work, later published in [Sédillot 1853] and analyzed in [Woepcke 1854a, 167–175]. [Aaboe 1954] is a clear and concise article describing the method as found in Chelebi (on which we rely here), and [Van Brummelen 1998] contains a popular account illustrating the connection to fixed-point iteration. [Rosenfeld/Hogendijk 2002–2003] contains a facsimile and translation into English of another treatise on the subject, by Qāḍī-Zādeh al-Rūmī or perhaps by Ulugh Beg himself. The argument for attribution of the treatise to the latter is made in [Ahmedov/Rosenfeld 1975].

of the sine triple-angle identities, easily derived from the sine summation formula, is

$$\mathrm{Sin}\,3\theta = 3\cdot \mathrm{Sin}\,\theta - 0;0,4\cdot (\mathrm{Sin}\,\theta)^3.^{36} \tag{4.10}$$

Substituting $\theta = 1°$ and $x = \mathrm{Sin}\,1°$, we arrive at the fundamental equation

$$\mathrm{Sin}\ 3° = 3x - 0;0,4x^3; \tag{4.11}$$

and since Sin 3° may be found by geometric means, we need only solve this equation.

The second idea is a way to find a solution to (4.11) without having to solve it analytically. Al-Kāshī begins by isolating the x term:

$$x = \frac{x^3 + 900\cdot \mathrm{Sin}\,3°}{2700} = \frac{x^3 + 47,6;8,29,53,37,3,45}{2700}. \tag{4.12}$$

We know from previous estimates that $x = 1;\_,\_,\_,\ldots$, which makes the x^3 term in the numerator on the right of (4.12) considerably smaller than 47,6;8, ... Thus, subtracting 1 from both sides,

$$0;\_,\_,\_,\ldots = \frac{(1;\_,\_,\ldots)^3 + 2,6;8,29,53,\ldots}{2700}. \tag{4.13}$$

Assuming the bracketed quantity in the numerator is just 1, we perform the division on the right side of (4.13) and get a result of 0;2, ... So, we know the next sexagesimal place is 2.

To arrive at the next sexagesimal place after that we simply repeat the process: set the x on the right side of the equation to the current estimate 1;2 and subtract 1;2 from both sides. This gives

$$0;0,\_,\_,\ldots = \frac{(1;2,\_,\_,\ldots)^3 + 36;8,\ldots}{2700}, \tag{4.14}$$

and dividing through the right side as before gives the next sexagesimal place as 49. Al-Kāshī continues the process to ten sexagesimal places, concluding with

$$\mathrm{Sin}\ 1° = 1;2,49,43,11,14,44,16,19,16. \tag{4.15}$$

This value is accurate to all but the last two places (which should be 26,18)—well beyond any practical astronomical need.

[36] The modern equivalent is $\sin 3\theta = 3\sin\theta - 4(\sin\theta)^3$.

Table of the Sine				
	0°	1°		4°
0′	0;0,0,0,0	1;2,49,43,11	…	4;11,7,23,54
1′	0;1,2,49,55	1;3,52,32,31	…	4;12,10,4,35
2′	0;2,5,39,49	1;4,55,21,50 [+1]	…	4;13,12,45,12
3′	0;3,8,29,44	1;5,58,11,7	…	4;14,15,25,44
4′	0;4,11,19,38	1;7,1,0,23	…	4;15,18,6,12
5′	0;5,14,9,33	1;8,3,49,38	…	4;16,20,46,35

Figure 4.7 An extract from the first page of the sine table in Ulugh Beg's *Sulṭānī Zīj*. (Columns of differences, used for interpolation, have been suppressed.)

The tools al-Kāshī employs here—the triple-angle identity and the idea of iteration—had been in place for centuries, for instance in al-Bīrūnī's work on the nonagon. But al-Kāshī was the first to bring these ideas together in this way, using the language of algebra. What we presume to be the result of al-Kāshī's labors, a monumental table of sines to five sexagesimal places for every minute of arc in Ulugh Beg's *Sulṭānī Zīj* (see figure 4.7),[37] was the last word in Islamic sines. The table was not outdone anywhere until Georg Rheticus's *Opus Palatinum* almost two centuries later.

Introducing the Tangent and Other Trigonometric Functions

The earliest trigonometric functions—the sine, versed sine, and chord—are closely related. The addition of the cosine function did not add much power, since the cosine of an arc is simply the sine of its complement; so no new table for the cosine is needed if a sine table is already handy. On the other hand we have seen that Ptolemy and Indian astronomers would have been aided by the tangent or cotangent. However, it was a surprisingly persistent conceptual barrier that kept the tangent in the world of gnomonics, and away from trigonometry proper.

Tables of the cotangent, known as the "shadow" function, are found in some of the earliest *zījes*, but the argument is consistently specified as an altitude rather than a general arc. Interest in shadows derives primarily from the study of sundials—the shadow cast by the gnomon tells us about the Sun's altitude—and its tradition in the Arabic world is long and detailed.[38] Al-Khwārizmī's *Zīj*, which contains one of the earliest instances of the shadow

[37] Six places for arcs between 87° and 90°, where the function increases slowly.

[38] For surveys of the history of gnomonics in the Arab world see [Schoy 1913] and [Schoy 1923b].

θ (arc of altitude)	Shadow (θ)
1°	687;29
2°	343;38
3°	228;58
⋮	⋮
30°	20;47
⋮	⋮
45°	12;0
⋮	⋮
60°	6;56
⋮	⋮
88°	0;25
89°	0;12 [−1]
90°	0;0

Figure 4.8
An extract from al-Battānī's shadow table (the cotangent)

function, seems to have had a table with R (the length of the gnomon) equal to 12;[39] al-Battānī's highly influential *Zīj al-Ṣābi'* (AD 900) and many others followed suit (see figure 4.8).[40] The function tabulated was usually

$$\text{Shadow}(\theta) = R\cot\theta = 12\cot\theta. \quad (4.16)$$

The parameter $R = 12$, immortalized on shadow squares on the backs of astrolabes, derives from Indian sources. According to al-Bīrūnī in his well-named *Exhaustive Treatise on Shadows*, this value came about as the ratio of a hand-span to the width of a finger:

> Many (times) I used to witness the Indians, if they wanted to ascertain the time for their operations . . . , they would pass their hands in the direction of the sun so that the hand, from the vicinity of the elbow, became parallel to the horizon, with the inside of the forearm and the arm toward the sky. Then they would erect half the middle

[39] See [al-Khwārizmī (Suter) 1914, 60], [al-Khwārizmī (Neugebauer) 1962, 105], [ibn al-Muthannā (Goldstein) 1967, 87–89], and the summary in [van Dalen 1996, 207]. A cotangent table with $R = 1$, probably due to Ḥabash al-Ḥāsib, is reproduced in [King 2004, 84–87].

[40] Al-Battānī's *Zīj* was foundational in promulgating Ptolemaic astronomy, and it had a large and consistent following. See the edition and Latin translation in [al-Battānī (Nallino) 1899/1903/1907]. The table of cotangents, although given to only two sexagesimal places, is remarkably accurate in the entries near the cotangent function's singularity at 0°. In this regard other eminent table-makers were not so fortunate.

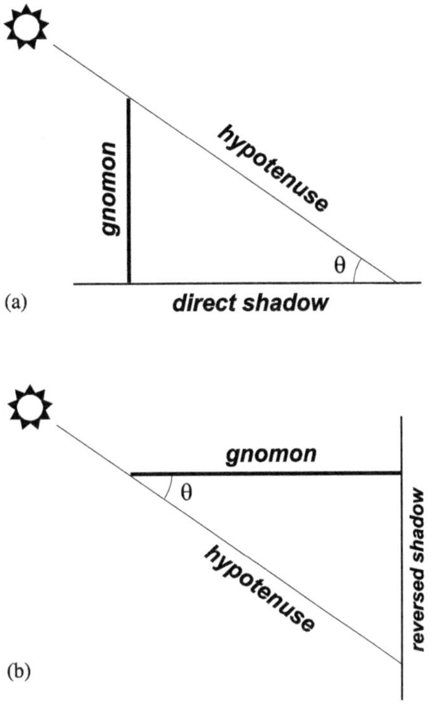

Figure 4.9
Al-Bīrūnī's definitions of the shadow functions

finger so that it became a gnomon, and its shadow extended itself along the inside of the palm and the forearm. Then they measured it in digits of the other hand.[41]

Other parameters were used, such as $R = 6\frac{1}{2}$ and $R = 7$ (from the ratio of a person's height to the length of his foot), $R = 60$, and of course Abū'l-Wafā''s and al-Bīrūnī's $R = 1$. But 12 was the most common choice.

Al-Bīrūnī's definitions of the shadow functions in the *Exhaustive Treatise on Shadows*, a standard within Arabic gnomonics, are as follows.[42] In figure 4.9(a) the gnomon is vertical; with argument θ the length of the "direct shadow" (*al-ẓill al-mustawī*) corresponds to our cotangent, while the "hypotenuse of the direct shadow" (*quṭr al-ẓill al-mustawī*) is the secant. Similarly, in figure 4.9(b) the gnomon is horizontal; the length of the "reversed shadow" (*al-ẓill maʿkūs*) is the tangent, and the "hypotenuse of the reversed shadow" (*quṭr al-ẓill maʿkūs*) is the cosecant.[43] So all six modern trigonometric functions are present—but only in a gnomonic context.

[41] [Al-Bīrūnī (Kennedy) 1976, vol. 1, 73].
[42] These definitions had already been used, for instance, by al-Battānī and Abū'l-Wafā'.
[43] [Al-Bīrūnī (Kennedy) 1976, vol. 1, 63–65; vol. 2, 23–24].

One nice example of al-Bīrūnī's masterful handling of gnomonics is his method to find the cardinal points of the compass during the summer months, after sunrise or before sunset (at a time when the North Star is not visible).

Text 4.2
Abū'l-Rayḥān al-Bīrūnī, Finding the Cardinal Points of the Compass
(from *Exhaustive Treatise on Shadows*)

[See figure 4.10.] Another [method] is that the graduated circle be *ABG*, with center *E* and *EB* perpendicular to the diameter *AG*, and we assume on it *BD* equal to the latitude of our locality and *GZ* equal to the declination of the Sun at the time [in question]. We join *D* [to] *E*, and we extend *ZH* parallel to *GE*, and we draw *ET* equal to *EH*. We extend *TK* parallel to *EB*, and *KLM* parallel to *GA*; and *MS* parallel to *BE*. And we join *L* [to] *S* and cut off *EO* to the amount of the gnomon set up at *E*. We extend *OF* parallel to *LS*, and we describe about *E* and at a distance *EF* a circle. We observe, in one of the two halves of the day, the entry of the shadow, and [its exit] from this small circle, and we pass from it a diameter in it which will be the east-west line, and the diameter perpendicular to it will be the meridian line.[44]

Explanation: Our problem is to determine the moment at which the shadow cast by a gnomon points due east (or west). Al-Bīrūnī tells us to mark off our terrestrial latitude ϕ and the Sun's declination δ in figure 4.10. Draw a line horizontally across from *Z* to *H*, and use a compass to draw $ET = EH$. Next draw straight up from *T* to *K*. Al-Bīrūnī asserts that *TK* is in fact the Sine of the Sun's altitude *h* when it is due east or west of us. If he is right, then $\overset{\frown}{MA}$ is *h* itself. Since $\angle LSE = \angle OFE$ is the angle corresponding to this arc and *OE* is the gnomon, we know that

$$EF = \text{Shadow}(h) = \text{Cot } h, \tag{4.17}$$

and the gnomon will cast a shadow of length *EF* at the moment that the shadow points due east or west.

Assuming that al-Bīrūnī is correct, all we have to do to determine the east/west line is to use figure 4.10 to determine *EF*, and draw a circle around the gnomon with radius *EF*. At the moment the shadow of the gnomon crosses the circle we will know that the shadow points due east/west, and the cardinal directions will have been determined.

[44] [Al-Bīrūnī (Kennedy) 1976, vol. 1, 168–169]. Kennedy's commentary is in vol. 2, pp. 95–96. See also [Kennedy 1959] and [Kennedy 1963]. This is an example of the use of an analemma, which we shall discuss in more detail later in this chapter.

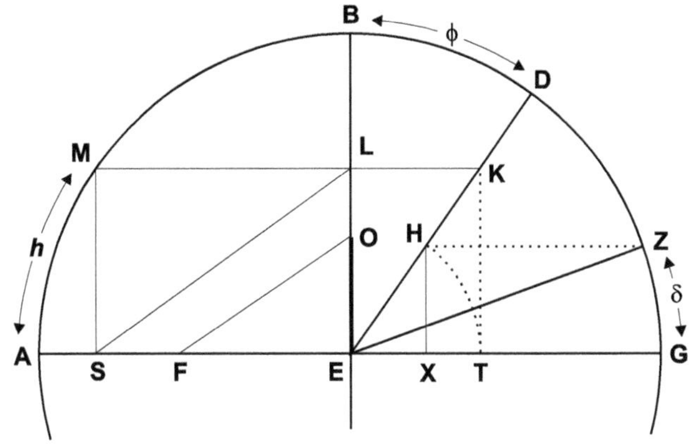

Figure 4.10 Al-Bīrūnī's construction of an east-west line using a gnomon

This is very clever indeed, but how do we know that TK is the Sine of h? Fortunately al-Bīrūnī goes on to prove it. In figure 4.11 the Sun rises at T and is due east of us at ☼; the arc ψ from the east point to the Sun's rising point T is called the ***ortive amplitude***. The Sun's path, or ***day-circle***, is separated from the celestial equator by its declination $\widehat{XB} = \delta$, so $DE = \text{Sin } \delta$.[45] Since $\angle DEG = \phi$,

$$\frac{ED}{EG}\left(= \frac{\text{Sin } \delta}{\text{Sin } \psi}\right) = \frac{\text{Cos } \phi}{R}. \quad (4.18)$$

But this latter quantity is also HX/EH in figure 4.10, and since $HX = \text{Sin } \delta$, $EH(=ET)$ must be equal to Sin ψ.

A similar argument gets us the second half of the proof. In figure 4.11 ZT/Z☼ $=$ Sin ϕ / Cos ϕ; in figure 4.10 ET/TK is equal to the same ratio. Since ZT and ET are both equal to Sin ψ, TK must be equal to Z☼, which is Sin h.

Al-Bīrūnī's unwillingness to simplify the above ratio Sin ϕ/Cos ϕ to a tangent reflects a general attitude to keep the gnomonic functions away from trigonometry proper.[46] This choice leads to the same problem that Ptolemy had experienced centuries before: using only a sine table, it is possible to compute a tangent—but not an arc tangent. In situations where an arc tangent

[45] One way to see this is to draw radii EX and EB; then $\angle BEX = \delta$, and the perpendicular from B to EX (which is equal to DE) is Sin δ.

[46] Converting the ratio of a sine and a cosine to a tangent changes the type of object from a ratio of two geometrical entities to a single entity; this would not have been natural.

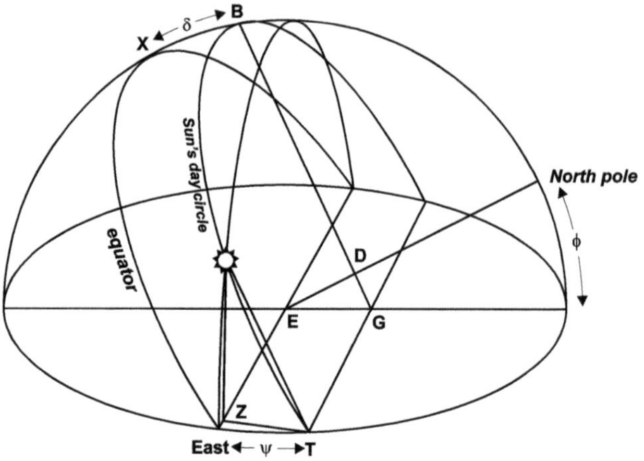

Figure 4.11
The proof of al-Bīrūnī's east-west line construction

was required, astronomers were forced to convert to an arc sine; i.e, if tan $\theta = a/b$, then

$$\phi = \text{arcSin}\left(\frac{R \cdot a}{a^2 + b^2}\right)^{47}. \tag{4.19}$$

The usefulness of dividing the sine of an arc by its cosine or vice versa beyond the confines of gnomonics was not entirely lost on early Islamic scientists. Ḥabash al-Ḥāsib, (fl. AD 850), one of the greatest of the mathematical astronomers (the name "al-Ḥāsib" actually means "the calculator"), tabulated and used the tangent in his *Jadwal al-Taqwīm* ("Table for Finding the Exact Positions").[48] These tables computed values of simple but useful combinations of the basic trigonometric building blocks, sines and cosines. Appropriately chosen, they could be combined to solve a wide variety of astronomical problems; we shall discuss them more systematically later in this chapter. Among Ḥabash's auxiliary functions was a tangent with $R = 60$, as well as two other functions using tangents:

$$\frac{\text{Tan }\theta \cdot \text{Sin }\varepsilon}{R} \text{ and Tan}^{-1}\left(\frac{\text{Tan }\varepsilon \cdot \text{Sin }\theta}{R}\right), \tag{4.20}$$

where ε is the obliquity of the ecliptic.

[47] See for instance Ibn Yūnus in [King 1972, 64–65].

[48] Detailed information on the *Jadwal al-Taqwīm* may be found in [Irani 1956]; on pp. 128–133 the case is made that Ḥabash was the first to use the tangent function to solve astronomical problems outside of gnomonics. The *zīj* in which this table appeared is summarized in [Debarnot 1987].

The tangent function was frequently useful in the solutions of problems in spherical astronomy; it arises especially when applying Menelaus's Theorem (or some equivalent) to a configuration wherein a 90° arc is broken into two parts. Often one needs to divide the sine of the first part by the sine of the other, which is equivalent to taking the tangent of the first part. Thus, although Ḥabash established the tangent only as an auxiliary function and not as a trigonometric function per se, he did find for it a role in the practice of spherical astronomy.

The tangent and the three minor functions (cotangent, secant, cosecant) were finally brought properly into the trigonometric fold with the sine and cosine in Abū'l-Wafā''s *Almagest*. Indeed, all six trigonometric functions may be defined by means of just one of his diagrams.[49] In figure 4.12 the argument is $\widehat{BZ}$; we have noted before that Abū'l-Wafā' introduces $R = 1$. Then the six functions are:

- sine $= TZ$;
- cosine $= ET$;
- tangent $= BH$;
- cotangent $= AY$;
- secant $= EH$; and
- cosecant $= EY$.

With all six functions in his arsenal Abū'l-Wafā' was able to solve astronomical problems more quickly than before. He also quickly demonstrated several basic identities involving the new functions, including

$$\sec\theta = \sqrt{1+(\tan\theta)^2} \text{ and } \csc\theta = \sqrt{1+(\cot\theta)^2}. \quad (4.21)$$ [50]

The advantages of including these functions in the trigonometric stable were not appreciated by all of Abū'l-Wafā''s colleagues. Ibn Yūnus, for example, either did not hear of the innovations or chose to continue under the old regime.[51] But with al-Bīrūnī's adoption, the continued life and acceptance of the new functions were assured. Their advantages in spherical astronomy, as Abū'l-Wafā' demonstrates dramatically in the section on spherical astronomy in his *Almagest*,[52] were too great to be ignored.

49 [Debarnot 1996, 513–514].

50 See either [von Braunmühl 1900/1903, vol. 1, 57] or [Cantor 1907/1913/1901/1908, vol. I, 748]. We have already noted that [von Braunmühl 1900/1903, vol. 1, 54–59] is the only available account of the trigonometry in Abū'l-Wafā''s *Almagest*.

51 [King 1972, 64].

52 Based on notes made by Benno van Dalen.

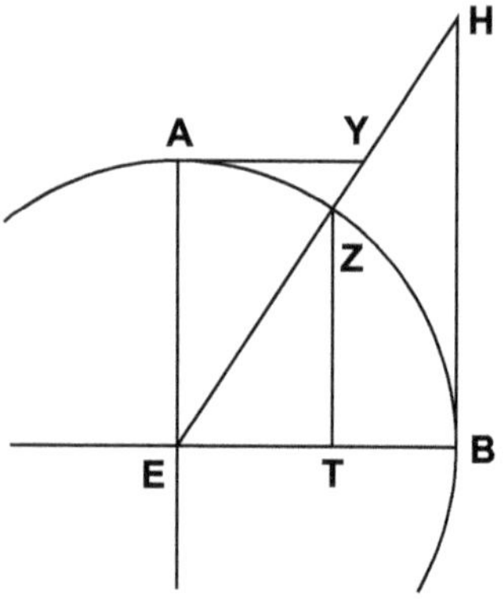

Figure 4.12
Abū'l-Wafā''s definitions of the tangent and other trigonometric functions

Streamlining Astronomical Calculation

As the sine and cosine had made the mathematics of astronomical work more convenient, the tangent and cotangent would also make their presence felt—but the transition was not immediate. One example of Islamic Ptolemaic astronomy should suffice for comparison: the solar equation, the key to the Sun's motion, which we saw handled with chords in Chapter 2. Recall that the Sun's mean position a_m in figure 4.13 is found using a mean motion table; the Sun's true position is determined by

$$a = a_m \pm q(a_m), \tag{4.22}$$

depending on whether a_m is less than or greater than 90°.

The following passage is by Kūshyār ibn Labbān, an Iranian scientist and Abū'l-Wafā''s colleague. His *Jāmiʿ Zīj* was an influential astronomical handbook, which we shall revisit later in this chapter.

Text 4.3
Kūshyār ibn Labbān, Finding the Solar Equation
(from the *Jāmiʿ Zīj*)

[See figure 4.13.] [Let] *ABG* [be] the circle of the eccentric orb with *E* as its center and *AG* as its diameter, and [let] *D* [be] the center of the orbit representing the ecliptic [i.e., parecliptic; *D* is the center of the Earth]. Then *DE* is the eccentricity. It has been found to be [equal to] 2 parts and 4 minutes plus half and a quarter [of a minute], based on [taking] *EA* [equal to] sixty parts. *A* is the position of the apogee, *B* is the body of the Sun, and $\overset{\frown}{AB}$ the solar mean anomaly. We drop *BH* perpendicular to *AE*. It is the Sine of the arc $\overset{\frown}{AB}$. [We drop] *DZ* perpendicular to *BZ*. The angle *ZED* is equal to the angle

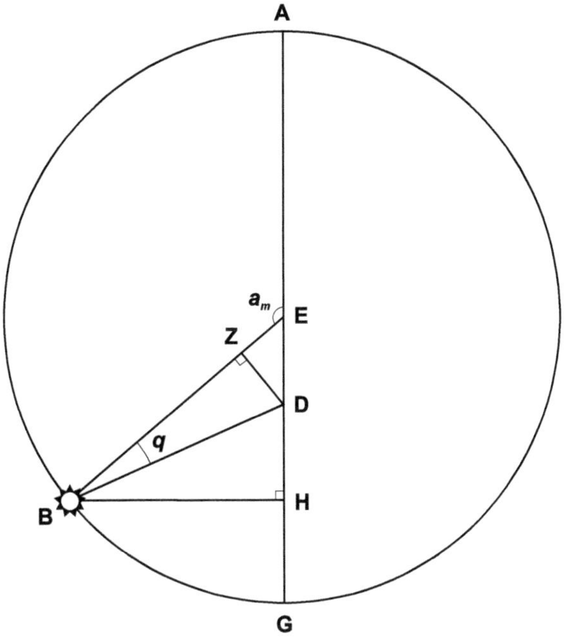

Figure 4.13 Kūshyār ibn Labbān on the solar equation

HEB, and the two angles *Z* and *H* are right. So the ratio of *EB* to *BH* is equal to the ratio of *ED* to *DZ*. *EB* is [equal to] sixty parts. *BH* and *ED* are known. So *DZ* is known, and *ZE* is known because *HE* is the Cosine of the mean anomaly [and *EB*:*HE* = *ED*:*ZE*]. So *BZ* is known. The [sum of the] squares of *BZ* and *ZD* is equal to the square of *BD*. So *BD* is known. The ratio of *BD* to *DZ*, which is known [i.e., which has been computed] based on [taking] *BE* as the radius, is equal to the ratio of sixty to *DZ* in the magnitude in which it [i.e., *DZ*] is desired [i.e., we want *DZ* for *BD* = 60]. So *DZ* based on [taking] *BD* as the radius is known. It is the Sine of the angle *ZBD*. So the angle *ZBD* is known, and it is the angle of the equation. That is what we wanted to demonstrate.[53]

Explanation: The goal is to find $q = \angle EBD$, given $a_m = \angle AEB$ and parameter $e = ED = 2;4,45$ (recalling that $R = BE = 60$). Since $\Delta DEZ \sim \Delta EBH$ we know that

$$\frac{EB}{BH} = \frac{ED}{DZ}, \text{ or } \frac{60}{\text{Sin}\, a_m} = \frac{e}{DZ},$$

and also that

$$\frac{EB}{HE} = \frac{ED}{ZE}, \text{ or } \frac{60}{\text{Cos}\, a_m} = \frac{e}{ZE}.$$

[53] From [Bagheri 2006, 140–141].

Therefore, since $BZ = 60 - ZE$,[54] we have two of the three sides of ΔBDZ. An arc tangent would finish the job, but Kūshyār instead applies Pythagoras to find BD and finishes with

$$q = \text{arcSin}\frac{DE}{BD}. \tag{4.23}$$

This method, like many of its kind, is very much modeled on Ptolemy's *Almagest*. The conservatism of finishing with an arc sine rather than an arc tangent is noteworthy, but it would not be long before other astronomers would move on. For instance, one of al-Bīrūnī's works contains no less than fifteen computations of the solar equation, two of which apply the arc tangent[55]—although in his masterwork the *Qānūn al-Masʿūdī*, he goes back to the arc sine.

Numerical Techniques: Approximation, Iteration, Interpolation

As effective as trigonometry powered by geometry can be, we have already seen several situations where it fails to provide practical results. The difficulty of computing Sin 1° is the most obvious instance of this shortcoming, but recall also the problems encountered by Indian astronomers in backtracking from true positions to mean positions of the Sun, Moon and planets. Finally, although the labor of composing a large astronomical table is not actually a theoretical barrier for trigonometry, it is a real problem that astronomers could not have overcome without occasionally abandoning direct computation in favor of the timesaving techniques of interpolation.

Occasionally, to avoid intense calculations certain cumbersome astronomical functions were just replaced with simpler ones. This happened for instance in the determination of the direction of Mecca, as we shall see later. The solar equation, relatively simple though it is, was also given this treatment. Rather than use some version of the correct formula

$$q(a_m) = \arctan\left(\frac{e\,\text{Sin}\,a_m}{R^2 + e\,\text{Cos}\,a_m}\right), \tag{4.24}$$

$q(a_m)$ was sometimes approximated by a multiple of the sine function:

$$q(a_m) = \max q \cdot \sin a_m. \tag{4.25}$$

[54] The subtraction assumes that the Cosine is not a signed quantity. In modern terms the Cosine in the diagram is actually negative and the sign in this formula should be positive.

[55] [Kennedy/Muruwwa 1958, 117]. According to al-Bīrūnī one of these two methods actually goes back to Ḥabash al-Ḥāsib!

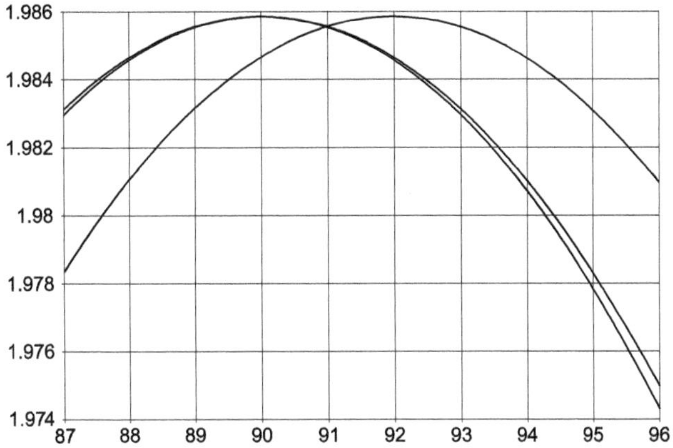

Figure 4.14 Graphs of the solar equation (shaded with its maximum at $a_m \approx 92°$) and approximations by the methods of sines and declinations (for $e = 2;4,45$ and $\epsilon = 23;51$). The method of sines is actually slightly closer to the solar equation when $a_m > 91°$.

Some *zījes* (for instance, al-Khwārizmī's[56]) used a multiple of the declination function:

$$q(a_m) = \max q \cdot \delta(a_m), \qquad (4.26)$$

presumably in an attempt to improve on the accuracy of (4.25).[57] The "method of sines," borrowed from Indian astronomy, is a reasonable approximation of $q(a_m)$ because q increases to a maximum not far from $a_m = 90°$ before decreasing to zero at 180°. But the peak of q occurs far enough after 90° to make the use of the method of sines easy to spot in a table. As can be seen from figure 4.14 neither method is very effective, and the "method of declinations" is really no better than the method of sines.

The solar equation is only one of a number of astronomical functions whose behavior resembles the sine; others include equations that arise in the motions of the Moon and the planets. Other astronomical functions resemble linear functions at key moments, such as the motions of the Sun and Moon near an eclipse. It is a natural idea, then, to begin with the simpler motion as an approximation to the more complicated one, and to find a means of improving the approximation step by step. This process leads naturally to the concept of iteration.

We have seen iteration used in Islam already, for instance by al-Bīrūnī and al-Kāshī in the determination of Sin 1°. The flourishing of iterative methods in

[56] [Kennedy/Muruwwa 1958, 118].

[57] See for instance [Salam/Kennedy 1967, 493].

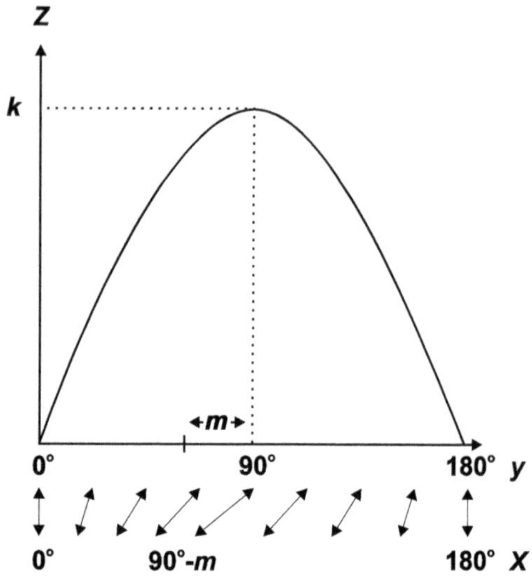

Figure 4.15
The first step of Ḥabash al-Ḥāsib's approximation process

India to solve astronomical problems similar to the ones faced by their Muslim colleagues may well have led to their transmission.[58] Indeed, one of the earliest and most sophisticated examples of iteration is found in Ḥabash al-Ḥāsib's *zīj*, which is known to rely heavily on Indian predecessors.[59]

The context in which Ḥabash's iteration occurs is parallax theory, the details of which would take us too far afield;[60] suffice it to say that Ḥabash is attempting to generate a function with its maximum some distance m to the *left* of 90°.[61] In figure 4.15, $z = f(y) = k \sin y$ is a "method of sines" approximation to a function that should peak at $90° - m$. Thus the y values must be altered continuously somehow to x values, as in the diagram, so that $y = 90°$ translates to $x = 90° - m$ while leaving the endpoints at 0° and 180° alone. Then the approximation

$$z = f(x) = k \sin(y(x)) \tag{4.27}$$

would give a function that looks like a sine wave, but with its maximum in the right place at $90° - m$. Now, the most obvious relation that does the trick is

[58] Indeed, there are a few traces even of Babylonian approximations (the step and zigzag functions that we saw in Chapter 1) in early Islamic science. See for instance [Lesley 1957, 136–139] and [Kennedy/Ukashah 1969].

[59] The method is described in both [Kennedy/Transue 1956] (mathematically) and [Kennedy 1969a] (historically). Our discussion is based on these two papers, but the presentation is altered in an attempt at clarity.

[60] The interested reader may pursue the parallactic context further in [Kennedy 1956b, 49–52].

[61] Of course the method could be adapted easily to work with a function whose peak is to the right of 90°, such as the solar equation.

$$x = y - m\,\sin y. \tag{4.28}$$

The problem with (4.28), known as Kepler's equation for its many appearances in celestial mechanics, is that we really want to get y in terms of x to be able to execute (4.27), not the other way around—and (4.28) is analytically unsolvable for y.

But as history has proven again and again, iteration is often able to solve the unsolvable. Suppose, given x_0, we want to find y so that $y = x_0 + m\sin y$ (rearranging (4.28)).[62] Thus, in figure 4.16 we need to determine y so that the distance one moves on the horizontal axis from x_0 to y corresponds to the height of the graph of $m\sin y$—visually, the place on the curve where a diagonal line rising rightward from x_0 intersects the graph of $z = m\sin x$.[63]

However, converting from the visual to the quantitative is not so easy here. Ḥabash describes the following solution: let

$$\begin{aligned} y_0 &= x_0 + m\sin x_0, \\ y_1 &= x_0 + m\sin y_0, \\ &\vdots \qquad \vdots \\ y_n &= x_0 + m\sin y_{n-1}. \end{aligned} \tag{4.29}$$

In figure 4.16 this sequence of computations corresponds to the following process: determine the height of the graph at x_0 (i.e., $m\sin x_0$) and move diagonally downward to the right onto the horizontal axis, thereby adding it to x_0; we end up at y_0. The height of the graph at y_0 is $m \sin y_0$; adding this quantity to x_0 corresponds to transferring the height of the graph at y_0 over to the left at x_0 and again moving diagonally down to the right. Repeat as necessary; it should become clear after a few iterations that the y_n's are converging to the desired value of y. Certainly neither Ḥabash nor anyone else explains why this convergence occurs; likely they were convinced not by any demonstration, but simply by experience.

Of course this visual description cannot conform to Ḥabash's or the originator's conception of the scheme—but the idea, in some other guise, may be the same. Since (as is typical) no explanatory writings exist, we may never know for certain how the method actually came about.

The purpose of schemes like Ḥabash's was to save the incredible amount of labor required to put together complete astronomical handbooks, which often contained dozens of tables computing a dizzying array of mathematically defined functions. Although iterative techniques had some currency, they were the exception rather than the rule. Much more common was

[62] The relation of this result to the basic equation for fixed-point iteration, $\theta = f(\theta)$, is clear.

[63] This matches our expectation that $x_0 = 90° - m$ should map to $y = 90°$, since x_0 is m units to the left of $y = 90°$, and the function at 90° has a value of $m \sin 90° = m$.

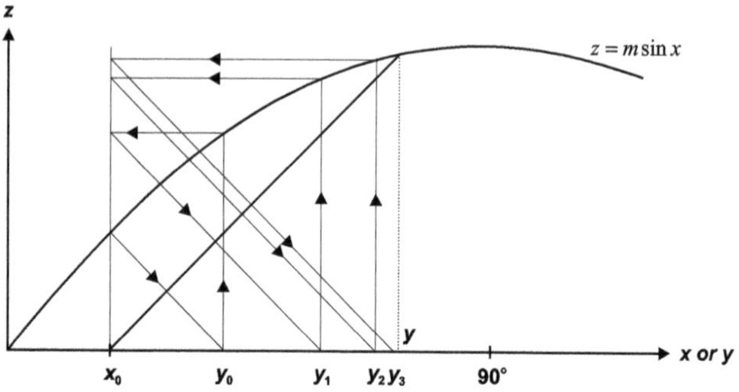

Figure 4.16
The second step of Ḥabash al-Ḥāsib's approximation process: iteration

the practice of computing only, say, every third or tenth entry of a table and filling in the rest of the entries by some form of interpolation. Appropriately applied, the results of these efforts could be virtually indistinguishable from a table computed entirely by direct methods, and it must have been a sore temptation to take this short cut often.

Although time was not taken to write about it, perhaps the most commonly used method was ***distributed linear interpolation***.[64] Already used by Ptolemy, this simple variant of linear interpolation had the advantage of being very fast to implement, with results in most cases almost as good as more sophisticated methods. The idea is to take advantage of the fact that the function differences to be inserted into the interval cannot usually be divided perfectly into equal parts. In figure 4.17, an extract from one of Kūshyār ibn Labbān's tables for Mars, Kūshyār has computed every third entry. In the gap from $a_v = 60° \ldots 63°$, for instance, the function increases by $1;5 = 65$ minutes. Breaking this into three parts, we find that the function must increase by 21, 22, and 22 minutes. Since the function's rate of increase is decreasing (i.e., its graph is concave down), the best course of action is to place the two 22-minute increments first, followed by the 21-minute increment.

Although distributed linear interpolation leads to a distinctive pattern of differences (see figure 4.17), the completed table does give the appearance of a smoothly changing function. Applied in the right situations (and it seems that the situations indeed were considered carefully), the method, albeit crude, is effective.

As prevalent as distributed linear interpolation seems to have been, the written literature of the tenth through the fifteenth centuries passed over it in favor of various second-order interpolation schemes. The earliest account of these more advanced methods that has been found so far is attributed to the

[64] Distributed linear interpolation has been described several times in the secondary literature; see for example [Rome 1939, 219], [van Dalen 1993, 11], [Van Brummelen 1991, 681–682], and the example on which our description is based, [Van Brummelen 1998, 278].

a_v	$p(a_v)$	*Differences*
60°	**23;13**	;22
61°	23;35	;22
62°	23;57	;21
63°	**24;18**	;22
64°	24;40	;21
65°	25;1	;21
66°	**25;22**	;22
67°	25;44	;21
68°	25;5	;21
69°	**25;26**	;21
70°	25;47	;21
71°	26;8	;21
72°	**26;29**	

Figure 4.17
Distributed linear interpolation practiced in an extract of one of Kūshyār ibn Labbān's tables for Mars. The bold entries have been computed directly; the others have been interpolated.

tenth-century astronomer al-Khāzin;[65] other authors include Ibn Yūnus (whom we shall visit in a moment), al-Bīrūnī,[66] Naṣīr al-Dīn al-Ṭūsī,[67] and Sayf-i-Munajjim (the *Ashrafī Zīj*).[68] The schemes were of course applied to trigonometric tables, but also to other astronomical tables and ephemerides. Oddly enough two of the best astronomers of medieval Islam, al-Bīrūnī and al-Kāshī,[69] produced interpolation methods that can only be described as misguided; but the others are equivalents of modern second-order formulas. It seems that many of the algorithms were inspired, either directly or indirectly, by Indian sources; al-Bīrūnī for instance had access to Brahmagupta's *Khaṇḍakhādyaka*, from which we have already seen a second-order scheme.[70]

[65] Described in [Hamadanizadeh 1978]. See also the survey paper [Hamadanizadeh 1987], which describes several of the schemes mentioned below. Ulugh Beg's sine table, for instance, employs al-Khāzin's method [van Dalen 2002, 351–353].

[66] Al-Bīrūnī's errant scheme has attracted some attention. See [Schoy 1927, 40–42] for a description, [Rosenfeld 1959] for a detailed account, [Kennedy 1978] for an attempt to understand its motivation, and [Rashed 1991] for a discussion of three different interpolation methods by al-Bīrūnī and a critique by al-Samaw'al.

[67] Originally described in [Kennedy 1962] and attributed to al-Kāshī, but identified with al-Ṭūsī in [Hamadanizadeh 1985]. In fact, as Hamadanizadeh points out, this method is the same as al-Khāzin's and may have been floating around the scientific community.

[68] Described in [Hamadanizadeh 1963].

[69] Al-Kāshī's system is described and analyzed in [Hamadanizadeh 1980, 41–44].

[70] A scheme which, for some reason, al-Bīrūnī chose to ignore. See especially [Kennedy 1978] and [Rashed 1991].

The modern understanding of second-order interpolation is that a parabola is being passed through three given points on the graph of the function, which is then used to generate other points on the curve. This is obviously not how the methods were conceived (recall Brahmagupta's description in chapter 3), but Ibn Yūnus's description is not all that far away from this interpretation.

Text 4.4
Ibn Yūnus, Interpolating Sine Values
(from the *Ḥākimī Zīj*)

Sines of arcs found by linear interpolation are always less than the actual values. . . . If you want to find the Sine of an arc other than the half-degrees whose Sines are given [in the table], and [you want to find] the difference between the Sine found by linear interpolation and the actual value, then first find the Sine of the arc in question by linear interpolation. . . . Also find the Sine of the half degree which is in between the two integral degrees less and greater than the arc in question, using the Sines calculated rigorously for each degree. Note the Sine of that half degree by linear interpolation, and note the difference between it and the Sine of the half degree in the *Zīj*, which was calculated rigorously. Multiply the difference by four, and the result will be the base for interpolation, so keep it in mind. Then look at the minutes of the arc in question over the integer degrees, and subtract them from 60. Multiply the remainder by the minutes which you subtracted, and then multiply the product by the interpolation base: the product will be the difference between the Sine of the degrees and minutes [of the arc in question found] by linear interpolation and its actual Sine, so add it to the value of the Sine which you found first by linear interpolation. The result will be the Sine of the degrees and minutes of arc, which you wanted to find exactly. Success is with God.[71]

Explanation: Although Ibn Yūnus does not work with an example, we shall. Our goal is to find Sin 9;20°, given that Sin 9°=9;23,9,51, Sin 9;30° =9;54,10,17, and Sin 10° = 10;25,8,1 [+1].[72] We first compute the "interpolation base" Δ, the difference between Sin 9;30° and the value for Sin 9;30° that one would arrive at by linear interpolation between Sin 9° and Sin 10° (see figure 4.18 for a visual interpretation); in our case $\Delta = 0;0,1,21$.[73] Where x is the

[71] From [King 1972, 82]. The passage has also been translated into German in [Schoy 1923a, 390–391], and the method is discussed in [King 1973b, 354–357]. The latter forms the basis of our explanation.

[72] These values are taken from the table of Sines in the *Ḥākimī Zīj*.

[73] The linearly interpolated value for Sin 9;30° is 9;54,8,56, which differs from the given value by 0;0,1,21.

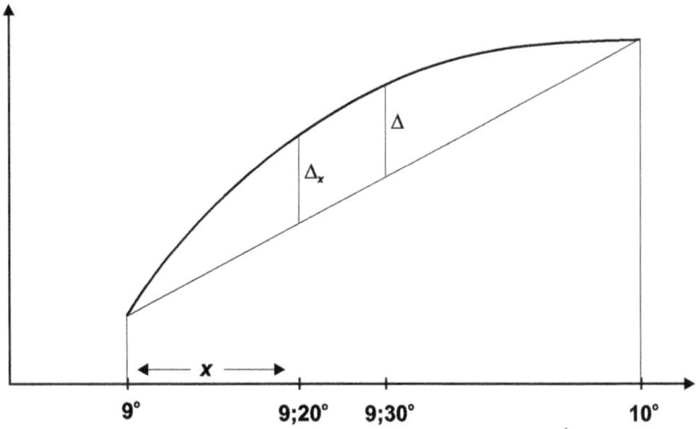

Figure 4.18 Ibn Yūnus's second-order interpolation scheme

amount by which the argument of the desired Sine exceeds 9° (in our case 20 minutes), Ibn Yūnus does the equivalent of assuming that the Sine curve exceeds the line connecting the endpoints by an amount equal to

$$4 \cdot x(1-x) \cdot \Delta .^{74} \quad (4.30)$$

The $x(1-x)$ term provides a nice quantity—a quadratic term—that begins and ends at zero (for $x = 0$ and $x = 1$ respectively), and reaches its largest value, $\frac{1}{4}$, in the middle when $x = \frac{1}{2}$. The multiplication by 4 and by Δ ensure that (4.30) has a maximum value of Δ, as required. Then the amount Δ_x that we must add to the linearly interpolated value of Sin 9;20° = 9;43,49,14[75] is simply(4.30):

$$\begin{aligned} \text{Sin } 9;20° &= 9;43,49,14 + 4 \cdot (0;20) \cdot (0;40) \cdot (0;0,1,21) \\ &= 9;43,49,14 + 0;\ 0,1,12 \quad (4.31) \\ &= 9;43,50,26. \end{aligned}$$

This value turns out to be accurate to all four sexagesimal places.

Muslim astronomers applied several other computational devices to their astronomical tables, mostly hoping to make arithmetical life easy on the user. To describe them all would take us too far afield; likely some remain undiscovered today. One, however, brings us back to familiar ground in Ptolemaic planetary theory. Recall that the planetary equation of anomaly p is a function of two variables, one of which (a_v) has a much greater influence on

[74] The use of 60 within the quotation stems from treating the excess above 9° as a number of minutes, rather than degrees.

[75] The linearly interpolated value is Sin 9;20° = 9;23,9,51 + (0;20)(10;25,8, 1 − 9;23,9,51) = 9;43,49,14.

p than the other (c_m). To compute a rectangular grid of entries of a double-argument function as complicated as p would be too time-consuming,[76] so Ptolemy invented an interpolation scheme to bypass the tabulation, which we described in chapter 2.

The tenth-century astronomer Kūshyār ibn Labbān, whose *Jāmiʿ Zīj* we have already encountered, implements a clever alternative to Ptolemy's scheme. Figure 2.34 is a graphical representation of $p(a_v, c_m)$; the goal is to compute p for given values of a_v and c_m using only single-argument tables—i.e., find the height of a randomly chosen point X using as data only a few known arcs on the surface. Ptolemy effectively tabulates function values along the four arcs AD, BE, CF, and $DYEF$, and assumes that $AX/AB = DY/DE$ (where a pair of letters represents the difference between two function values; i.e., the difference between their heights). Since X is the only point in this ratio not on a tabulated curve, it may be computed. Kūshyār, on the other hand, tabulates only BE and $DYEF$ and assumes that $X/B = Y/E$ (where, again, letters represent function values). And again in this ratio only X is not tabulated, so it may be computed. Although Kūshyār's method holds the advantage over Ptolemy of requiring the tabulation of only two functions instead of four,[77] it does lead to significant errors for certain values of the arguments, for the planets with large epicycles (Venus and Mars). So far, only one other table has been found that adopts this approach.[78]

Early Spherical Astronomy: Graphical Methods and Analemmas

Outlining the beginnings of spherical trigonometry and astronomy in medieval Islam is a vexed task, not well served by the limited historical evidence. At least three different mathematical approaches to the subject are possible. First, there is the style of the Greeks exemplified in Ptolemy's *Almagest*, which relies on Menelaus's Theorem (or, as the Muslims called it, the Transversal Figure). Once this result is established, Greek spherical trigonometry operates solely on the surface of the sphere. Second, we have the Indian style, which builds its theory primarily through the use of similar

[76] This is not to say that it was not done, in Islam or surrounding cultures. See for example (among others) [Tichenor 1967], [Saliba 1976], [van Dalen 1999, 321–328], and [Goldstein 2003].

[77] To be precise, Kūshyār tabulates three functions, but one is simply a constant multiple of the other. Kūshyār also adopts two other computational efficiencies in his tables, making them easier to use but further obscuring their mathematical structure. See [Van Brummelen 1998] for an account.

[78] The anonymous *Rīqānī Zīj* (ca. 1090), in the lunar tables (personal communication, Benno van Dalen).

plane triangles within the sphere and the Pythagorean Theorem. Third, we have the tradition of the analemma, involving the rotation of one circle onto the plane of another. We have seen already that the analemma began in Greece, although few examples remain. We have also argued that its use in India might have been overstated; however, it quickly became a prominent tool in Islamic astronomy. So, we shall begin with the analemma and some related constructions.

The problem of identifying the sources of Islamic spherical trigonometry is not aided by the fact that the earliest extant writings often contain rules for calculation, but no theoretical descriptions. For instance, al-Khwārizmī's *zīj* describes how to solve problems of spherical astronomy such as rising times and related astrological matters using tables,[79] but like his Indian predecessors he does not say how these rules arose in the first place. Finally, the earliest texts do not distinguish clearly between spherical trigonometry and its application in spherical astronomy, again echoing the Indian approach, but also not that far removed from the Greeks.

Indeed, it is possible to find a great deal of spherical trigonometry buried within early Muslim solutions to astronomical problems.[80] Al-Khwārizmī gives a geometric procedure to determine the ortive amplitude[81] ψ in terms of the Sun's declination δ and terrestrial latitude ϕ that translates to

$$\sin\psi = \frac{\sin\delta}{\cos\phi}, \tag{4.32}$$

which turns out to be equivalent to the Rule of Four Quantities applied to a right triangle. His calculation of the azimuth α of the Sun in terms of its altitude h as well as ψ and ϕ,

$$\cos\alpha = \frac{\sin\psi - \tan\phi\sin h}{\cos h}, \tag{4.33}$$

is equivalent to the other fundamental identity of spherical trigonometry, the Law of Cosines.[82] Indeed, as we shall see later, a very similar problem solved

[79] [Al-Khwārizmī (Neugebauer) 1962, 46–55]. An analemma is used on pp. 50–51 to explain the *Sūrya Siddhantā*'s approach, but as Neugebauer would have realized, this is simply his derivation of the pattern of calculation. The paper he quotes, [Schmidt 1944], concludes that "the methods that were used . . . for the computation of the length of daylight thus seem to be essentially independent of Greek and Babylonian methods."

[80] [King 2004, 26–38] is a useful compilation of the essential formulas of spherical astronomy in medieval Islam.

[81] The distance from the east point of the horizon to the point where the Sun rises, see figure 4.11.

[82] These two rules are discussed in [Rosenfeld 1993, 305–306]. The spherical Law of Cosines may be arrived at using either Indian or Ptolemaic methods.

by al-Battānī a few decades later[83]—also identical to the spherical Law of Cosines—may have been the path that led Regiomontanus to the purely trigonometric result. So the early Islamic scientists were certainly capable of finding and using rules of the sophistication and importance of the laws of sines and cosines, but until later the rules remained in their astronomical contexts, and were not posited independently.

Al-Khwārizmī's description of (4.32) takes a rather novel form, below.

Text 4.5
Al-Khwārizmī, Determining the Ortive Amplitude Geometrically
(Geometric Construction of the Ortive Amplitude
for Every [Zodiacal] Sign and for Every Latitude)

[See figure 4.19.] Draw a circle *ABCD* and divide its arc *AD* into 90 parts. From the point *D* on the circle lay off the arc *DG* having the magnitude of the latitude of [your] town and draw a line from this [point] *G* to the center [of the circle], that is the line *EG*. Then from the point *A* on the circle lay off the arc *AH* having the magnitude of the declination of that zodiacal sign for which you want [to make] the construction. From [the end of] this [arc] draw a line parallel to the line *AE*. It is the line *HF*, which reaches the line *BD* which is a diameter. Then look where it cuts the line *EG*; let it be the point *M*. Then take the pair of compasses with opening *ME*, put its leg in the center point, that is in the point *E*, and find the place where the other leg cuts the line *DE*. Let this be the point *N*. From the point *N* raise the perpendicular reaching the arc *AG*, it is the perpendicular *ON*. From [the point *O*] where [the perpendicular] reaches count out [the arc] to the point *A*. The thing you get is the ortive amplitude for the sign you wanted.[84]

Explanation: One will look in vain for a derivation in this passage; this is simply a description of a method to construct the arc of the solution geometrically. To see that it produces the correct result, note that the sine of the purported solution, $\psi = \widehat{AO}$, satisfies the following relation:

$$\sin\psi = EN = ME = \frac{ME}{EF} \cdot EF = \frac{1}{\cos\phi} \cdot \sin\delta, \tag{4.34}$$

which we know to be true by (4.18).

[83] See [Delambre 1819, 20], [von Braunmühl 1900/1903, vol. 1, 53–54] and [al-Battānī (Nallino) 1899/1903/1907, vol. 1, 185].
[84] From [Rosenfeld 1993, 306].

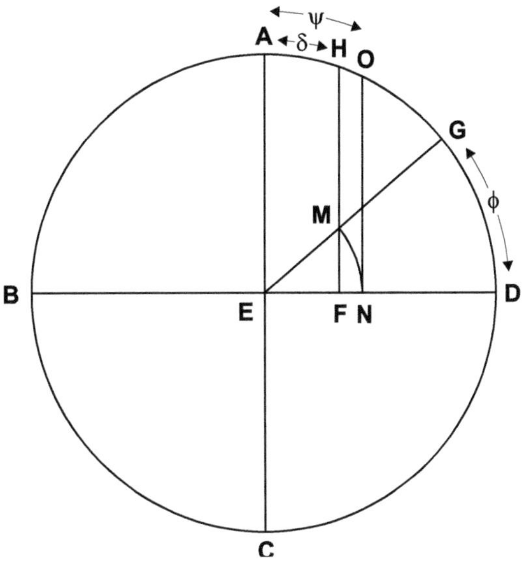

Figure 4.19
Al-Khwārizmī's graphical solution to the ortive amplitude problem

This sort of technique has been given names like "graphical methods" or "geometric trigonometry," the latter since it is a geometric analogue of our computational trigonometry in the same sense that Book II of Euclid's *Elements* (sometimes called "geometric algebra") is a geometric analogue of algebra. We have seen a similar diagram earlier in this chapter, in Text 4.3 from al-Bīrūnī's *Exhaustive Treatise on Shadows*, although there it was accompanied by a proof. Similar diagrams may be found, for instance, in the works of Ḥabash al-Ḥāsib and al-Māhānī.[85] Although procedures like this existed in India, these diagrams went much further.[86]

The graphical methods are closely related to what have become known as analemmas in the Arabic tradition. In Greek use, the analemma involved the rotation of one circle in the celestial sphere onto the plane of another. In Islam we find examples of analemmas similar to the Greek applications, but the tradition is considerably richer. Analemmas abound to solve a wide variety of problems, including the usual topics in spherical astronomy, but also the design of astronomical instruments (especially the astrolabe), mathematical geography, and astronomical timekeeping.[87]

[85] Several examples are in [Kennedy/Kunitzsch/Lorch 1999]; on the same topic see also [Lorch 1998].

[86] [Kennedy/Kunitzsch/Lorch 1999, 11–12]. Occasionally, such as in the decomposition of the ratio in the middle of (4.34), similar triangles are used in a way that evokes an essentially arithmetic procedure rather than a geometric one. See also [Lorch 1998, 222–223].

[87] The literature on analemma methods in Islam is extensive. For examples of its use in spherical astronomy see for example [Kennedy/Sharkas 1962] and [ʿId 1969]. On instruments see [Berggren 1991/92], [Lorch 1998] and [Kennedy/Kunitzsch/Lorch 1999]. Applications in

Sometimes analemmas are accompanied by the equivalent trigonometric calculations but often they are not, as in the case in al-Bīrūnī's report of Ḥabash al- Ḥāsib's (fl. 850) determination of the direction of Mecca, known as the *qibla*.[88] One of the earliest in the Islamic world to compose analemmas, Ḥabash was one of the most remarkable figures of early Islamic mathematical astronomy. His analemmas are among his most creative contributions.[89]

Ḥabash's analemma for the *qibla* gives a direct visual marker for the appropriate direction, which may have been more useful to architects and worshippers than a number. The trigonometric analogue would eventually become known as the "method of the *zījes*," on which we shall say more later. For now we follow the analemma itself, as a particularly clever example of the genre. A moment's thought should confirm that the direction of Mecca is the same as the direction of Mecca's zenith N in the observer's sky (line LN in figure 4.20), and that the needed data are the latitudes of the observer and Mecca (ϕ_o and ϕ_m respectively) and the difference in longitude ΔL between the two locations.

Al-Bīrūnī describes only the analemma itself with the overly hopeful statement that "the proof of what Ḥabash said is evident"—dubious, since the analemma's circle changes its role no less than three times. Our explanation, adapted from a modern reconstruction,[90] is as follows.

The analemma's main circle (figure 4.21) begins its life as the meridian circle, with the observer's zenith A drawn at left. The diameters of the equator ZEB and of the parallel circle of Mecca HMY are drawn appropriately according to the values of ϕ_o and ϕ_M. The circle now becomes the equator; since diameter ZEB is shared by both the equator and the meridian circle, we may use ZEB as if it is on the equator. We draw $\Delta L = \widehat{ZT}$. In figure 4.20 dropping a perpendicular from Mecca's zenith N to the equator produces S; since ES is the radius of the parallel of Mecca, we may find S on the analemma by drawing $ES = MY$. Now if we drop perpendicular SU onto ZE, the configuration $ZUEST$ will be identical in the two figures.

But since diameter $ZUEB$ is also on the meridian, the analemma's circle may become the meridian again. This allows us to project U onto Mecca's

---

mathematical geography are [Kennedy 1959], [Kennedy 1963], [al-Bīrūnī (Kennedy) 1976, vol. 1, 168–169; vol. 2, 95–96] which we have already seen, [Kennedy 1984] and [Kennedy 1989]. On the analemma in spherical astronomy and timekeeping see [King 1972] and [King 1999].

[88] The method attributed here to Ḥabash may actually predate him; see [Berggren 1985, 5]. The account which follows is based partly on [Kennedy/ʿId 1974, 6–10]. See also [Debarnot, in al-Bīrūnī (Debarnot) 1985, 49–51] for a corresponding calculation of the *qibla* for Samarra by Ḥabash.

[89] [Debarnot, in al-Bīrūnī (Debarnot) 1985, 42–63] is a detailed survey of Ḥabash's contributions to spherical astronomy, including analemmas on pp. 48–51. On analemmas for the *qibla* see also [Schoy 1921], [King 1975, 115–118], [Berggren 1980], [Carandell 1984], and [Suzuki 1987/88].

[90] [Kennedy/ʿId 1974].

Figure 4.20
Deriving Ḥabash's analemma for the *qibla*. Dotted lines are within the celestial sphere; thin dashed lines and curves will be used later in this chapter. Thickly dotted lines follow the path of projections: *N* drops perpendicularly onto *TE* to define *S*, then *S* onto *ZE* to define *U*, then *U* onto *HY* to define *O*, and *O* onto *AE* to define *L*. The sought direction of the *qibla* is *LN*.

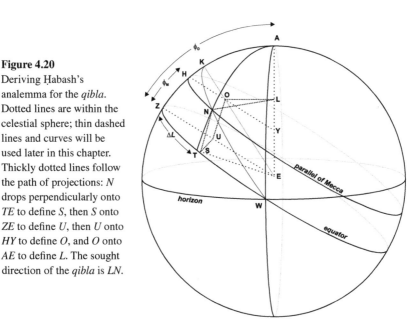

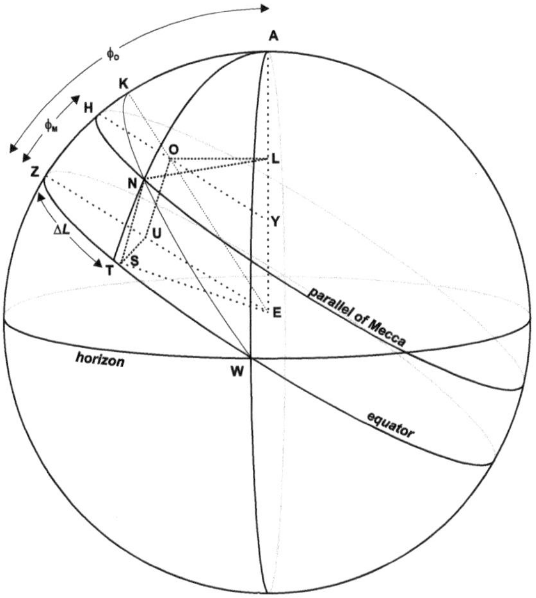

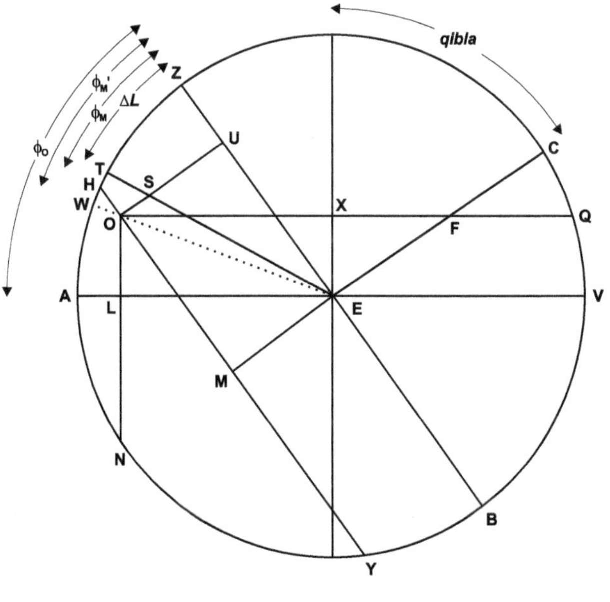

Figure 4.21
Ḥabash's analemma for the *qibla*

parallel circle, defining O in both diagrams. In fact we may take another step: project O onto AE, defining L. Now, LN in figure 4.20 is the direction line that we want. Since ALE is also on the plane of the great circle through the two zeniths A and N, our analemma circle becomes the latter circle, and we may draw LN easily enough in both diagrams.

This procedure gives LN as the correct length in the analemma, but not the correct direction. However, this problem is easily solved: draw horizontal OXQ, so that $OL = EX$; then draw EFC so that EF is equal in length to LN. Then ΔEFX in the analemma will be the same as ΔLNO in figure 4.20, and the *qibla* has been established.

No doubt this method is an extremely clever and perfectly correct solution to a difficult problem. And, as with most analemmas, it is possible to construct a conventional trigonometric solution that corresponds to the geometry of figure 4.21.[91] Indeed, it was a popular exercise to convert analemmas to trigonometric formulas, and we shall see later that this one was the source of the most common accurate method of solving the *qibla* problem. But is the analemma itself properly part of medieval trigonometry? According to the definition with which we began this book the answer is "no"; the method is purely geometric and there is no need to invoke the quantitative aspect of the subject at all. The exercise of converting an analemma into a trigonometric solution, as common as it was, is not usually a trivial affair. The concerns of and tools available to the solver in the two contexts, related though they may be, differ somewhat.[92] Since the analemma comes first in the process, it seems fair to consider it not as part of trigonometry as such, but rather as a mentoring older sibling.[93]

[91] The following modern solution differs from what we shall see later. We begin by noting that $ES = MY = \cos\phi_M$; therefore $SU = ES \sin\Delta L = \cos\phi_M \sin\Delta L$. Also $UO = EM = \sin\phi_M$, so SO may be found by subtracting SU from UO. Join EO; then SO, ES and the enclosed angle $\angle ESO = \Delta L + 90°$ are known and Euclid II.12 (the Law of Cosines for obtuse angles) may be invoked to determine EO. From EO and UO we can find $\angle UOE$, and since $\angle SOF = 90° - \phi_O$ we have $\angle FOE = \angle OEL$.. Thus all of ΔOEL is known. But $LN^2 = AL \cdot LV = (R - LE) \cdot (R + LE)$, and so the *qibla* = arc cos(OL/LN).

[92] For instance, in the above footnote OL is easily found by the analemma simply by dropping the perpendicular to AE. To find the length of OL trigonometrically requires an argument, including adding a line segment to the diagram (EO), an invocation of the Law of Cosines, and the determination of a couple of angles. Conversely, at the end of the argument the trigonometry does not even require the construction of EF. The distinction between analemmatic and trigonometric reasoning is revealed dramatically in the comparison of four analemmas for the *qibla* in [Berggren 1980].

[93] In a review of [al-Bīrūnī (Debarnot) 1985] Jan Hogendijk points out that Debarnot demonstrates the link between analemmas and spherical trigonometry, and calls for "a systematic study of analemma methods in [the] Arabic tradition . . . as a preliminary to writing the history of Arabic trigonometry" [Hogendijk 1988]. This is not that study; it certainly would be welcomed.

Menelaus in Islam

Spherical trigonometry proper, with trigonometric functions and numeric computation, quickly became a significant part of Arabic astronomy. The translation movement especially in the ninth century brought in a great deal of Greek science, and along with it came Greek mathematical techniques. The *Elements* was of course a fundamental text for any mathematical enterprise, but for the study of spherical astronomy the *Almagest* was the role model—even more than Menelaus's *Spherics*, which is the fundamental work on the trigonometry of the sphere.

As we have seen, both the *Spherics* and the *Almagest* contain treatments of Menelaus's Theorem, although the *Spherics* is (on the face of it) a mathematical work while the *Almagest* concentrates on the astronomy. Readers of chapter 2 will recall that the theorem as stated in the *Almagest* asserts two slightly different results with respect to the ***spherical quadrilateral*** of figure 4.22. Both are stated using the chord function, asserting that a certain ratio of chords is equal to the composition of two other ratios of chords. However, converting to sines is child's play; indeed, a leading translation of the *Spherics* in the Arab world actually uses sines.[94] The first of the two theorems expressed in sines (known as "disjunction"[95] because the ratio on the left refers to two separate arcs) is

$$\frac{\mathrm{Sin}\,a}{\mathrm{Sin}\,b}=\frac{\mathrm{Sin}(c+d)}{\mathrm{Sin}\,d}\cdot\frac{\mathrm{Sin}\,g}{\mathrm{Sin}\,h}; \tag{4.35}$$

the second (known as "conjunction" because one of the terms in the ratio on the left is a combination of two arcs) is

$$\frac{\mathrm{Sin}(a+b)}{\mathrm{Sin}\,a}=\frac{\mathrm{Sin}(g+h)}{\mathrm{Sin}\,g}\cdot\frac{\mathrm{Sin}\,f}{\mathrm{Sin}(e+f)}.^{96} \tag{4.36}$$

Arabic spherical astronomers seized on the "Sector" or "Transversal Figure" (as they called it) to solve an extremely diverse set of problems, including those of spherical astronomy found in the *Almagest* but also the *qibla*, the design of astronomical instruments, and geography, to name a few. Generally the arcs were chosen so that some of the combined arcs ($a+b$, $c+d$, $e+f$, and $g+f$ in figure 4.22) are 90°. This strategy usually led to

[94] Abū Naṣr Manṣūr's translation, available in [Krause 1936].

[95] The terms "disjunction" and "conjunction" derive from Ptolemy's *Almagest*. See [Bond 1921, 300].

[96] Of course both results appear identical using modern rather than medieval sines, since the base circle radius simply cancels out in each of the three ratios.

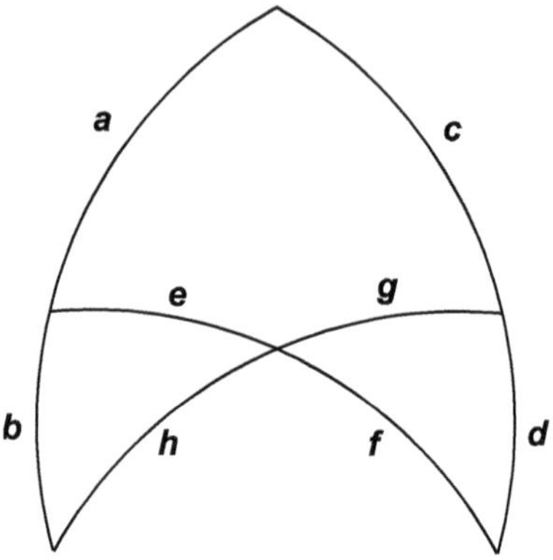

Figure 4.22
Menelaus's (the Transversal or Sector) Theorem

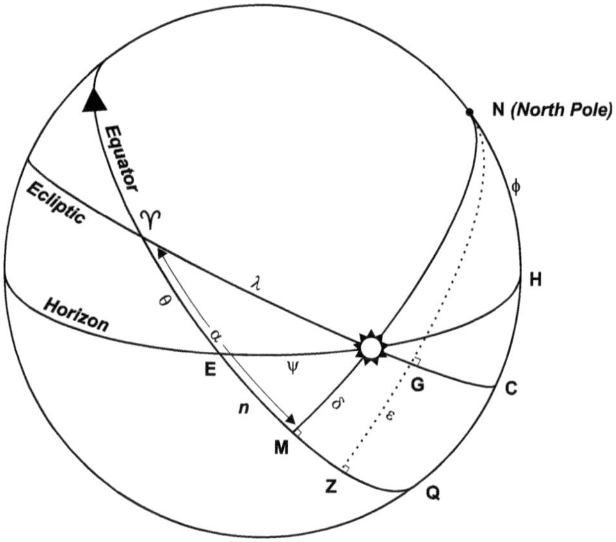

Figure 4.23
Al-Kūhī's solution to the rising times problem

simplifications, and in fact (as we shall see) to solutions to some problems that are equivalent to other important theorems of spherical trigonometry.

One typical example of the use of the Transversal Figure, of interest for more than just its artistry, is in a short treatise by the great tenth-century geometer Abū Sahl al-Kūhī. The goal is to solve the familiar problem of rising times that we saw first in the *Almagest*: in figure 4.23, given the local latitude $\phi = \widehat{NH}$ and the Sun's position on the ecliptic $\lambda = \widehat{♈☼}$, find the arc on the equator $\theta = \widehat{♈E}$ (the ***oblique ascension***) that rises above the horizon as arc

λ rises. (Since the equator rotates uniformly it is the circle on which time may be measured, hence the use of the term "rising time" for equatorial arcs.)

When we saw this problem before in Ptolemy's *Almagest*, we assumed knowledge of the Sun's equatorial coordinates α and δ, but al-Kūhī solves it (and a few other problems along the way) from scratch:

Text 4.6
Al-Kūhī, Finding Rising Times Using the Transversal Theorem
(from Rising Times of Arcs of a Known Arc of the Ecliptic)

Some of our colleagues who are well-advanced in this art of ours asked us at the Royal Palace . . . about finding the rising time of a known arc of the ecliptic. . . . And he requested us to do that for him using [only] our knowledge of the Transversal Figure, which is in Ptolemy's *Almagest*, and no other theorems. And he claimed that he can derive that by a way that is shorter, easier and involves less work than that of the people who know [just] the Transversal Figure, and that that is not only because of his acuity in this art, but because of another theorem not known as "The Transversal." . . . And he claimed that he and others were freed by it [the new theorem] from knowing the Transversal Figure in these operations, and from looking into it. But it is my opinion that, although his judgment may be allowable for himself, it is not so for others. But [here is] an investigation of it.

If we multiply the sine of a known arc of the ecliptic by the sine of the whole declination, the result is the sine, expressed in minutes, of the inclination of that degree. And if that is divided by the cosine of the latitude of the town there results the sine of the ortive amplitude of that degree in that town. And if the cosines, expressed in minutes, of each of the ortive amplitude and of the known arc of the ecliptic area divided by the cosine of the latitude of that degree there result two sines. The difference between the arc of one of them and the other is the rising time of that degree in that town. The complement of one of the two arcs is the rising time in sphaera recta for that degree, and the complement of the other is the equation of daylight for that degree in that town. That is what we wanted to prove.[97]

Explanation: Clearly al-Kūhī was on the cusp of a conflict between old (the Transversal Figure) and new; we shall return to this debate and its likely combatants shortly.

Although al-Kūhī does not explain his procedure thoroughly, he applies the Transversal Figure four times to two different spherical quadrilaterals,

[97] From [Berggren/Van Brummelen 2001, 42].

which we reconstruct below. The advantage of choosing many of the arcs to be 90° ($\overset{\frown}{NZ}, \overset{\frown}{NM}, \overset{\frown}{♈G}, \overset{\frown}{NQ}, \overset{\frown}{EH}, \overset{\frown}{EQ}$ and $\overset{\frown}{♈Z}$) is obvious; one important point in this regard is that it allows the conversion of angle ε between the equator and the ecliptic into arc $\overset{\frown}{GZ}$.

- On $NZ♈☼$ apply (4.36); this gives $\dfrac{\sin 90°}{\sin\varepsilon} = \dfrac{\sin 90°}{\sin\delta} \cdot \dfrac{\sin\lambda}{\sin 90°}$ or simply $\sin\delta = \sin\lambda \sin\varepsilon$, and we have the declination δ.
- On $E☼NQ$ apply (4.36); this gives $\dfrac{\sin 90°}{\sin\phi} = \dfrac{\sin 90°}{\sin\delta} \cdot \dfrac{\sin\psi}{\sin 90°}$,[98] which gives in turn the ortive amplitude ψ.
- On $E☼NQ$ apply (4.36) but assign the arcs so that EM and MQ are in the ratio on the left side of the equation; this gives $\dfrac{\sin 90°}{\sin MQ} = \dfrac{\sin 90°}{\cos\psi} \cdot \dfrac{\cos\delta}{\sin 90°}$, which gives MQ. Its complement is the ***equation of daylight***, or ascensional difference, $n = \overset{\frown}{ME}$.
- On $NZ♈☼$ apply (4.36) but assign the arcs so that $♈M$ and MZ are in the ratio on the left side; this gives $\dfrac{\sin 90°}{\sin MZ} = \dfrac{\sin 90°}{\cos\lambda} \cdot \dfrac{\cos\delta}{\sin 90°}$, and we have MZ.

Finally, $\theta = ♈E = ZQ = MQ - MZ$.[99]

If ever proof were needed of the power of the Transversal Figure, this is it—four successive applications to only two configurations solve the problem, and produce several important quantities along the way. However, as we shall see shortly, this confirmation of its efficacy did not prevent its eventual replacement at the foundation of spherical astronomy.

The Transversal Figure led scientists in another surprising direction, exemplified in a text by Thābit ibn Qurra (824–901). He was instrumental in bringing Greek science to the West; he was involved in the production of excellent Arabic translations of many Greek works, especially those of Archimedes, and he wrote commentaries on both the *Elements* and the *Almagest*. His mathematics and astronomy were certainly inspired by his Greek ancestors, and in this respect he was a leader for many who came after him. Nevertheless his own mathematical and astronomical output was exceptionally creative, and extended the Greek ideas in new directions.

[98] The cosine arises, here and elsewhere, by taking the sine of 90° minus the given arc.
[99] $♈Z = EQ = 90°$, so both $♈E$ and ZQ are equal to $90° - EZ$.

This characterization certainly applies to Thābit's *On the Transversal Figure*.[100] This book may be described as the earliest ever written that deals specifically with a trigonometric topic, but its content is quite unexpected. It begins ordinarily with an account of Ptolemy's proof of Menelaus's Theorem, and proceeds to describe a simpler, more elegant proof that avoids the lemmas used in the *Almagest*, back in chapter 2. Thābit does introduce his proof with his own lemma, one that has attracted some attention of its own:

Lemma: (See figure 4.24.) Given two great circles with common diameter *AG*, drop perpendiculars from any two points *E* and *Z* on one circle onto the plane of the other, producing *K* and *L*. Then

$$\frac{\text{Crd}\, 2AE}{\text{Crd}\, 2AZ} = \frac{EK}{ZL}. \tag{4.37}$$

At first glance this lemma is innocuous, but if one drops perpendicular arcs from *E* and *Z* along the sphere to *X* and *Y*, with almost no effort (4.37) becomes

$$\frac{\text{Sin}\,\widehat{AE}}{\text{Sin}\,\widehat{AZ}} = \frac{\text{Sin}\,\widehat{EX}}{\text{Sin}\,\widehat{ZY}}, \tag{4.38}$$

none other than the Rule of Four Quantities applied to the configuration *AXYZE*. Although we do not have an explicit statement from Thābit to this effect, those who have noticed the connection agree that it could hardly have escaped Thābit's notice that this useful result is an obvious consequence of his lemma.[101] There is a passage in Nāsir al-Dīn al-Ṭūsī's treatise on the Transversal Figure that quotes Abū Naṣr Manṣūr stating that Thābit wrote on the Rule of Four Quantities as a replacement to the Transversal Figure in spherical astronomy; but if that is true, Thābit's work is now lost.[102] If Abū Naṣr was correct, this work would be the first explicit statement of the Rule of Four Quantities with astronomical intent.[103]

[100] *On the Transversal Figure* is available in Latin in [Björnbo 1924] and [Carmody 1960, 150–164], and in English in the lamentably rare [Lorch 2001b]; see also the discussion [Bellosta 2004]. On the combinatorial aspects of the treatise see also [Koelblen 1993] and [Koelblen 1994].

[101] See for instance [von Braunmühl 1900/1903, vol. 1, 46–47] or [Lorch 2001b, 396]. A French translation of the lemma, in al-Ṭūsī's hand, is in [Naṣīr al-Dīn al-Ṭūsī (Caratheodory) 1891, 200–202].

[102] [Lorch 2001b, 6].

[103] Since the Rule of Four Quantities is a trivial consequence of the Transversal Figure it is not hard to find astronomical problems solved earlier in a manner that appears to reflect the Rule, but we have made the point before that this is not the same thing as stating it explicitly as a theorem.

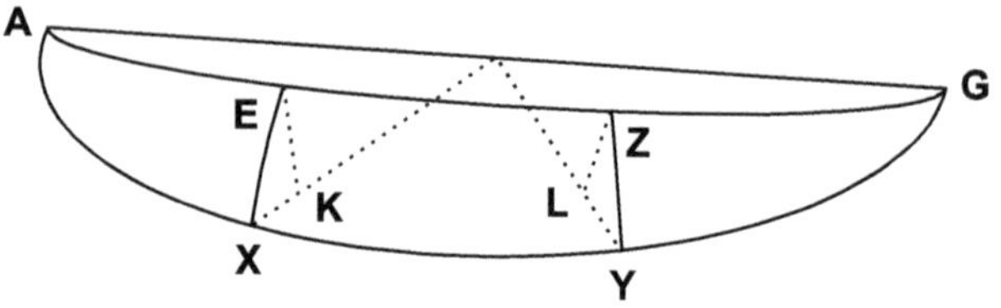

Figure 4.24
Thābit ibn Qurra's lemma to his proof of the Transversal Figure

Thābit continues *On the Transversal Figure* with what turns out to be a combinatorial problem. Both theorems of the Transversal Figure assert a composition of ratios in the form

$$\frac{a}{b} = \frac{c}{d} \cdot \frac{e}{f}. \tag{4.39}$$

Some particular application of the Transversal Figure may require the ratio of, say, c to f, which may be expressed in two ways as

$$\frac{c}{f} = \frac{a}{b} \cdot \frac{d}{e} \text{ or } \frac{c}{f} = \frac{a}{e} \cdot \frac{d}{b}. \tag{4.40}$$

There are altogether eighteen ways in which the original expression can be rearranged, two for each of the nine pairs of quantities that can be placed in ratio by (4.39). (Certain ratios, such as a/d, cannot be formed from (4.39) at all.) Thābit lists all eighteen "modes," as he calls them, and gives a separate demonstration for each one.

All of this may appear very strange to the modern reader, who might simply write $adf = bce$ and be done with it, and consider the eighteen variants merely as trivial rearrangements of the original proposition. However, ancient and medieval conceptions of ratio[104] tie the quantities in ratio to each other more tightly than a modern reader would. The equivalence of the right sides of the two equations in (4.40), for instance, is for a medieval mathematician a non-trivial proposition that requires a formal proof.

Thābit seems to have initiated a tradition of interest in this subject, since we find a number of authors discussing it afterward, including such luminaries as al-Nayrīzī, al-Khāzin, Ibn al-Haytham, al-Sijzī, Maslama al-Majrīṭī, and Jābir ibn Aflaḥ. These later treatises, some depending on Thābit but others not, generally worked toward covering what each author considered to be a complete treatment, either mathematical or astronomical, of the Transversal Figure.[105]

[104] Thābit actually proposes a new conception of ratios and their composition in a sequel to *On the Transversal Figure*, entitled *On the Composition of Ratios*. See the English translation in [Lorch 2001b, 167–326] and the discussion in [Crozet 2004].

[105] [Lorch 2001b, 353ff.]; on al-Sijzī's work see [Berggren 1981], and on Maslama al-Majrīṭī see [Lorch 1996].

Menelaus's Replacements

As powerful as the Transversal Figure is, it is not the simplest tool with which to operate on the sphere. It usually requires that the solver add certain arcs to the given problem in order to generate one or more appropriate spherical quadrilaterals to hit upon the right version of the theorem to apply, and then to juggle the six different quantities that emerge until some useful statement appears—as we saw in al-Kūhī's solution of the rising times problem. Spherical astronomical applications of the Transversal Figure tended to follow certain patterns: for instance, arcs 90° in length were sought after; and when angles were included among the givens of a problem, the great circle for which the angle is at the pole was added to the diagram, so that the angle could be converted easily to an arc.

These similarities in approach suggest that some simplified form of the Transversal Figure might turn out to be useful in astronomical work. For instance consider figure 4.25, where $\overset{\frown}{ABC}$ is the equator, $\overset{\frown}{ADE}$ is the ecliptic, and the angle between them at A is the fixed value ε. Suppose the Sun is at D (so $\lambda = \overset{\frown}{AD}$) and we wish to determine its declination $\delta = \overset{\frown}{BD}$. The obvious solution is to draw $\overset{\frown}{CEF}$ with its pole at A so that $\overset{\frown}{CE} = \varepsilon$; then extend BD and CE up to the pole F of $\overset{\frown}{ABC}$. We now have a transversal figure, to which we apply conjunction (4.36). Luckily three of the arcs are 90° in length, so the result

$$\frac{\text{Sin } 90^\circ}{\text{Sin } \varepsilon} = \frac{\text{Sin } 90^\circ}{\text{Sin } \delta} \cdot \frac{\text{Sin } \lambda}{\text{Sin } 90^\circ} \qquad (4.41)$$

simplifies nicely to the standard formula $\text{Sin } \delta = (\text{Sin } \lambda \cdot \text{Sin } \varepsilon / R)$. This is basically what Ptolemy does in the *Almagest*.

But this is also how to prove the Rule of Four Quantities from the Transversal Figure, for if we take the arcs $\overset{\frown}{DF}$ and $\overset{\frown}{EF}$ to be auxiliary to the main diagram, (4.41) becomes simply

$$\frac{\text{Sin } \overset{\frown}{AD}}{\text{Sin } \overset{\frown}{AE}} = \frac{\text{Sin } \overset{\frown}{BD}}{\text{Sin } \overset{\frown}{CE}}, \qquad (4.42)$$

where $\overset{\frown}{AE} = 90^\circ$. In this sense at least Ptolemy may be said to have known the Rule of Four Quantities—and many others after him, since this sort of situation arises very frequently, especially when changing coordinates.[106] But it

[106] [Berggren 1985] includes an example involving the "method of the *zījes*" to determine the *qibla* in which the Rule of Four Quantities is applied four times.

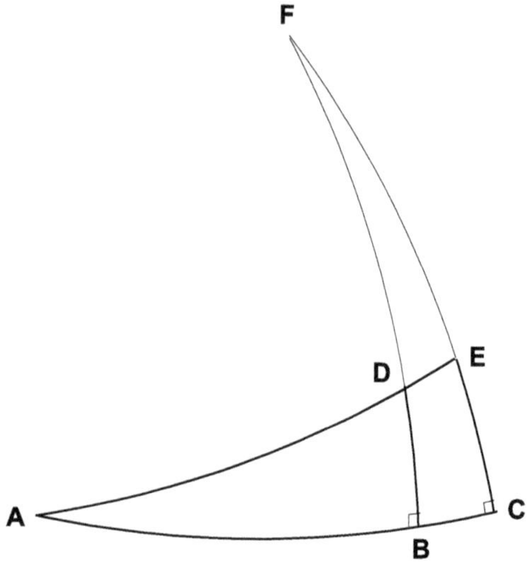

Figure 4.25
Deriving the Rule of Four Quantities from the Transversal Figure

is a very different thing to prove the result explicitly as a mathematical theorem (as we have seen Thābit ibn Qurra do, almost), and to assert a role for it as a starting point for spherical astronomy, independently of the Transversal Figure.

We are extremely fortunate to have a first-hand account of the emergence of the Rule of Four Quantities and other theorems in al-Bīrūnī's *Maqālīd ʿilm al-hay'a* (*Keys to Astronomy*), written only a decade or so after these discipline-changing events in the last few years of the tenth century.[107] The story reconstructed partly from al-Bīrūnī's work is full of intrigue. In the last several years of the first millennium, Abū Naṣr Manṣūr ibn ʿIrāq[108] (whom we have met before, with his use of $R = 1$ in a set of astronomical tables) received a request from a colleague to prove various theorems in spherical astronomy, including rising times. His response, the *Book of Azimuths*, used the Transversal Figure extensively but included two results derived from what he called the "figure that frees" one from needing the Transversal Figure. The passage in al-Kūhī's memo in Text 4.6 may well refer to this.

Unfortunately the *Book of Azimuths* is lost, and we cannot say for certain what this figure was. We do know that Abū Naṣr's colleague

[107] The *Keys to Astronomy* has been translated into French in [al-Bīrūnī (Debarnot) 1985] (mostly superceding the summary in [Kennedy 1971a]), which also includes a valuable and extensive account of spherical trigonometry and astronomy up to the eleventh century, in considerably more detail than we have space for here.

[108] On Abū Naṣr Manṣūr see the studies collected in [Samsó 1969]; on spherical trigonometry in particular see pp. 38–45, 66–70, and 134–152.

Abū'l-Wafā' (who had also used $R = 1$, among other achievements) replied to it, criticizing Abū Naṣr's use of the old-fashioned Transversal Figure. Now Abū'l-Wafā' died in 998, so by then his new-fangled *Almagest* with its new theorems and use of tangents would have been already or nearly completed. Abū Naṣr's defense against Abū'l-Wafā', the *Treatise on the Determination of Spherical Arcs*, luckily is available;[109] it contains a proof of a form of the Rule of Four Quantities[110] and a great deal more.

Abū Naṣr's path to the solution of rising times appeals to another, related theorem: in the same diagram (figure 4.25)

$$\frac{\operatorname{Sin}\widehat{AD}}{\operatorname{Sin}\widehat{AB}} = \frac{\operatorname{Cos}\widehat{DB}}{\operatorname{Cos}\widehat{CE}},^{111} \tag{4.43}$$

a result closely related to what was eventually called Geber's Theorem.[112] We proceed, following Abū Naṣr, as follows (figure 4.23):[113]

- From the Rule of Four Quantities applied to ♈☼GZM we find $\frac{\operatorname{Sin}\lambda}{\operatorname{Sin}\delta} = \frac{\operatorname{Sin}90°}{\operatorname{Sin}\varepsilon}$, which gives us the declination δ.
- From (4.43) applied to the same figure we find $\frac{\operatorname{Sin}\lambda}{\operatorname{Sin}\alpha} = \frac{\operatorname{Cos}\delta}{\operatorname{Cos}\varepsilon}$, which gives us the right ascension α.
- From the Rule of Four Quantities applied to E☼HQM we find $\frac{\operatorname{Sin}\psi}{\operatorname{Sin}\delta} = \frac{\operatorname{Sin}90°}{\operatorname{Sin}(90° - \phi)}$, which gives us the ortive amplitude ψ.
- From (4.43) applied to the same figure we find $\frac{\operatorname{Sin}\psi}{\operatorname{Sin}n} = \frac{\operatorname{Cos}\delta}{\operatorname{Cos}(90° - \phi)}$, which gives us the equation of daylight (ascensional difference) n.

[109] See [Luckey 1941] for a German translation and [Matvievskaya/Tllashev 1983] for a Russian translation. See also [Kunitzsch/Lorch 1994], which contains translations of two passages from Abū Naṣr solving the problem of the "ascension of the azimuth"—the earlier treatise using the Transversal Figure, and the later one using the new methods described in this section.

[110] As we shall see, Abū Naṣr's approach to this theorem makes it more likely that he thought of it as a special case of the Law of Sines. The result is the same either way.

[111] This is easily demonstrated from the Transversal Figure. Assuming that the four long arcs are all 90°, solve for Sin a/Sin g (according to the assignment of arcs in figure 4.22) in both the disjunction and conjunction versions of the Theorem; then set the results equal to each other. Abū Naṣr establishes the result using the Law of Sines.

[112] In a spherical triangle with right angle at C, $\cos A = \cos a \sin B$. The theorem is named after Jābir ibn Aflaḥ, a twelfth-century Andalusian astronomer. We shall encounter the man, and the result, again.

[113] From [Luckey 1941, 437–440].

Finally, $\theta = \alpha - n$.

In fact, Abū Naṣr refers to several of the arcs in the above figures by their associated angles, an important point to which we shall return in a moment. But applications in spherical astronomy were often considerably smoother to solve when dealing strictly with arcs, and when the Rule of Four Quantities later became a dominant tool it was typically used with arcs as above, not with angles.

Indeed, several centuries later the greatest of Muslim computational astronomers, Jamshīd al-Kāshī, solved spherical astronomical problems relying on little else than the Rule of Four Quantities as his trigonometric apparatus, almost always choosing $\widehat{AE} = 90°$ in figure 4.25.[114] In fact in his *Khāqānī Zīj* he even extended its application to the determination of planetary positions, an unprecedented move. The models for planetary motion that we have seen thus far take place entirely on the plane of the ecliptic, but in fact the planets move above and below the ecliptic. Since Ptolemy this problem had been dealt with somewhat crudely using approximations, but al-Kāshī devised a scheme in which the epicycle is enclosed in a sphere, and the motions dictated in the *Almagest* are modeled by the movements of several great circles on this sphere. The problem of determining the planet's position suddenly requires spherical trigonometry, and al-Kāshī returns again and again to the Rule of Four Quantities to solve it.[115]

That the Rule of Four Quantities was in the air at the end of the tenth century is underlined by the fact that there was a third claimant to the priority crown, Abū Maḥmūd al-Khujandī, who wrote a treatise on its use in solving astronomical problems. Al-Khujandī may well have been the first to use it, and his work on this topic would be picked up by Kūshyār ibn Labbān (whose name we have encountered in the context of interpolation with double-argument functions).

Although al-Khujandī disputed priority with Abū'l-Wafā' over the Rule of Four Quantities, it was clear from the start that Abū'l-Wafā''s contribution was much deeper and more meaningful than a single theorem. His *Almagest* contains a completely new approach to spherical trigonometry and astronomy, and includes several new theorems—each of which might serve, on its own, as an adequate replacement for the Transversal Figure. As the first person to use the tangent function on a par with the sine and cosine, one example that must have come naturally to his attention was the Law of Tangents: in figure 4.25,

[114] [Kennedy 1985] is a survey of al-Kāshī's approach to spherical astronomy; 23 of 26 problems are solved with the Rule of Four Quantities.

[115] For a full account see [Van Brummelen 2006].

$$\frac{\operatorname{Sin}\widehat{AB}}{\operatorname{Sin}\widehat{AC}} = \frac{\operatorname{Tan}\widehat{BD}}{\operatorname{Tan}\widehat{CE}}. \tag{4.44}$$

As we have found time and again with respect to other theorems, hints of this relation appear as early as Ptolemy's *Almagest*, but it is not until Abū'l-Wafā''s *Almagest* that it is stated explicitly.[116] He uses it early and often; for instance, to find the right ascension of a given arc of the ecliptic from its declination he applies it to ♈☼*GZM* (figure 4.23), and arrives immediately at

$$\frac{\operatorname{Sin}\alpha}{R} = \frac{\operatorname{Tan}\delta}{\operatorname{Tan}\varepsilon}.^{117} \tag{4.45}$$

Curiously the Law of Tangents was rejected by both al-Khujandī and his follower Kūshyār ibn Labbān, on the grounds that the use of a table of tangents might lead to numerical problems for arcs approaching 90°, where the tangent function increases very rapidly.[118] However, Kūshyār seems later to have changed his mind, for the Law of Tangents is found as one of the theorems in his *Jāmiʿ Zīj*.[119]

In his *Keys to Astronomy* al-Bīrūnī credits Abū'l-Wafā' with the Law of Tangents, but Bīrūnī's general approach is to discredit Abū'l-Wafā' in favor of Abū Naṣr Manṣūr—which is perhaps to be expected, since the latter had been his teacher. Bīrūnī's main bone of contention is the last of our great theorems of spherical trigonometry, the spherical Law of Sines, and he leaves no doubt in the reader's mind that Abū Naṣr Manṣūr deserves credit. Its statement is breathtakingly elegant: in an arbitrary spherical triangle (figure 4.26),

$$\frac{\operatorname{Sin}a}{\operatorname{Sin}A} = \frac{\operatorname{Sin}b}{\operatorname{Sin}B} = \frac{\operatorname{Sin}c}{\operatorname{Sin}C}. \tag{4.46}$$

Abū Naṣr's *Treatise on the Determination of Spherical Arcs* begins with the Law of Sines, and many of the book's results are derived from the Law's right-angled version.[120] The Law of Sines also appears in Abū'l-Wafā''s *Almagest*, but it does not have as central a place; it is derived from the Rule of Four Quantities.[121]

[116] See [Khaĭretdinova 1985a] on this point, and for a more general history of the Law of Tangents.
[117] Based on notes made by Benno van Dalen.
[118] [Debarnot 1996, 513].
[119] [Berggren 1987, 25–26].
[120] See also [Suter 1910] for a German translation, and [Matvievskaya/Tllashev 1983] for a Russian translation, of another document by al-Bīrūnī containing a proof of the Law of Sines by Abū Naṣr.
[121] See [Berggren 1986, 174–176] for an account of Abū'l-Wafā''s proof.

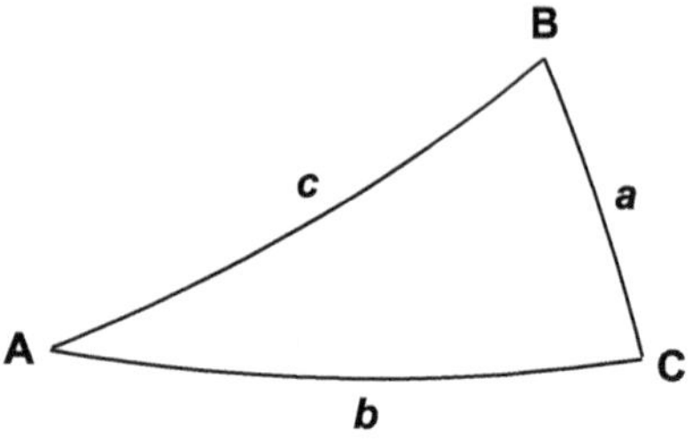

Figure 4.26
The spherical Law of Sines

Generally, Abū Naṣr seems more willing to allow angles to play an independent role in spherical trigonometry than does Abū'l-Wafā'. For instance, Abū Naṣr's statement of the Rule of Four Quantities confines itself to ΔADB in figure 4.25:

$$\frac{\text{Sin}\,\widehat{BD}}{\text{Sin}\,\angle A} = \frac{\text{Sin}\,\widehat{AD}}{\text{Sin}\,90^\circ}.^{122} \tag{4.47}$$

The inclusion of $\angle A$ allows the removal of references to C and E, allowing Abū Naṣr to focus solely on the triangle of interest with no distracting external constructions. In fact, in a very real sense this episode may be said to herald the birth of spherical trigonometry as the study of triangles.[123]

While Abū Naṣr's focus on the spherical triangle does not immediately provide much extra power in astronomy, it does lead to further mathematical insights—in particular, the notion of the polar triangle. In figure 4.27, from one of Abū Naṣr's own treatises, consider ΔABG.[124] Draw great circles with poles A, B, and G, which intersect at H, K, and S. The resulting ***polar triangle***, ΔHKS, has some elegant properties that recommend it to the mathematical reader. For one, ΔABG turns out to be the polar triangle of ΔHKS, so that polar triangles are dual to each other.[125]

The main advantage of a polar triangle is its transformation of angles in the original triangle into sides, and vice versa: the sides of the polar

[122] In this form the Rule of Four Quantities may be seen as a special case of the Law of Sines, which is in fact how Abū Naṣr proves it.

[123] This interpretation of the shift in spherical trigonometry in the late tenth century is one of the many significant contributions of Debarnot's several studies of this period; see for instance [Debarnot 1996, 506–507].

[124] Abū Naṣr's work on polar triangles has been analyzed in [Debarnot 1978], which includes a translation into French of the relevant passage.

[125] A duality relation between a pair of objects is a transformation that returns the original object when applied, then applied again to the transformed object. Often the once-transformed object behaves in one way differently, but in another way analogously to, the original object. In this case, as we are about to see, angles and sides of polar triangles play reversed roles. Abū Naṣr

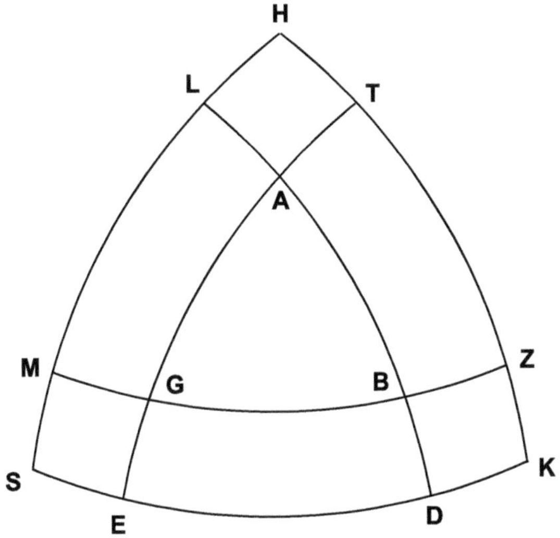

Figure 4.27
Abū Naṣr Manṣūr and the polar triangle

triangle are equal to the supplements of the angles in the original triangle (and the angles are supplements of the original sides). As Abū Naṣr shows,

$$\begin{aligned}\overset{\frown}{KS} &= \overset{\frown}{KD} + \overset{\frown}{DE} + \overset{\frown}{ES} = (90^\circ - \overset{\frown}{DE}) + \overset{\frown}{DE} + (90^\circ - \overset{\frown}{DE}) \\ &= 180^\circ - \overset{\frown}{DE} = 180^\circ - \angle A.\end{aligned} \tag{4.48}$$[126]

This simple correspondence between sides and angles allows one to simplify considerably the exposition of solving triangles. For instance, by the time he reached figure 4.27 Abū Naṣr had already dealt with the case where three sides of the triangle are known. He handles the case of three known angles simply by converting to the polar triangle: its three sides are known, so its angles may be found using the earlier case. They may then be converted back to the sides of the original triangle using (4.48).

---

demonstrates the duality of polar triangles as follows: since $\overset{\frown}{AG}$ passes through the poles of $\overset{\frown}{DE}$ and $\overset{\frown}{TZ}$, both of the latter circles must pass through the poles of $\overset{\frown}{AG}$. Since K is the intersection of these two circles, it must be a pole of $\overset{\frown}{AG}$. Similarly for H and S.

[126] Since K is the pole of $\overset{\frown}{EGAT}$, $\overset{\frown}{KE} = 90^\circ$ and hence $\overset{\frown}{KD} = 90^\circ - \overset{\frown}{DE}$; similarly, $\overset{\frown}{ES} = 90^\circ - \overset{\frown}{DE}$. Abū Naṣr does not actually go through the equations here; he simply points out the arcs that are 90° in magnitude and asserts that since $\overset{\frown}{DE}$ is known, $\overset{\frown}{KS}$ is known.

Systematizing Spherical Trigonometry: Ibn Muʿādh's *Determination of the Magnitudes* and Naṣīr al-Dīn al-Ṭūsī's *Transversal Figure*

One might expect that the fundamental developments that rendered Menelaus's Theorem obsolete would trigger someone to write a comprehensive treatise on the reworked field, and one would not be disappointed. However, the very quick arrival of the first such work, and especially its appearance not in eastern Islam but far to the west in Spain, should raise eyebrows. The book in question is *Determination of the Magnitudes of the Arcs on the Surface of a Sphere*[127] by one Abū ʿAbd Allāh Muḥammad ibn Muʿādh al-Jayyānī (ca. 989–after 1079), who seems to have spent most of his life in Cordoba.[128] Not much is known about him; he wrote several treatises in astronomy and astrology, a couple of which were translated by Gerard of Cremona, and a defense of Euclid's conception of ratio. Most tantalizing to us is the fact that he spent several of his younger years in Cairo, where he may have studied with Ibn al-Haytham and learned of the new spherical trigonometry of the East.

Determination of the Magnitudes is the earliest surviving account of spherical trigonometry independent of astronomical application. It begins with a statement of the Transversal Figure (without proof), and follows up with problems of the form: given the sum or difference of two arcs and the ratio of their sines, find the two arcs. This result turns out to be useful in solving certain spherical triangles and leads Ibn Muʿādh to construct a table of tangents—not shadow lengths, but explicitly defined as the "division of the sine by the cosine."[129]

Many of Ibn Muʿādh's theorems have a familiar ring: the Law of Sines, the Rule of Four Quantities, an equivalent of Abū Naṣr's (4.43), Geber's Theorem, the Law of Tangents, and several corollaries. In the section where Ibn Muʿādh goes on to solve triangles there is a hint that he may have had a predecessor: the order of the triangle problems seems to have been rearranged to correspond with the givens of each triangle, rather than an ordering according to which theorems are called upon to solve them.[130] Even the polar triangle makes an appearance, to solve the case of three known angles.[131]

[127] Edited and translated into Spanish in [Villuendas 1979]; see also the summary of the text in [Samsó 1980].

[128] What is known of Ibn Muʿādh's life may be found in [Sabra 1967, 84–85].

[129] The table, which gives the tangent for every 1° as well as a few values for arcs between 89° and 90°, is reproduced in [Villuendas 1979, 118]. [Samsó 1980, 62–64] speculates that it may derive partly from the sine tables of al-Khwārizmī-Maslama.

[130] [Samsó 1980, 65].

[131] See [Khaĭretdinova 1985b] for the connection between Ibn Muʿādh and Naṣīr al-Dīn al-Ṭūsī on the polar triangle, and [Debarnot 1979] on its beginning with Abū Naṣr Manṣūr.

This evidence suggests very strongly that Ibn Muʿādh had extensive access to our Eastern writers Abū Naṣr Manṣūr, Abū'l-Wafā', and al-Bīrūnī, although there may have been an intermediary source. If the link is to be believed—and how could it not be?—then we have the beginning of a case that late tenth- and early eleventh-century Eastern scientific authors were known, at least to some extent, in Muslim Spain.[132]

Ibn Muʿādh's book may or may not have been the first work of its kind; it was followed by several similar treatises on spherical trigonometry more or less independent of astronomy in the East in the eleventh through thirteenth centuries.[133] A case can be made that al-Bīrūnī's *Keys to Astronomy* falls into this category, although the astronomical connection is never very far from al-Bīrūnī's mind and astronomical problems are solved within its pages. Beyond doubt, however, the ultimate systematic treatment came with the thirteenth-century Iranian scientist Naṣīr al-Dīn al-Ṭūsī and his *Treatise on the Quadrilateral*.[134]

Few would argue against the claim that al-Ṭūsī was one of the greatest scientists of medieval Islam.[135] His accomplishments, encompassing over 150 treatises, extended beyond the exact sciences to philosophy and theology, even poetry. Within the sciences he contributed to mineralogy, mathematics (on the parallel postulate) and medicine, but his central focus was astronomy. His *Ilkhānī Zīj* (named after the Mongol dynasty that ruled thirteenth- and fourteenth-century Iraq and Iran) was one of the most influential of the astronomical handbooks, and directly inspired al-Kāshī's previously-mentioned *Khāqānī Zīj*. Among al-Ṭūsī's other astronomical accomplishments was a commentary on Ptolemy's *Almagest*[136] and his supervision of the famous astronomical observatory of Maragha. One of his greatest feats was the *al-Tadhkira fī ʿilm al-hay'a* (*Memoir on Astronomy*).[137] Among other things, this book describes the "Ṭūsī couple," a pair of circles moving uniformly so as to produce linear oscillations. This invention became the foundation of efforts to reconstruct Ptolemaic astronomy to fit more closely the requirements of Aristotelian physics.

On the other hand, the *Treatise on the Quadrilateral* is an unabashedly mathematical book. Its topics are motivated by mathematical ideas, and it

[132] The evidence, particularly with respect to al-Bīrūnī, is gathered together in [Samsó 1996].

[133] Some of these are listed in [Samsó 1980, 61].

[134] The *Treatise on the Quadrilateral* is available in Arabic and French translation in [Naṣīr al-Dīn al-Ṭūsī (Caratheodory) 1891], and in Russian in [Naṣīr al-Dīn al-Ṭūsī (Mamedbeyli/Riznichenko/Rosenfeld) 1952]. A good account of its contents is [von Braunmühl 1900/1903, vol. 1, 65–71].

[135] On al-Ṭūsī's life see [Ragep, in Naṣīr al-Dīn al-Ṭūsī (Ragep) 1993, 3–23].

[136] See [Saliba 1987].

[137] The *Tadhkira* is available in Arabic edition and translation in [Naṣīr al-Dīn al-Ṭūsī (Ragep) 1993].

strives for mathematical completeness in a way that an astronomer might find off-putting. As its title indicates, the *Treatise* goes back to the Transversal Figure as its foundational theorem—but it does so in a more powerful and thorough manner than had ever been attempted before.

The first two of the *Treatise*'s five books build toward a theory of plane trigonometry. Book I deals with the theory of ratios in an "uncompromisingly arithmetical" manner,[138] preparing the reader to be able to manipulate ratios easily in following chapters. Book II handles the surprisingly many theorems that can be asserted and cases that arise with respect to the plane version of the Transversal Figure. As we have seen before, this subject can be as much about combinatorics as about geometry.

These topics set up Book III, where plane trigonometry arises and triangles are solved. Al-Ṭūsī begins in the usual way by solving right-angled triangles when certain elements are given: for instance, given two sides, the third may be found using the Pythagorean Theorem. Then the angles are found in a rather old-fashioned way. For instance, to find $\angle C$ in figure 4.28 circumscribe a circle, and rescale the units so that $AC = 120$. Then the length of AB may be converted to its corresponding arc in a chord table (such as Ptolemy's), and this arc is twice $\angle C$. Next al-Ṭūsī deals with arbitrary triangles, often by splitting them into right triangles.

The heart of Book III comes next, where two important theorems are demonstrated. The first is an explicit statement of the plane version of the Law of Sines, accompanied by two proofs and explanations on how to apply it to solve triangles. Since the first proof has appeared in several places[139] we give a brief account of the second, in the case where all angles are acute. In figure 4.29 consider ΔABC; extend AB and AC to AD and AF respectively, both of which are equal to 60. Drop AE perpendicularly onto BC and extend to H; also drop perpendiculars DK and FT. Then

$$\frac{AB}{AE} = \frac{AD}{AK} = \frac{R}{\text{Sin}\angle B} \text{ and } \frac{AE}{AC} = \frac{AT}{AF} = \frac{\text{Sin}\angle C}{R}.$$

Combining the two ratios, we arrive at

$$\frac{AB}{AC} = \frac{\text{Sin}\angle C}{\text{Sin}\angle B}. \tag{4.49}$$

Today this appearance of the Law of Sines would be more prominent, but al-Ṭūsī passes quickly on to his next result, which he describes as "very

[138] [Lorch 2001b, 391].

[139] For instance in [von Braunmühl 1900/1903, vol. 1, 66–67] and [Berggren 1986, 138–139].

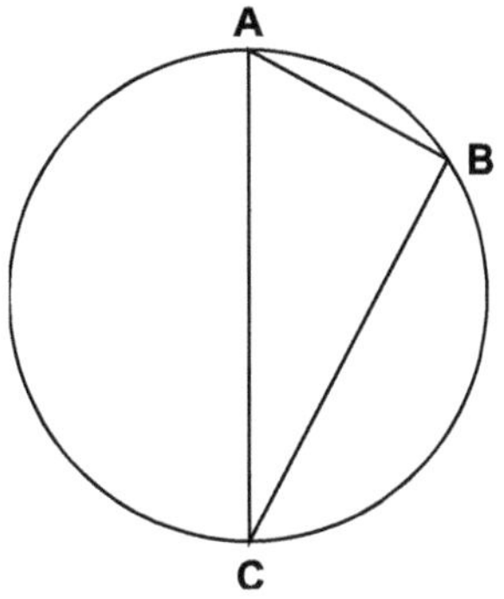

Figure 4.28
Finding angles from sides in a right triangle, in al-Ṭūsī's *Treatise on the Quadrilateral*

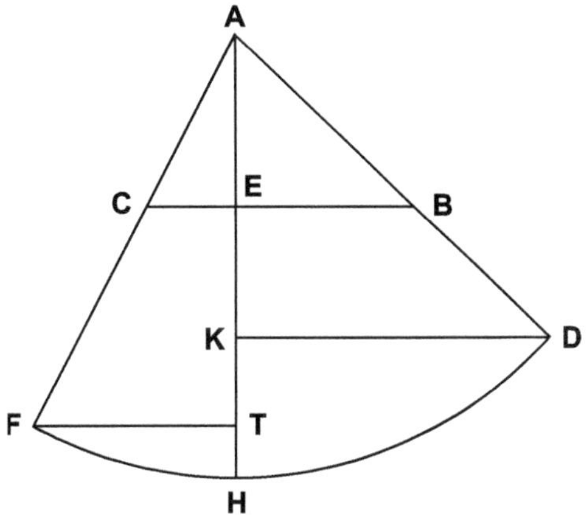

Figure 4.29
One of al-Ṭūsī's proofs of the plane Law of Sines

useful in the theory of the plane quadrilateral": given $\alpha \pm \beta$ and the ratio of Sin α to Sin β, determine the two arcs α and β—the very problem we have just seen mentioned in Ibn Muʿādh's *Treatise.* We also recognize (4.49) as the sine equivalent of the two lemmas used by Ptolemy for his proof of the spherical Transversal Figure;[140] presumably al-Ṭūsī includes the result here at the end of Book III since it is still within the domain of plane trigonometry.

Now that his preparations are complete, al-Ṭūsī commences his study of the spherical Transversal Figure in Book IV. He begins with Ptolemy's treatment, but extends well beyond it to a thorough study and tabulation of all of the possible cases that arise when using this theorem to solve problems. We skip

[140] For instance in Lemma A, figure 2.17(a) in chapter 2, $\alpha = \overset{\frown}{AZ}$ and $\beta = \overset{\frown}{ZB}$. The chord of $\alpha + \beta$, AB, is divided into the ratio Sin α/Sin β. Then the radius HKZ divides $\alpha + \beta$ into the individual arcs. Al-Ṭūsī goes on to mention [Naṣīr al-Dīn al-Ṭūsī (Caratheodory) 1891, 80] that this theorem is very frequent in astronomical operations when $\alpha \pm \beta = 90°$. Indeed, in the $\alpha + \beta$ case this corresponds simply to finding α by means of an arc tangent.

over Book IV lightly; what is of most interest to us is the chapter's conclusion, where al-Ṭūsī introduces the most substantial, and concluding, fifth book:

> The ancients did not fail to . . . make use of [the spherical quadrilateral] with confidence, as is seen in Menelaus's book on *Spherics* and in the beginning of Ptolemy's *Almagest*. But the moderns, either because they were unwilling to engage in the examination of various ratios and their varieties or to avoid the lengths that the use of composed ratios involves in practice, imagined and studied other figures intended to take the place of the quadrilateral and to gain the utility that they afford, without needing to resort to many distinctions and composed ratios. Also we believed it useful, having engaged in this study, to speak of the methods used by the moderns, in order to complete with the help of God all that has been done in this branch of the science.[141]

Al-Ṭūsī does not disappoint; Book V contains a thorough reexamination of spherical trigonometry with the triangle as its focus rather than the quadrilateral. Thus a study of the properties of a spherical triangle recalling Book I of Menelaus's *Spherics*, as well as a discussion of angles and their relation to arcs on the sphere, are al-Ṭūsī's starting points. This latter point is significant; it had only been with Abū Naṣr Manṣūr and his colleagues that angles had begun to be taken seriously on the celestial sphere.[142] Al-Ṭūsī reflects the views of his tenth-century predecessors by allowing angles to stand on their own, but he still feels it necessary to describe quite carefully how they are defined in terms of the corresponding arcs.

Al-Ṭūsī's account of modern spherical trigonometry revolves around two theorems: the Law of Sines and the Law of Tangents. With the Law of Sines he begins as one might expect, with the right-angled case. He provides a string of demonstrations, by such familiar names as Abū Naṣr Manṣūr, al-Bīrūnī, Abū'l-Wafā' and al-Khujandī, and concludes (as one might anticipate) with a proof of his own based on the Transversal Figure. Several important results follow as corollaries: in a spherical triangle (such as figure 4.26) with a right angle at C,

- $\cos c = \cos a \cos b$ (which he attributes to tenth-century scientists al-Nayrīzī and al-Khāzin);
- $\cos A = \cos a \sin B$ (Geber's Theorem); and
- the less useful $\dfrac{\cos A}{\cos a} = \dfrac{\sin b}{\sin c}$.

141 [Naṣīr al-Dīn al-Ṭūsī (Caratheodory) 1891, 114]; translated from French.

142 See for instance [Luckey 1937, 124–132], and [Luckey 1941] on Abū Naṣr Manṣūr.

On the Law of Tangents, al-Ṭūsī goes beyond theorem-proving, and makes a spirited case for the use of tangents and related functions in general. The fact that he defines them carefully, in a modern purely geometric spirit essentially as Abū'l-Wafā' had done, indicates that he felt some of his readers would not necessarily have accepted them implicitly, or perhaps even been familiar with them. As with the Law of Sines, the Law of Tangents is proved several times. One of these proofs comes directly from the Transversal Figure for the case where one angle is right, thereby looking somewhat old-fashioned; it is nevertheless elegant.

In figure 4.30 let $\angle B$ in ΔABC be right, and extend arcs so that $\overset{\frown}{AD} = \overset{\frown}{AE} = \overset{\frown}{EF} = \overset{\frown}{BF} = 90°$. Then, by one of al-Ṭūsī's varieties of the disjunctive statement of the Transversal Figure, we have

$$\frac{\operatorname{Sin}\overset{\frown}{BC}}{\operatorname{Sin}\overset{\frown}{CF}} = \frac{\operatorname{Sin}\overset{\frown}{AB}}{\operatorname{Sin}\overset{\frown}{AE}} \cdot \frac{\operatorname{Sin}\overset{\frown}{DE}}{\operatorname{Sin}\overset{\frown}{DF}}. \tag{4.50}$$

The first and third of the ratios in this expression simplify to tangents (divided by R); in the second ratio, Sin $\overset{\frown}{AE} = R$. This gives the desired result

$$\frac{\operatorname{Tan}\overset{\frown}{BC}}{\operatorname{Sin}\overset{\frown}{AB}} = \frac{\operatorname{Tan}\overset{\frown}{DE}}{R}. \tag{4.51}$$

Again as before, al-Ṭūsī goes on to establish several corollaries to the Law of Tangents, this time using the new functions, including the following: in a triangle right-angled at C,

- $\cos A \cot b = \cot c$;
- $\cot A = \cos C \tan B$; and
- the less useful $\cot A \sin a = \tan b \cos c$.

From here, al-Ṭūsī addresses systematically the solution of all spherical triangles using both the Law of Sines and the Law of Tangents, applying the polar triangle to the case of three known angles as Abū Naṣr had done before him. But al-Ṭūsī begins by defending charges made by unnamed astronomers (perhaps al-Khujandī and Kūshyār ibn Labbān) that the tangent function is not useful when the arcs exceed 45°, due to its rapid increase and the large errors that result when interpolating within the tables.

Al-Ṭūsī considers five cases. In the first, he imagines that we are to use $\sin a = \tan b \cdot \tan c$ to solve for a.[143] If $b > 45°$ (i.e., $\tan b > 1$) then we shall

[143] Of course al-Ṭūsī uses the medieval functions rather than the modern ones; to avoid having to describe repeated multiplications and divisions by R we convert to modern formulas.

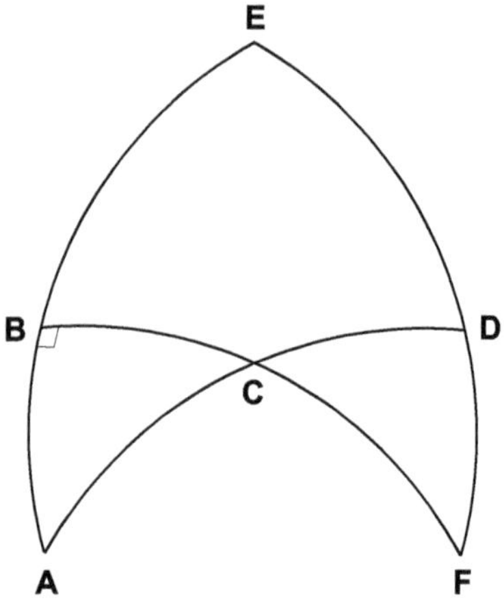

Figure 4.30
Al-Ṭūsī's demonstration of the right-angled case of the Law of Tangents

need to have c less than 45° (i.e., tan $c < 1$), since their product must be a sine and hence less than 1. To avoid using the dangerous part of the tangent table where $\theta > 45°$, al-Ṭūsī recommends replacing tan $b \cdot$ tan c with tan c/tan $(90° - b)$. And indeed this conversion works. If we wish to find, for instance, tan 88.6° · tan 0.25°, the use of an accurate tangent tabulated for each degree and one fractional sexagesimal place gives the result 0;12,1,48, as opposed to the correct value 0;10,42,44. Using Ṭūsī's conversion we arrive at the much better value 0;10,42,51.

It is a matter of (probably futile) debate to pinpoint a moment when trigonometry became its own subject. The events of the late tenth century certainly signal the beginnings of an emergence of sorts from astronomy; al-Bīrūnī's *Keys to Astronomy* deals extensively (but not entirely) with spherical trigonometry on its own. The systematic works by Ibn Muʿādh and others are definitively trigonometric, but in terms of their quality they cannot compare to al-Ṭūsī's *Treatise on the Quadrilateral*, which has traditionally been considered the formative work. We leave the case as it stands, and conclude only with the remark that spherical trigonometry seems to have attracted more attention as a lively mathematical subject than plane trigonometry ever did. It is thus an irony that modern students of mathematics include plane trigonometry as part of their basic tool set, but most have never have heard of its spherical sibling.

Applications to Religious Practice: The *Qibla* and Other Ritual Needs

We have seen trigonometry mature considerably as it grew within medieval Islam. The expansion to all six modern functions and the discovery of associated theorems, in both plane and spherical varieties, transformed trigonometry into a powerful systematic tool for use within mathematical astronomy, its parent discipline. Astrological applications were also not long in coming; indeed, astrology would not be completely separated from astronomy for some centuries. However, trigonometry was to make itself useful much further afield.

Perhaps trigonometry's earliest new venture, still relying heavily on astronomy, was its support of various aspects of Islamic ritual practice. Among the five pillars of Islam (the religious duties of every practicing Muslim) is the requirement to fast during daylight hours throughout the sacred month of Ramadan. On the face of it this does not sound astronomical; however, it becomes so when one considers what one means by a "month."

In the Muslim calendar, the months depend on the relative speeds of the Sun and Moon. The Moon travels about 13 times as fast through the zodiac as the Sun, passing it by roughly every 29½ days (see figure 4.31). When the Moon moves too close to the Sun it vanishes from sight for a couple of days, eventually reappearing on the Sun's other side as a thin crescent. When the crescent becomes visible, a new month is said to begin. So, the dates of the fast depend rather delicately on a variety of factors, some of them meteorological; but what one mostly needs to know are the precise positions of the Sun and Moon.[144] We have seen the Sun's motion modeled trigonometrically; although the Moon is more complicated, similar methods were used to model its motion. Thus, predicting the emergence of the lunar crescent was a mathematical question, and a difficult one at that. As early as the eighth century a number of scientists attempted to come up with an answer, some of their efforts borrowed from Indian astronomy. They included in their output various collections of trigonometrically defined tables to help the faithful determine as accurately as possible the beginning of Ramadan.[145]

Another of the pillars of Islam, the daily prayers, requires mathematics for two reasons. The five daily times dictated for these prayers are defined according to lengths of shadows, hence by the altitude of the Sun.[146] Also crucial is the direction in which the Muslim faithful must face while praying: toward the Kaʿba in Mecca, an edifice that represents the geographical center

[144] Since the first visibility of the lunar crescent must take place just after sunrise, the relative positions of the Sun and Moon with respect to the horizon are also important.

[145] Studies of a number of these tables have been analyzed recently. While no full survey exists, [King 2004, 666–668] gives an introduction to the topic, while [King 1987] gives some background and describes the relation to the earlier Indian theory.

[146] For a survey of prayer times see [King 2004, 529–622], and on tables for regulating them see [King 2004, 191–456].

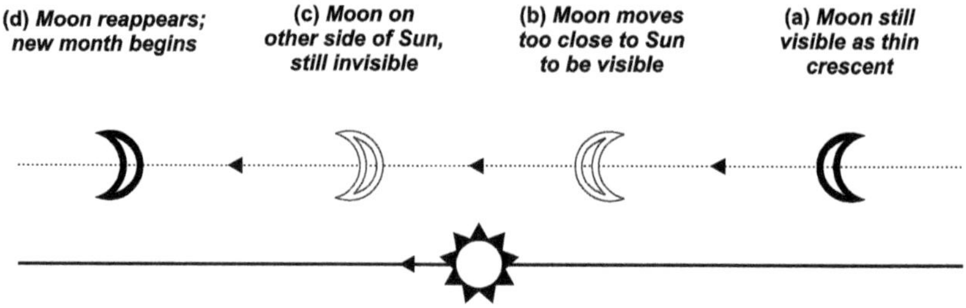
(d) Moon reappears; new month begins
(c) Moon on other side of Sun, still invisible
(b) Moon moves too close to Sun to be visible
(a) Moon still visible as thin crescent

Figure 4.31 The lunar crescent and the new month

of the religion. This sacred direction, the *qibla*, has a broader significance: various activities such as recitation of the Qur'an, ritual slaughter of animals, and bodily functions, are to be performed either in the direction of the *qibla* or perpendicular to it.[147] Since the *qibla* generated a great deal of attention among Muslim scientists, we also shall focus on it.

Now in practice, mathematics was not always used to define the direction of the *qibla*. Prayer walls of mosques were oriented according to various different traditions, and often pointed only vaguely in the direction of Mecca. But this situation did not discourage substantial scientific interest in the question of determining the *qibla* precisely, and trigonometry—both plane and spherical—was available to lend a hand.

The problem is not exclusively mathematical, since it requires knowledge of the coordinates both of Mecca and of the observer's location. In figure 4.32, O is the observer with latitude ϕ_O and longitude L_O, and M is Mecca with latitude ϕ_M and longitude L_M; the goal is to find q, the *qibla*. Perhaps the most difficult part of this enterprise is to find the difference in longitude $\Delta L = L_O - L_M$. In theory, this could be done by observing lunar eclipses simultaneously in the two locations and noting the difference in local times;[148] but in practice, problems in accurately measuring local time and determining the moments at which eclipse phases begin could cause errors. Of course this procedure only has to be done once; most users simply looked up the geographical coordinates in a table.

Some of the earliest solutions to the *qibla* problem were approximate. The most common of these, reported by al-Battānī, became the most popular technique and was used as late as the nineteenth century. The identification of al-Battānī's name with this simple method was to leave him open to

[147] For general summaries of the *qibla* and its connection to Islamic society see [King 1985] and especially [King 2004, 741–771]; [King 1986] and [King 1996, 128–157] delve into the mathematics in greater detail. On the *qibla* in the Islamic West see [Rius 2000].

[148] [King 1996, 147].

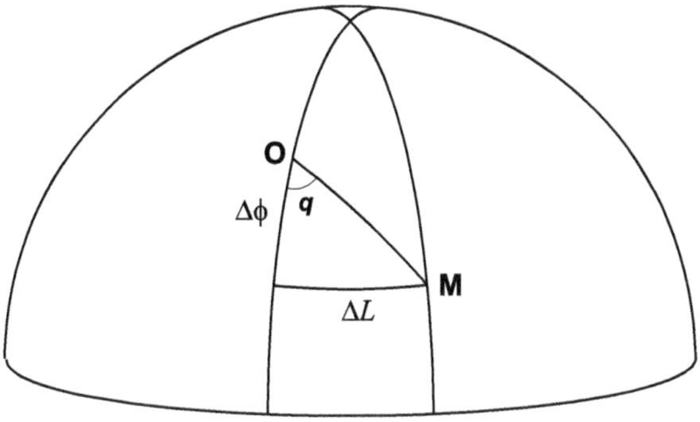

Figure 4.32
The *qibla q*

(perhaps unfair) criticism by later astronomers who were more interested in precise solutions.[149]

Text 4.7
Al-Battānī, a Simple Approximation to the Qibla
(from the *Ṣābi' Zīj*)

[See figure 4.33.] If you want to know the azimuth of Mecca, which is the azimuth of the *qibla* for prayer, by this method, you extend on [the sundial] a line from the center of the circle and that line will be the azimuth of the *qibla* in that locality. . . . [P]lace one end of a ruler on the latitude difference measured from the east line in the direction of Mecca with respect to latitude [that is, north or south] and do likewise from the west line in that direction on the circumference of the circle until the side of the ruler falls on the same latitude difference. Then draw a line with the side of the ruler to connect the eastern mark and the western one. Then take the longitude difference and count it on the circumference of the circle from the meridian in the direction of Mecca with respect to longitude in the south on the circumference, and do likewise in the north. Place the side of the ruler on the two points and use it to draw a straight line. Where these two lines intersect represents the position of Mecca with respect to azimuth from that locality. Place the edge of the ruler on the center of the circle and on the point of intersection and draw a straight line extending it on the sundial to the circumference of the circle in the south [for localities north of Mecca]. This line is the azimuth of the *qibla* in that locality.

If you want to calculate the value of the azimuth of the *qibla*, take the Sine of the longitude difference between the two localities and the Sine of their latitude difference. Multiply each of these by itself and add [the squares],

[149] Among those who took shots at al-Battānī were al-Sijzī, Abū Naṣr Manṣūr, al-Khujandī, and al-Bīrūnī [Berggren 1982, 50–51]. See also Ibn Yūnus in [King 1972, 265–266].

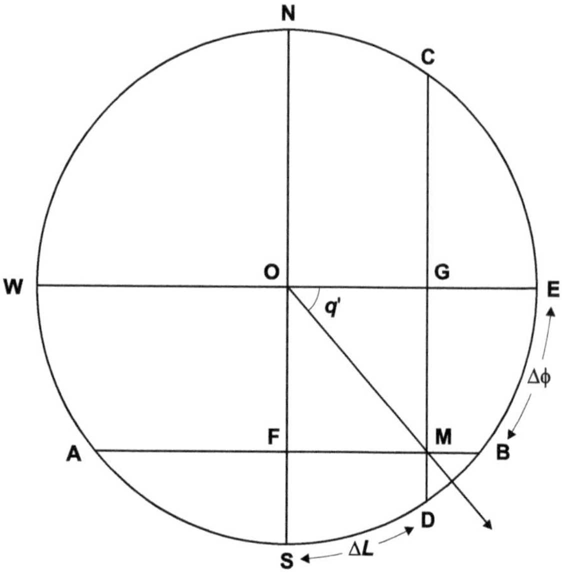

Figure 4.33 Al-Battānī's approximate solution to the *qibla*

then take the square root of the sum. The result is the hypotenuse of the triangle which subtends the right angle, and this is the distance between the center of the circle and the point of intersection of the longitude and latitude lines on the circumference of the circle, so keep it in mind. Then go back to the Sine of the latitude difference and multiply it by the radius [of the base circle, i.e. 60], and divide [the product] by the hypotenuse of the triangle. Take the arc Sine of the result and the arc will be the azimuth of Mecca. Measure it on the circumference from the east point or the west point according to the direction of Mecca for that locality with respect to longitude and latitude. Make a mark on the circumference at that point and draw a straight line from the center of the circle to that mark: this line represents the azimuth of Mecca from that locality.[150]

Explanation: This rather prolix text actually describes a simple, almost obvious method. Measure the difference in latitude $\Delta\phi$ along the edge of the circle from E and W (defining A and B), and draw AB. Similarly, measure ΔL from N and S (defining C and D) and draw CD. The two new lines intersect at M, and OM is asserted to be the direction of the *qibla*.

Al-Battānī's calculation works as follows. Since the radius of the circle is 60, $FM = \text{Sin}\,\Delta L$ and $GM = \text{Sin}\,\Delta\phi$; thus $OM = \sqrt{(\text{Sin}\Delta L)^2 + (\text{Sin}\,\Delta\phi)^2}$. The *qibla* is found using ΔOGM:

[150] From [King 1986, 103–104].

$$q' = \text{arc Sin}\left(\frac{GM}{OM}\right) = \text{arc Sin}\left(\frac{R\,\text{Sin}\,\Delta\phi}{\sqrt{(\text{Sin}\,\Delta L)^2 + (\text{Sin}\,\Delta\phi)^2}}\right).^{151} \quad (4.52)$$

Any approximation is a compromise between accuracy and simplicity; this one tends to the latter. It leads to errors of up to several degrees in the regions in which it was used (although it works quite well in the central lands of Islam), but it was extremely easy to use. We might think of it as approximating the arc from the observer to Mecca as the hypotenuse of a plane right triangle with sides Sin ΔL and Sin $\Delta\phi$.

Readers of previous chapters will recognize in al-Battānī's solution a common situation: a simpler path was open to him by calculating $q' =$ arc Tan (GM/FM). But he did not yet have full command of a tangent function, and so he needed to convert to an arc Sine.

Perhaps not surprisingly, the earliest precise solution of the *qibla* problem complete with proof comes directly from the Transversal Figure, by Baghdadi mathematician and astronomer al-Nayrīzī (fl. ca. 900), a contemporary of al-Battānī.[152] Al-Nayrīzī considers the problem using the celestial sphere (figure 4.34) rather than the Earth's surface; then the goal becomes to determine the direction one must face to point toward Z_M, the zenith of Mecca, given the observer's and Mecca's latitudes ϕ_O and ϕ_M and the longitude difference ΔL. So we need to determine $q = \widehat{SQ}$, the azimuth of the *qibla*.[153] Al-Nayrīzī applies the same sort of economy we saw earlier in al-Kūhī's approach to rising times: he uses only two transversal figures, but applies the theorem twice to each configuration. First he looks at figure $SCWBN_{eq}A$, from which he deduces by conjunction

$$\frac{\text{Sin}\,\widehat{N_{eq}S}}{\text{Sin}\,\widehat{SA}} = \frac{\text{Sin}\,\widehat{N_{eq}C}}{\text{Sin}\,\widehat{CB}} \cdot \frac{\text{Sin}\,\widehat{BW}}{\text{Sin}\,\widehat{WA}}. \quad (4.53)$$

[151] q' is the complement of the angle usually used to represent the *qibla*, $q = \angle SOM$.

[152] Al-Nayrīzī's treatise was translated into German in [Schoy 1922]; an edition and translation have appeared recently in [Hogendijk 2000]. See also [Debarnot, in al-Bīrūnī (Debarnot) 1985, 62–63].

[153] The azimuth of a point on the celestial sphere is the position of the point projected onto the horizon, i.e., the direction one must turn to face the point. We saw in al-Battānī's case that the *qibla* was measured with respect to the East point; more typically it was measured, as here, from the South point. Other points on this diagram are N_{eq}, the north pole of the celestial equator; and N, E, S, and W, the cardinal points on the horizon. A, B, C and Q are intersections of the various great circle arcs.

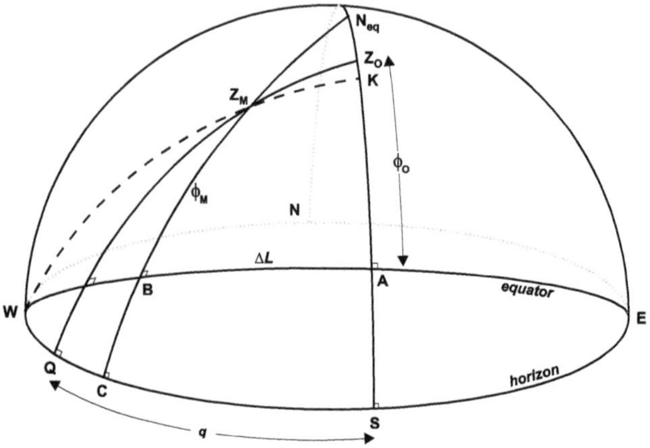

Figure 4.34
Al-Nayrīzī's *qibla* solution, and the method of the *zījes*

This simplifies to $\phi_O = \cot\widehat{BC}\cdot\cos\Delta L$ (in modern terms), so $\widehat{BC}$ is known.[154] Using the same figure but applying disjunction, we have

$$\frac{\mathrm{Sin}\,\widehat{N_{eq}A}}{\mathrm{Sin}\,\widehat{AS}} = \frac{\mathrm{Sin}\,\widehat{N_{eq}B}}{\mathrm{Sin}\,\widehat{BC}}\cdot\frac{\mathrm{Sin}\,\widehat{CW}}{\mathrm{Sin}\,\widehat{WS}} \tag{4.54}$$

which corresponds to $\cos\phi_O = \sin\widehat{BC}/\cos\widehat{CS}$, so $\widehat{CS}$ is known.

Al-Nayrīzī turns next to $SCQZ_MN_{eq}Z_O$. Applying conjunction gives

$$\frac{\mathrm{Sin}\,\widehat{SN_{eq}}}{\mathrm{Sin}\,\widehat{Z_OS}} = \frac{\mathrm{Sin}\,\widehat{N_{eq}C}}{\mathrm{Sin}\,\widehat{CZ_M}}\cdot\frac{\mathrm{Sin}\,\widehat{QZ_M}}{\mathrm{Sin}\,\widehat{QZ_O}} \tag{4.55}$$

or $\sin\phi_O = \cos\widehat{BC}\cdot\cos\widehat{Z_OZ_M}\big/\sin(\widehat{BC}+\phi_M)$, giving the distance between the two zeniths $\widehat{Z_OZ_M}$. Finally, applying conjunction again but assigning the arcs in the reverse order gets us the *qibla*, $q = \widehat{SQ}$:

$$\frac{\mathrm{Sin}\,\widehat{QS}}{\mathrm{Sin}\,\widehat{CS}} = \frac{\mathrm{Sin}\,\widehat{QZ_O}}{\mathrm{Sin}\,\widehat{Z_OZ_M}}\cdot\frac{\mathrm{Sin}\,\widehat{N_{eq}Z_M}}{\mathrm{Sin}\,\widehat{CN_{eq}}} \tag{4.56}$$

or $\sin q\big/\sin\widehat{CS} = \cos\phi_M\big/(\sin\widehat{Z_OZ_M}\cdot\cos\widehat{BC})$.

[154] Since al-Nayrīzī does not have a tangent function he must convert to a sine before finding $\widehat{BC}$, as we have seen before.

Al-Nayrīzī's method is closely related to the most popular precise solution to the *qibla* problem, the "method of the *zījes*"—so named by al-Bīrūnī because a number of *zīj* authors describe it in their works. This latter method may be derived from Ḥabash al-Ḥāsib's analemma for the *qibla*, which we saw earlier in this chapter. Some of the many authors who employed this technique are familiar names like Ibn Yūnus, Abū'l-Wafā', Kūshyār ibn Labbān, al-Bīrūnī, Ibn al-Haytham, Ibn Muʿādh, Ibn Isḥāq, al-Marrākushī, Jamshīd al-Kāshī, and Ulugh Beg.[155] We shall follow the method as outlined in al-Bīrūnī's *Taḥdīd al-Amākin*, where he uses the Rule of Four Quantities to full effect.

The best way to understand the method of the *zījes* is to consider it as a pair of coordinate transformations. We begin by transforming the coordinates of the zenith of Mecca, from those with respect to the equator (latitude ϕ_M and longitude difference ΔL) to a new set with respect to the observer's meridian $SN_{eq}N$. Thus, in figure 4.34 we must introduce a new great circle through W and Z_M, projecting Z_M perpendicularly onto the meridian at K. The new coordinates are the "modified longitude difference" $\Delta L' = \widehat{Z_M K}$, and the "modified latitude" $\phi'_M = \widehat{AK}$. Once we have $\Delta L'$ and ϕ'_M, they may be transferred to a third set of coordinates measured with respect to the local horizon, $q = \widehat{SQ}$ and the altitude of the zenith of Mecca $h_M = \widehat{QZ_M}$. Of course, q is the quantity we want.

Our first coordinate transformation begins with an application of the Rule of Four Quantities to figure $N_{eq}Z_M BAK$. This gives

$$\frac{\operatorname{Sin} \widehat{N_{eq}Z_M}}{\operatorname{Sin} \widehat{Z_M K}} = \frac{\operatorname{Sin} \widehat{N_{eq}B}}{\operatorname{Sin} \widehat{AB}}, \qquad (4.57)$$

which simplifies to $\sin \Delta L' = \sin \Delta L \cos \phi_M$. Next, on figure $WZ_M KAB$ we have

[155] The literature of the method of the *zījes* is substantial. A general description and a text containing perhaps its earliest appearance (although without proof) may be found in [King 1986, 112–117]. On al-Bīrūnī, al-Kāshī and Ulugh Beg see [Berggren 1980]. Al-Bīrūnī's method as described in his geographical treatise the *Taḥdīd al-Amākin*, along with the accompanying analemma, is in English translation in [al-Bīrūnī (Ali) 1967, 252–255] with commentary in [Kennedy 1973, 209–214]. See also [Berggren 1985], a follow-up to [Berggren 1980] in which the origin of the method is traced back to the theory of sundials in Ptolemy's *Analemma* (on this work see [Luckey 1927]). [Berggren 1985, 5–10] also discusses Ibn Yūnus, Abū'l-Wafā', and Kūshyār ibn Labbān on the subject. Also on Ibn Yūnus see [King 1972, 256–265]. On al-Bīrūnī in the *Keys to Astronomy* see [al-Bīrūnī (Debarnot) 1985, 50–51, 260–264]. On Ibn Muʿādh and Ibn Isḥāq see [Samsó/Mielgo 1994]. On Ibn al-Haytham see [King 1986, 116], [Dallal 1995], and another treatment of the *qibla* in [Schoy 1921].

$$\frac{\operatorname{Sin}\widehat{WZ_M}}{\operatorname{Sin}\widehat{Z_MB}} = \frac{\operatorname{Sin}\widehat{WK}}{\operatorname{Sin}\widehat{KA}}, \tag{4.58}$$

or $\sin\phi'_M = \sin\phi_M/\cos\Delta L'$. The first transformation is now complete.

The second coordinate conversion begins with the definition of an intermediate quantity, the "modified latitude difference" $\Delta\phi' = \widehat{KZ_O} = \phi_O - \phi'_M$. The altitude of the zenith of Mecca h_M may now be found by applying the Rule of Four Quantities to figure WZ_MKSQ:[156]

$$\frac{\operatorname{Sin}\widehat{WZ_M}}{\operatorname{Sin}\widehat{Z_MQ}} = \frac{\operatorname{Sin}\widehat{WK}}{\operatorname{Sin}\widehat{KS}}, \tag{4.59}$$

or $\sin h_M = \cos\Delta L' \cos\Delta\phi'$.[157] Finally the *qibla* itself emerges from figure Z_OZ_MQSK:

$$\frac{\operatorname{Sin}\widehat{Z_OZ_M}}{\operatorname{Sin}\widehat{Z_MK}} = \frac{\operatorname{Sin}\widehat{Z_OQ}}{\operatorname{Sin}\widehat{QS}}, \tag{4.60}$$

or $\sin q = \sin\Delta L'/\cos h_M$.

Al-Nayrīzī also pursues a pair of transformations; however, his first step determines $\widehat{BC}$ and $\widehat{CS}$, before he turns to h_M and q. This procedure is actually slightly simpler than the method of the *zījes* since it does not require great circle $\widehat{WZ_MK}$, but unfortunately it produces numerical instabilities in the calculations, and al-Nayrīzī ended up with a severely flawed value for the *qibla* of Baghdad.[158]

To the modern reader the relation of these computations to Ḥabash's analemma for the *qibla* discussed earlier in this chapter must seem somewhat elusive, so it is worth an explanation here. The modified longitude difference $\Delta L' = \widehat{Z_MK}$ in figure 4.34 corresponds to $\widehat{NK}$ in figure 4.20, which displays the derivation of the analemma. Thus, on the plane of that great circle, it

[156] [Kennedy 1973, 213] appeals to the Transversal Figure here.

[157] $\sin h_m$ can be seen alternately as an expression for the cosine of the distance between the two zeniths $\widehat{Z_OZ_M} = 90° - \widehat{Z_MQ}$.

[158] This occurs in the calculation of $\widehat{CS}$ [Hogendijk 2000, 50, 53]; see also pp. 54–55 on the comparison with the method of the *zījes*. Al-Nayrīzī's *qibla* of Baghdad is 29;7°, as opposed to the correct value 13;29°.

follows immediately that $NO = SU = \sin \Delta L'$ and $EO = \cos \Delta L'$. Now $\Delta L'$ can be determined by observing that $\sin \Delta L = SU/ES = \sin \Delta L'/\cos\phi_M$. The modified latitude $\phi'_M, \widehat{KA}$ in figure 4.34, is $\angle ZEO$ in figure 4.20, so $\sin\phi'_M = UO/EO = \sin\phi_M/\cos\Delta L'$. The modified latitude difference, then, is $\Delta\phi' = \angle OEL = \phi_O - \phi'_M$.

Turning to the second coordinate conversion, in figure 4.20 EL is equal to the vertical height of N (the zenith of Mecca) above the horizon, so it is the sine of the altitude of Mecca, h_M. So $EL = \sin h_M = EO \cdot (EL/EO) = \cos \Delta L' \cdot \cos \Delta\phi'$. Finally, for the *qibla* itself we need $q = \angle OLN$. We have $\sin q = ON/LN = SU/LN = \sin \Delta L'/\cos h_M$. Obviously $\angle OLN$ is not properly represented on the analemma itself (figure 4.21), so it is transferred to the middle of the diagram.

Both the approximate and the exact solutions of the *qibla* formed the basis for tables from as early as the ninth century, varying dramatically in size and sophistication. Since q is a function of both the difference in longitude ΔL and in latitude $\Delta\phi$, the tables had to be constructed with two arguments. And since q is sensitive to changes in both arguments, existing streamlining procedures such as Ptolemaic interpolation would not have been effective. As a result *qibla* tables required a full rectangular grid of entries, and some of them were truly massive. The best of the known tables (based on an exact formula) was composed by Shams al-Dīn al-Khalīlī, an astronomical timekeeper in late fourteenth-century Damascus; it contains over 2,800 entries, extracted in figure 4.35.[159] In addition to these computational solutions various instruments were designed to determine the *qibla*, most remarkable among them two brass world-maps that have been the subject of a recent study.[160]

Astronomical Timekeeping: Approximating the Time of Day Using the Height of the Sun

Just as important as knowing the direction of the *qibla* was determining the time of day. It would seem obvious that we should be able to find the time from the altitude of the Sun in the sky, but it turns out that this apparently simple problem requires a certain amount of astronomical information and insight. In fact, the problem led to the entire discipline (until recently understudied) of astronomical timekeeping. If we can solve the problem, this would give the five

[159] The complete table has been published in [King 1975]. Surveys of *qibla* tables may be found in [King 1986, 133–141], [King 1999, 64–70], and [King/Samsó 2001, 90–91]. On specific tables based on approximate formulas see also [King 1975, 120–122], [Lorch 1980], and [King 1986, 118–126]; on a specific table based on an exact formula see [Hogendijk 1994].

[160] [King 1999].

$L \backslash \varphi_o$		10°		11°		30°		56°	
7°	127°	70;50		71;26	[+1]	84;0		*77;22*	[+1]
8°	126°	70;49	[+1]	71;26	[+2]	84;26		*76;26*	
9°	125°	70;47	[+1]	71;25	[+2]	84;53	[+1]	*75;30*	[−1]
⋮	⋮	⋮	⋮	⋮	⋮	⋮	⋮	⋮	⋮
30°	104°	67;31	[+1]	68;43	[−1]	*84;29*		*53;44*	[+1]
⋮	⋮	⋮	⋮	⋮	⋮	⋮	⋮	⋮	⋮
50°	84°	52;46	[−2]	55;1	[−3]	*64;54*	[+1]	*27;4*	[+1]
⋮	⋮	⋮	⋮		⋮	⋮	⋮	⋮	⋮
64°	70°	13;41	[−2]	14;57	[+1]	*18;17*	[−1]	*4;55*	
65°	69°	9;15		10;5	[−1]	*12;26*	[+1]	*3;16*	[−1]
66°	68°	4;42	[+3]	5;8	[+3]	*6;20*	[+4]	*1;38*	[−1]

Figure 4.35 An extract from al-Khalīlī's *qibla* table. For ease of use longitudes and latitudes have been given directly, rather than as differences from Mecca's longitude and latitude. Italicized entries (given in red in the original table) indicate that the *qibla* direction is southerly rather than northerly.

daily prayer times, since they are defined by shadow lengths. Although (just as with the *qibla*) religious practice did not often follow the astronomers' responses to this ritual requirement, the scientific community continued to show interest and develop new approaches to problems of timekeeping.

And indeed, there was some cultural and financial support for these practices. Timekeeping, originally the domain of the *muezzins* who mostly practiced indigenous folk astronomy, found a new group of practitioners in the thirteenth century: the *muwaqqits* (who included al-Khalīlī among their number), who were supported by mosques. Others not associated with religious institutions practiced their craft under the title *mīqātī*.[161] The first mention of *muwaqqits* was in Cairo; interest soon spread to Yemen and Syria, and later to Tunis and Ottoman Turkey. Clearly the niche of interest in the science of the subject was enough to keep at least some of its specialists in business.[162]

A couple of quantities must be observed before the problem can be solved, but fortunately they are within the reach of almost anyone, virtually regardless of astronomical knowledge. These are the maximum altitude of the Sun over the course of the day ($H = \widehat{DF}$ in figure 4.36) and the semi-diurnal

[161] See the description of the *muwaqqits* and *mīqātī* in [King 1990].

[162] For a survey of the history of astronomical timekeeping see [King 1990]. Readers with a more voracious appetite for the subject will find it satisfied by the two comprehensive volumes [King 2004] and [King 2005].

arc (the Sun's journey from sunrise to noon $D = \widehat{JD}$). Along with the Sun's declination δ determined from the time of year and the local terrestrial latitude ϕ, we have enough data to turn to the mathematics.

We first need a preliminary result. In ΔACD we know that $\angle ADC = \phi$, $CD = \sin H$, and $AD = \text{vers}\ D \cos\delta$.[163] So $\cos\phi = CD/AD = \sin H/(\text{vers}\ D \cos\delta)$, which we rearrange to

$$\frac{\text{vers}\ D}{\sin H} = \frac{1}{\cos\delta\cos\phi}\ . \tag{4.61}$$

This ratio will be useful to us in a moment, but it also turns out to play a role in a number of other timekeeping applications.

We are ready to proceed with the original problem. Recall that $t = \widehat{☼D}$ is the hour-angle, or time until noon. So, if the day-circle had a radius equal to 1, then AS would be equal to $AD - SD = (\text{vers}\ D - \text{vers}\ t)$. But since the radius of the day-circle is $\cos\delta$, $AS = (\text{vers}\ D - \text{vers}\ t)\cdot\cos\delta$. Therefore ST, which equals $\sin h$ (where h is the Sun's altitude), is also equal to

$$AS\cdot\frac{ST}{AS} = ((\text{vers}\,D - \text{vers}\ t)\cdot\cos\delta)\cdot\cos\phi\ .$$

Setting the two expressions for ST equal to each other and solving for vers t, we apply (4.61) and arrive at an expression for t in terms of our givens h, D, and H:

$$\text{vers}\,t = \text{vers}\,D - \frac{\sin H}{\cos\delta\cos\phi} = \text{vers}\,D - \sin\,h\cdot\frac{\text{vers}\,D}{\sin H}. \tag{4.62}$$

This equation makes t a function of the three variables D, h, and H—essentially impossible to tabulate for all but the most obsessed. Fortunately, obsessed scientists are not very hard to find. In this case the challenge was met by one thirteenth-century Cairo astronomer, Najm al-Dīn al-Miṣrī; his triple-argument table included over 400,000 entries.[164] In Najm al-Dīn's defense we should note that he found ways to use $t(D,h,H)$ to solve a startling variety of spherical astronomical problems.[165]

[163] Assuming that circle $\widehat{JD}$ has radius 1 or R, the versed sine of $\widehat{JD}$ is defined to be the distance from D to the perpendicular projection JA of J onto diameter DA. JA is of course the sine of the arc.

[164] On this table see [Charette 1998]. Najm al-Dīn was not satisfied even with this table and had designs to build an even larger one that would avoid requiring the user to interpolate between entries. He calculated that the prospective table would hold 2.65 billion entries, filling over 5,300 books. The fact that it has not been found is a good indication that he never composed it; it would have been rather easy to find.

[165] See [Charette 1998, 31–41].

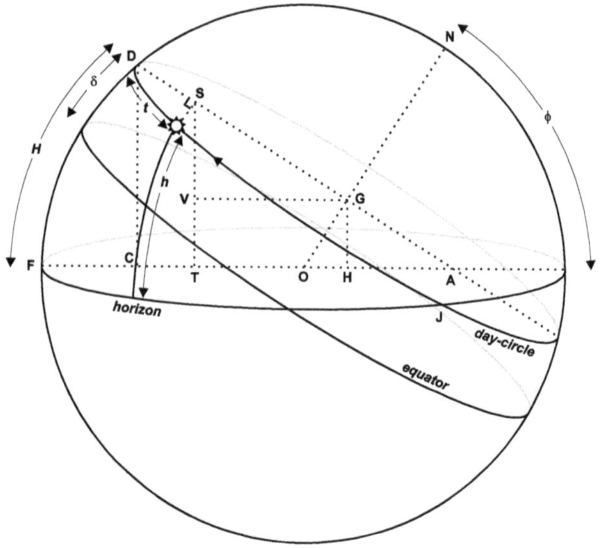

Figure 4.36
Diagram for solving problems of astronomical timekeeping

But more commonly, the formidable computational challenge posed by (4.62) was bypassed by tabulating t (or $T = D - t$) for a particular latitude ϕ, which dropped D from the list of arguments. A number of other varieties exist: sometimes the argument H was replaced by the Sun's longitude λ; there are also a number of tables that rearrange (4.62) to give the Sun's altitude h as a function of t. These sorts of tables, foundational as they are, may be found everywhere that timekeeping was studied and applied—as early as the thirteenth century, and as late as the nineteenth.[166]

The problem with powerful tables (or indeed powerful tools in general) is that they can be difficult to build and difficult to use. It is thus not surprising that many Muslim astronomers adopted two simpler Indian approximate procedures for finding the time of day. The first, an arithmetic scheme, need not concern us.[167] The second, which had been used in certain Greek sundials (although the Muslims learned about it from Indian sources), is trigonometric. One way to think of it is as follows. At the equinox, the semi-diurnal arc D is equal to exactly six hours or 90°. If we substitute this value into (4.62) (and convert to $T = D - t$), we get the much simpler formula

[166] [King 2004, 43–74] is a summary of over 30 sets of tables of this type. One is attributed to the early giant of astronomical timekeeping Ibn Yūnus (see [King 1973b]), but the form we find in the manuscripts likely is due to someone else.

[167] See [King 2004, 557–558] on the scheme itself, and following pages for its implementation in legal texts.

$$T = \frac{1}{15}\arcsin\left(\frac{\sin h}{\sin H}\right), \tag{4.63}$$

where the $\frac{1}{15}$ arises simply by the switch from angular to time measurement.[168]

But (4.63) works only at the equinox; what does one do at other times of year? Suppose it is summertime and the day circle is above the equator, as in figure 4.36. Then there will be more than six hours between sunrise and noon (i.e., $\widehat{JD} > 90°$), and (4.63) will be incorrect. But we might have reason to hope that the approximation does a reasonable job even when the Sun is not at the equinox. If we measure times T and t in ***seasonal day hours***—that is, divide the longer summer day into twelve equal periods—then the time from sunrise to noon will be six hours by definition, and (4.63) just might give a decent approximation to the time of day without having to deal with the fact that the semi-diurnal arc D varies. And this hope turns out to be justified: in the lower terrestrial latitudes the error in the approximation is on the order of only a few minutes of time, and even at the northern limits of civilization the formula is never wrong by more than about twenty minutes.[169]

The approximate formula is found in the literature much earlier than the exact one, in fact almost as early as Arabic science itself. It first appears in the eighth century, including in the works of Yaʿqūb ibn Ṭāriq, famous for his introduction of Indian astronomy to Islam.[170] The approximate formula is also found in the works of such luminaries as Ḥabash al-Ḥāsib and the Andalusian astronomer al-Zarqāllu; it is partly from the latter's *Toledan Tables* that the formula also gained some currency in Europe. Tables for timekeeping of this type, designed either for specific latitudes or for all, were composed as early as the ninth century and as late as the sixteenth.[171]

New Functions from Old: Auxiliary Tables

By now the reader may have formed the impression that the methods used to deal with problems in spherical astronomy often relied on similar notions

168 The Rule of Four Quantities also lurks in this. In figure 4.36, if we are at the equinox then $\widehat{JD} = 90°$, and the approximation rule follows by considering the figure $J☼DF$.

169 See the error analyses in [Puig 1986, 67–72] and [King 2005, 131–132]. [King 2005, 111–197] is the fundamental study of this approximate formula, and much of this section and some of the following one is based on it.

170 See [Pingree 1968] for a survey of Yaʿqūb ibn Ṭāriq's works and his relation to the Indian sources.

171 [King 2005, 133–147].

and techniques. Of course trigonometric tools such as the Transversal Figure and the Rule of Four Quantities were used repeatedly, but also the ideas behind the constructions—projections, transformations of coordinates, and so on—appeared in different guises in different situations. Inevitably, solutions that come from the same collection of tools often share properties. In Islamic astronomy these similarities were recognized quite early, and a number of scientists were led naturally to the idea of tabulating ***auxiliary functions***.

An auxiliary function has no astronomical import of its own, but appears frequently as a trigonometric unit within solutions to a variety of astronomical problems. We have already seen a few examples, including in Ḥabash al-Ḥāsib's ninth-century *Jadwal al-Taqwīm*, which contains an early appearance of the tangent. Al-Nayrīzī also tabulated a set of fourteen auxiliary functions, which included among their number the tangent and cosecant.[172] Another example mentioned earlier in these pages is Abū Naṣr Manṣūr's "Table of Minutes," particularly noteworthy for its introduction of $R = 1$ for the trigonometric base circle.

Both Ḥabash's and Abū Naṣr's auxiliary functions had only a single argument, so their tables were relatively small. However, much more substantial double- and even triple-argument tables were on the way. The 400,000-entry behemoth by Najm al-Dīn al-Miṣrī which we discussed in the previous section was by far the largest of these tables. Najm al-Dīn went far beyond Ḥabash's goal of converting celestial coordinates: his table was designed to solve, on its own, all major problems of spherical astronomy.

Perhaps the best-executed of all sets of auxiliary tables are those by Shams al-Dīn al-Khalīlī, who computed the *qibla* tables we saw in figure 4.35.[173] His three tables, all of double-argument functions, comprise a mere 13,000 entries, and are remarkably accurate given the computational task involved; they were to be used extensively for centuries. Like al-Miṣrī, al-Khalīlī set himself the task of building tables to solve all problems of spherical astronomy. The functions he chose are

$$f(\phi,\theta) = \frac{R\,\mathrm{Sin}\,\theta}{\mathrm{Cos}\,\phi}, \tag{4.64}$$

$$g(\phi,\theta) = \frac{\mathrm{Sin}\,\theta \cdot \mathrm{Tan}\,\phi}{R}, \tag{4.65}$$

and

[172] See [Schoy 1923, 392] and [King 2004, 164–165].

[173] On al-Khalīlī's auxiliary tables see [King 1973a] and [King 2004, 378–401].

$$G(x,y) = \text{arc Cos}\left(\frac{R \cdot x}{\text{Cos}\, y}\right). \tag{4.66}$$

They are tabulated for every degree of argument, and the values retain reasonable accuracy even in certain places in the table for *G* where the function becomes numerically unstable.[174]

Al-Khalīlī's motive for choosing these three functions in particular might come from his work on the *qibla*. One correct formula is

$$q(\phi_O, L) = \text{arc Cos}\left[\frac{R\left[\frac{\text{Sin}\, h(\phi_O, L) \cdot \text{Tan}\, \phi_O}{R} - \frac{R \cdot \text{Sin}\, \phi_M}{\text{Cos}\, \phi_O}\right]}{\text{Cos}\, h(\phi_O, L)}\right], \tag{4.67}$$

where ϕ_O and ϕ_M are the latitudes of the observer and Mecca respectively, *L* is the observer's longitude, and $h(\phi_O, L)$ is the altitude of Mecca's zenith above the horizon. At first glance it seems obvious that al-Khalīlī's auxiliary functions are simply taken from portions of this formula, since

$$q(\phi_O, L) = G(g(\phi_O, h) - f(\phi_O, \phi_M), h).^{175} \tag{4.68}$$

However al-Khalīlī lists a series of astronomical problems that may be solved with the same auxiliary functions, some of them put together just as naturally; consider the following example.

Text 4.8
Al-Khalīlī, Using Auxiliary Tables to Find the Hour-angle

Section on the hour-angle. Find the value of the first function [*f*] on the page corresponding to the terrestrial latitude with the altitude as [vertical] argument and the value of the second function [*g*] as before in the section on the half arc. If the declination is southerly add the two values, otherwise take the difference. The result is the auxiliary sine. Enter it in the table of the third

[174] It has been found that al-Khalīlī generated the table for *f* from the values in the table for *g* using the relation $f(\phi, 90° - \phi \pm n) = \text{Cos}\, n \pm g(\phi, n)$, anticipating the later technique of prosthaphaeresis (see chapter 5) by replacing the time-consuming task of multiplication with a much easier addition. See [Van Brummelen 1991], [Van Brummelen/Butler 1997, 46–48], and a later confirmation in [King 2004, 380].

[175] Oddly enough it does not appear that al-Khalīlī's *qibla* table was computed from his auxiliary tables; perhaps his *qibla* table was computed first. See [Van Brummelen 1991, 688–689] and [King 2004, 390].

> function [G] with the solar or stellar declination as [vertical] argument, and you will find the hour-angle. If the declination is northerly and the first function is less than the second, subtract the result from 180° and the remainder will be the hour-angle. If the two functions are equal then the hour-angle will be 90°.[176]

Explanation: We shall take advantage of the modern conveniences of negative numbers and unit-circle trigonometric functions for the following derivation.

Figure 4.36 portrays late morning on a summer day. The Sun, participating in the daily rotation of the celestial sphere, is approaching its highest point D, which it will reach at local noon. The path traced by the Sun is called the ***day-circle*** (with center G), and the time remaining until noon is represented by the ***hour-angle*** $t = \widehat{☼D}$. Our goal is to find t given the Sun's altitude h, its declination δ, and the local latitude ϕ.

From δ we may quickly determine $GD = \cos\delta$. Drop a perpendicular from the Sun to GD, defining S; then $GS = GD \cdot \cos t = \cos\delta \cdot \cos t$. Therefore, since in ΔSVG $\angle S = \phi$, we have $SV = GS \cdot \cos\phi = \cos\delta \cdot \cos t \cdot \cos\phi$. But SV is also equal to $ST - GH = \sin h - OG \cdot \sin\phi = \sin h - \sin\delta \cdot \sin\phi$. So

$$\cos\delta \cos t \cos\phi = \sin h - \sin\delta \sin\phi,$$

and solving for t, we arrive at the formula

$$t = \arccos\left(\frac{\sin h - \sin\delta \sin\phi}{\cos\delta \cos\phi}\right) = \arccos\left(\frac{\dfrac{\sin h}{\cos\phi} - \sin\delta \tan\phi}{\cos\delta}\right). \quad (4.69)$$

And again the formula decomposes nicely into the language of al-Khalīlī's auxiliary functions, almost as if they were designed for this use:

$$t = G(f(\phi,h) - g(\phi,\delta),\delta). \quad (4.70)$$

The best auxiliary functions are those that combine simplicity with the capacity to solve many important problems. By this standard al-Khalīlī's functions are first-rate; they were not surpassed until the modern period.

[176] From [King 2004, 381].

Trigonometric and Astronomical Instruments

Until now we have considered the application of trigonometry to astronomical problems to be the domain of pencil-and-paper; that is, the solution is derived using geometric diagrams, with the aid of trigonometric values usually obtained from numerical tables. However, in medieval Islam there arose a healthy tradition of an entirely different sort: the construction of physical, hand-held instruments made of brass and other metals, etched with curves and attached to rulers in various ways. These devices, many of which were ingenious in their mathematical design, allowed the user to solve the underlying trigonometric problem simply by adjusting the instrument appropriately, and reading off the answer (usually) as a length measured on a scale.

These instruments[177] may not fit what we usually mean by trigonometric computation. However, they usually handled the problems they were designed to solve with as much accuracy as users required of them, and generally with considerably more facility and reliability than solutions by means of tables. Thus they are entitled to an honored place in the history of trigonometry.[178]

Among the most common of such instruments, the horary quadrant was designed to solve precisely the problem we have been studying: determining the time of day by the altitude of the Sun.[179] Some horary quadrants are called "universal" since they work well for all the terrestrial latitudes for which they were used (in the Islamic world at least, if not in Europe) and for any time of year. Sometimes found on the back of an astrolabe, this type of quadrant was the next most popular astronomical instrument of the medieval period. It went through a number of revisions starting as early as the ninth century; one variety found its way to Europe under the name *quadrans vetus*.[180]

The most popular type of universal horary quadrant, in its simplest manifestation, works as follows. The circular edge of the quadrant (figure 4.37(a))

[177] See [King 2005], where they are wryly dubbed "instruments of mass calculation."

[178] A number of astronomical instruments, especially astrolabes, sundials, and apparatuses designed to predict planetary positions, employed physical geometrical embodiments of the astronomical situation to generate their results. Since they do not employ trigonometry directly, we pass over them here—although this is admittedly a fine distinction.

[179] The most comprehensive modern study of the universal horary quadrant is [King 2005, 199–258]. [Charette 2003], which contains an edition and translation of a fourteenth-century treatise dealing with a number of mathematical instruments, discusses the universal horary quadrant on pp. 211–215. See also [King 1995] for a concise history of quadrants in general, [King 2005, 71–80] for descriptions and photographs of the various types of quadrants, and the outdated but still useful [Schmalzl 1929], also on the history of various types of quadrants.

[180] See [Lorch 1981, 118–119]. The literature on the *quadrans vetus* in Latin is considerable; but [King 2002] traces three varieties of it back to a ninth- or tenth-century treatise from Baghdad.

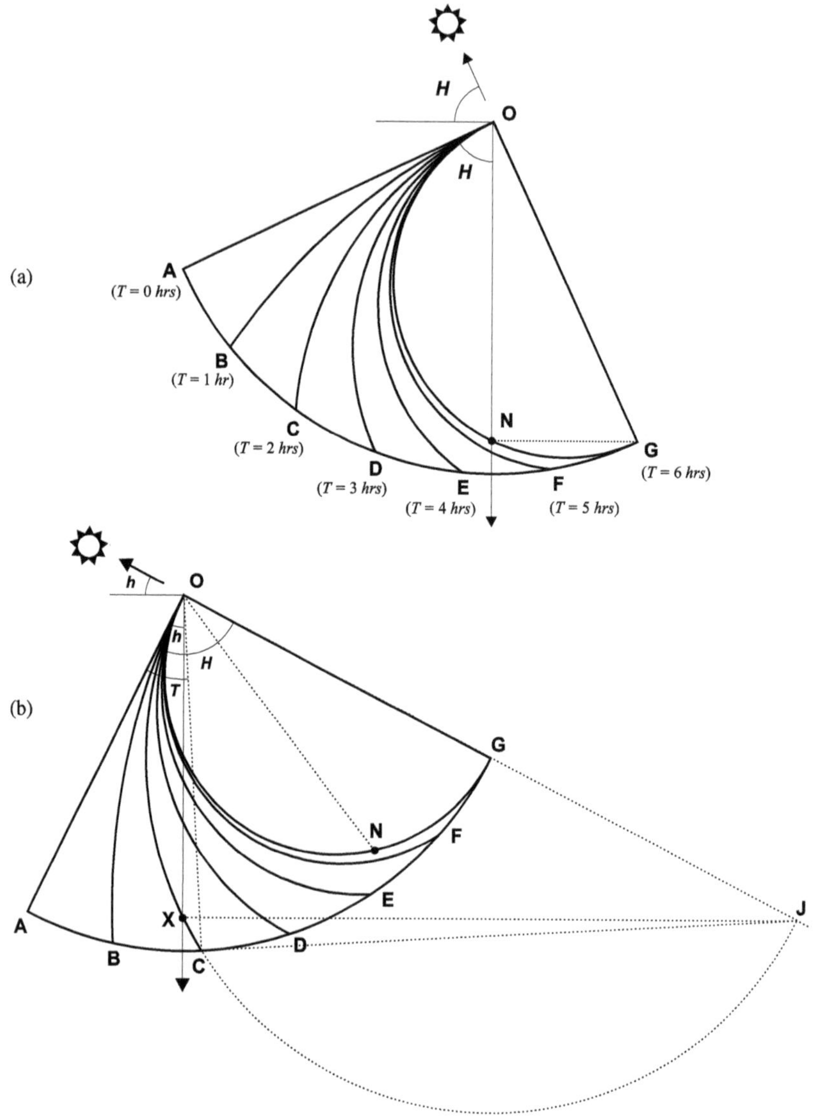

Figure 4.37 The universal horary quadrant. In (a) above the Sun is sighted at noon, and a bead N on the plumb line is fixed to lie on semicircle $\widehat{ONG}$. In (b) below, the Sun is sighted at some other time of day, and the bead moves to X. Since in our configuration X lies on the circular arc corresponding to $T = 2$ hours, it is two hours after sunrise or before sunset.

is divided in six equal parts corresponding to six seasonal day hours, and circular arcs are drawn from O through B, C, D, E, F, and G so that their diameters lie on OG (or its extension). The Sun is sighted along OG at noon, and a bead is fixed at N on the plumb line along the semicircle corresponding to noon ($T = 6$ seasonal day hours);[181] thus $\angle AON = H$, the Sun's altitude at noon. At any other time (figure 4.37(b)) the user sights the Sun and notes the position of the bead X with respect to the arcs on the plate. In our figure X happens to lie on the arc for $T = 2$ hours, so it is 2 seasonal day hours after sunrise or before sunset.

Why does this method work? It relies heavily on the fact that a triangle inscribed in a semicircle is right-angled. In ΔOGN, $\angle G = H$, and so (assuming the quadrant has radius 1) $ON = \sin H$. In the circle corresponding to the current hour ($OXCJ$ in figure 4.37(b)) we have two inscribed triangles. In ΔOXJ $\angle J$ is equal to h, the Sun's current altitude,[182] so $OX(= ON) = OJ \sin h$; whereas in ΔOCJ $\angle J$ is equal to T, the time of day measured in hours, so $OC(= 1) = OJ \sin T$. Hence $ON = \sin h / \sin T$. So, setting the two expressions for ON equal to each other, we arrive at

$$\sin h = \sin H \sin T, \tag{4.71}$$

the same approximate timekeeping formula that we saw in the previous section with (4.63)![183]

So, although the universal horary quadrant was designed specifically to find the time of day, it may be seen more generally as an analog computer to solve equations of the form (4.71)—a handy alternative to using the auxiliary tables for this equation which were common in both medieval Islam and the European Renaissance.

As old as the study of Islamic science is, it is still changing all the time. One example of this is an instrument discovered by David King while this chapter was being written. Constructed by the famous astrolabist Nastulus (fl. ca. 900), this device is a unique solution to the timekeeping formula (4.71)—using polar coordinates! In figure 4.38, the user moves the ruler to the Sun's longitude λ measured around the outside of the instrument, known from the time of year. The Sun's altitude is then found on the ruler, i.e., the distance from the center.[184] The time of day is determined using the six

[181] If the user does not wish to wait until noon to find the time and happens to know ϕ and δ, he can simulate noon by positioning the plumb line so that $\angle AOX = 90° - \phi \pm \delta$, which is equal to H, the Sun's altitude at noon.

[182] This is true because $h = \angle AOX$ and $\angle OJX$ are both complements of $\angle JOX$. A similar argument applies to establish that $\angle OJC = T$.

[183] The factor of 15 in (4.63) arises because we are measuring in hours rather than degrees.

[184] To be completely accurate, the distance is measured from the edge of the center disk that holds the ruler in place.

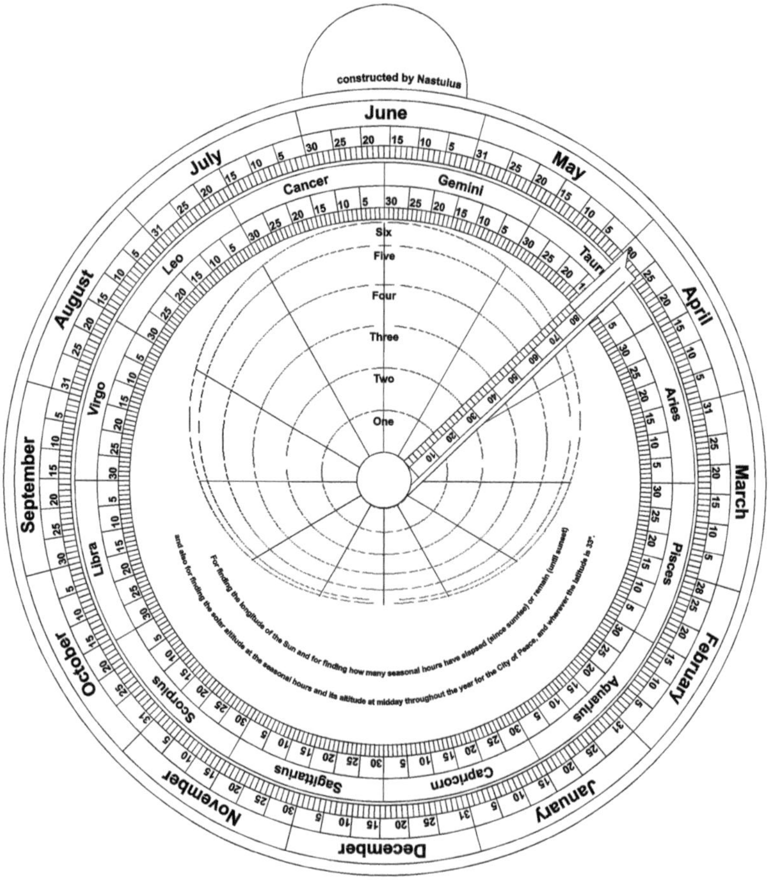

Figure 4.38 A computer-generated replica of a recently discovered instrument for timekeeping by Nastulus

oval-shaped curves, which in the morning represent hours since sunrise working from the center outward, and in the afternoon hours before sunset working inward.

Although Nastulus's instrument *determines* T from h and H (the latter is found from λ and the local latitude—Baghdad's, in Nastulus's case), it is *constructed* differently. For a fixed value of T, and a set of values of λ (measured radially), Nastulus would have computed or looked up in a table a set of values of h (measured from the center), which he then plotted on the instrument. Somehow, he joined these points in a smooth curve. So, in effect Nastulus was plotting contour curves of a double-argument function in polar coordinates!

Another instrument related to the horary quadrants but with a more explicitly mathematical purpose, the sine quadrant also developed in the ninth and tenth centuries. Both devices, in fact, may be found in the literature as early as treatises probably by al-Khwārizmī.[185] While the initial motivation for these devices was to find the time of day from the altitude of the Sun, they soon became universal in another sense. If the user was clever enough, they could be manipulated to solve virtually any problem in spherical astronomy—including most of the spherical topics discussed in this book, even the difficult problem of the *qibla*.[186]

Perhaps the most startling impression that the sine quadrant leaves on the modern reader is the appearance of a Cartesian coordinate grid (see for instance figure 4.39), which we are accustomed to thinking of as a more recent European invention. These coordinates make it easy to look up the sine or cosine of any arc: simply set the thread (or ruler) to the appropriate angle on the curved edge of the quadrant, and follow the horizontal and vertical lines to the scales on the other two edges. But the sine quadrant is much more versatile than that.

Text 4.9
Al-Sijzī (?), On an Application of the Sine Quadrant

To determine the ascensions of the signs at sphaera recta.[187]

[See figure 4.40.] When *AD* is the declination of the beginning of Taurus, and we draw *DE* parallel to *AB*, and *AZ* is the maximum declination,[188] and we draw *BZ*, and we draw from its intersection [with] *DE*, which is point *T*, line *TH* parallel to *BG* and draw *BD* cutting *TH* at *K*, and when *BD* is divided into 90, [then] *BK* is the ascension of the beginning of Taurus.

Explanation: The goal is to determine the right ascension α, given the declination δ and the "maximum declination" (the obliquity of the ecliptic) ε. Since

$$\frac{BK}{BD(=1)} = \frac{KH}{TH},$$

[185] On the sine quadrant see [Schmalzl 1929, 83–99], [Lorch 2000], and [King 2005, 71–74]. On the contribution of al-Khwārizmī see [King 1983, 28–30], [Charette/Schmidl 2004, 179–181], and [King 2005, 162–164].

[186] [King 1975, 109–120] describes several methods of finding the *qibla* using a sine quadrant.

[187] From [Lorch 2000, 259].

[188] We have removed the phrase "the complement of" before "the maximum declination" to render the argument correct. See [Lorch 2000, 264].

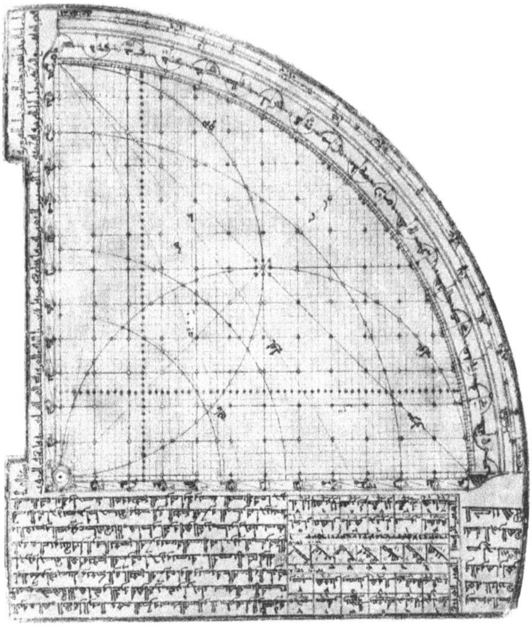

Figure 4.39
A sine quadrant constructed by Muhammad al-Ṣakāsī al-Jarqasī in Damascus around AD 1800 (from the archive of the Institut für Gechichte der Naturwissenschaften, Frankfurt, courtesy of Prof. David A. King)

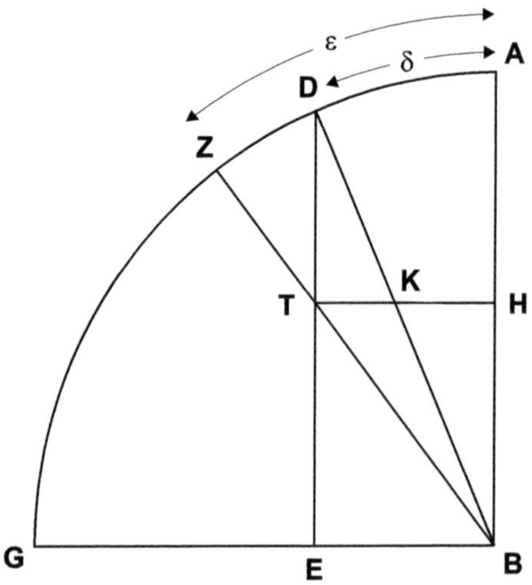

Figure 4.40
Using a sine quadrant to find the right ascension

$$BK = \frac{KH}{HB} \cdot \frac{HB}{TH} = \tan\delta \cdot \cot\varepsilon.^{189} \qquad (4.72)$$

If we grant the author the liberty of interpreting BK as $\sin\alpha$ rather than just α (and this sort of thing was rather common), we arrive at

[189] Of course the author would have expressed, for instance, $\tan\delta$ as $\sin\delta/\cos\delta$.

sin α = tan δ·cot ε, the same formula we found in Ptolemy's *Almagest* in chapter 2.

There is nothing particularly special about the astronomical context of this procedure; it is simply a means of composing ratios in order to multiply two trigonometric quantities. Similar processes are used by our author to solve other problems involving trigonometric products, which we have seen time and again in spherical astronomy. With the right construction on the quadrant's face, virtually any problem could be solved in this way.

Trigonometry in Geography

One of the most critical innovations in trigonometry in the Islamic world was the opening up of its applications well beyond astronomy. Although there are earlier isolated instances of such uses, it is in Islam that we really see trigonometry applied systematically in another scientific field, namely, mathematical geography.[190] It is no surprise that this discipline is where trigonometry first branches out, since the techniques used in spherical astronomy are transferred easily from the celestial sphere to the Earth's surface. Thus trigonometry was still being practiced on a grand scale, if not an astronomical one.

Mathematical geography divides roughly into two parts. Cartography, the production of maps from projections of the spherical surface of the Earth onto a flat surface, continued its development from Hellenistic times. As we saw from Ptolemy's *Geography* in Text 2.5, trigonometry needs to be employed occasionally to locate key points on the maps.[191] However, in geodesy, the study of the dimensions of the Earth, trigonometry—especially the spherical variety—becomes indispensable.

And on geodesy, one treatise stands out above all: the *Taḥdīd Nihāyāt al-Amākin li-Taṣḥīḥ Masāfāt al-Masākin* ("Determination of Coordinates of Positions for the Correction of Distances Between Cities") by al-Bīrūnī.[192] This work sets out to determine the relative positions of Ghazna, Baghdad and Mecca on the Earth's surface: in other words, its task (in part) is to find the *qibla* at Ghazna. In fact, our earlier account of the method of the *zījes* is taken from this work. But al-Bīrūnī goes beyond the *qibla* into a variety of

[190] See the surveys of Islamic mathematical geography in [Schoy 1924] and [Kennedy 1996], the latter mostly superceding the former.

[191] See [Berggren 1982] for an edition and translation of a treatise by al-Bīrūnī, containing no less than eight map projections. Trigonometric methods are used on pp. 58–59.

[192] The *Taḥdīd* is available in an Arabic edition in [al-Bīrūnī (Bulgakov) 1962] and an English translation in [al-Bīrūnī (Ali) 1967]; [Kennedy 1973] is an extensive scientific commentary.

related topics, and he ends up with a treatise covering most of the topics in geodesy: the determination of terrestrial latitudes and longitudes, the size of the Earth,[193] and great circle distances between cities, to name a few.

The problem of the distance between cities is one instance where the fundamental role played by the *qibla* in mathematical geography becomes clear. In figure 4.41 the two cities are A with coordinates $\phi_A = \overset{\frown}{AH}$ and L_A, and B with coordinates $\phi_B = \overset{\frown}{BT}$ and L_B; then $\Delta L = L_A - L_B = \overset{\frown}{HT}$. The goal is to find $\overset{\frown}{AB}$.

One way to proceed is simply to follow the method of the *zījes*; although we are now operating on the surface of the Earth rather than the celestial sphere, the situation is analogous. Only a few years earlier, Abū'l-Wafā' had used this method to find the distance from Baghdad to Mecca.[194] In fact Abū'l-Wafā' had given two methods, the second applying his new tangent function.[195] But al-Bīrūnī chose an entirely different course, bringing back a theorem we have not seen in some time.[196]

Firstly, notice that ΔAZW, ΔDBX, and ΔHTY (where W, X, and Y are points on the axis through E) are similar isosceles triangles. Thus

$$\frac{AW}{AZ} = \frac{DX}{BD} = \frac{HY}{HT}, \text{ or } \frac{\operatorname{Cos}\phi_A}{AZ} = \frac{\operatorname{Cos}\phi_B}{BD} = \frac{R}{\operatorname{Crd}\Delta L}; \qquad (4.73)$$

so $AZ = \operatorname{Cos}\phi_A \operatorname{Crd}\Delta L/R$ and $BD = \operatorname{Cos}\phi_B \operatorname{Crd}\Delta L/R$. But $AZBD$ is a cyclic quadrilateral (since $AD = BZ$ and $AZ \parallel BD$), so Ptolemy's Theorem applies to it:

$$AB \cdot ZD = AZ \cdot BD + AD \cdot BZ,$$

or

$$AB^2 = \left(\frac{\operatorname{Cos}\phi_A \operatorname{Crd}\Delta L}{R}\right) \cdot \left(\frac{\operatorname{Cos}\phi_B \operatorname{Crd}\Delta L}{R}\right) + \left(\operatorname{Crd}\Delta\phi\right)^2. \qquad (4.74)$$

This gives AB in terms of the radius of the earth, which al-Bīrūnī had already determined earlier.

[193] See [Berggren 1986, 141–143] for an account.

[194] To be precise, the analogy between figure 4.34 and figure 4.41 is as follows: in the latter diagram let Q be the pole of the meridian through A. Then $N_{eq} \leftrightarrow E$, $Z_O \leftrightarrow A$, $K \leftrightarrow P$, $A \leftrightarrow H$, $Z_M \leftrightarrow B$, and $W \leftrightarrow Q$.

[195] [Kennedy 1984] describes both methods (although not drawing the analogy to the method of the *zījes*), and includes an analemma and demonstration by Debarnot of the second.

[196] The following material is taken from [al-Bīrūnī (Ali) 1967, 192–193]; see also the commentary in [Kennedy 1973, 144–146].

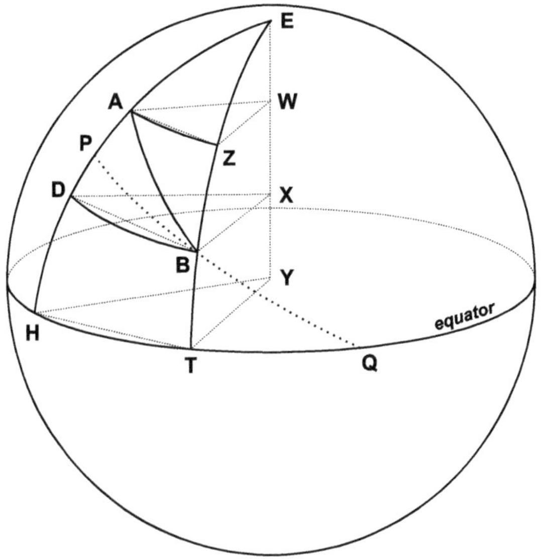

Figure 4.41
Al-Bīrūnī on the distance between cities

This appearance of Ptolemy's Theorem is not typical of the *Taḥdīd*; usually al-Bīrūnī applies the Rule of Four Quantities whenever he can. In general, the *Taḥdīd* contains few of the trigonometric innovations circulating at the time, even those in al-Bīrūnī's other works such as the use of $R = 1$ and the tangent function. Perhaps he chose the traditional methods here since trigonometry was not the main point of this book.

Trigonometry in al-Andalus

Until now we have not spent much time distinguishing between different geographical regions and local situations in the medieval Muslim world, which stretched all the way from the borders of India to Spain and covered at least eight centuries. This shortcoming, which has always afflicted the study of Muslim science, is beginning to be addressed in the research literature.[197] The case of Muslim Spain ("al-Andalus," not the current Andalusia) from the eighth to the fifteenth centuries, is a strong example of a scientific subculture with its own concerns and style, one that has been clearly recognized by the history of science community.[198]

From the middle of the ninth to the early eleventh century, scientific knowledge seems to have been exchanged somewhat freely between eastern

[197] The importance of concentrating on local situations in understanding Arabic (or any other culture's) science, guided by but distinguished from the overarching tradition, was stated strongly in [Sabra 1996].

[198] See [Vernet/Samsó 1996] for a summary of the sciences in al-Andalus.

and western Islam. We have already seen a striking example of this in Ibn Muʿādh's early eleventh-century treatise on the new spherical trigonometry, which appeared in al-Andalus not long after its creation in the east. Starting around this time, however, al-Andalus became gradually less dependent on eastern Islamic science and developed its own methods and tastes. This movement might have something to do with why Ibn Muʿādh's spherical trigonometry does not seem to have instigated further developments.

Perhaps the most striking characteristic of the Andalusian scientific community was the rich interaction of different geographical and religious cultures that occurred there. Muslim, Jewish and Christian scholars worked side by side and exchanged information with each other, and the knowledge they generated spread well beyond Spain. Indeed, we shall see soon that, for all the significant contributions made in eastern Islam, it was Andalusian trigonometry that transmitted to Europe and provoked advancements there. However, it is a historical mistake to consider al-Andalus only for its role in transmission to another culture, and we shall discuss this later.

Two important Andalusian contributions to trigonometry other than those already raised elsewhere in this chapter deserve some space: one just before 1100, and the other just after. The first accomplishment is a collection of astronomical tables developed in Toledo by a group of scientists, most prominent among them Abū Isḥāq ibn al-Zarqāllu (also known as Azarquiel). These tables, lost in the original Arabic but available in two Latin translations (one by Gerard of Cremona), were not especially original; most of their content was borrowed from the *zījes* of al-Khwārizmī (in the Maslama version) and al-Battānī. From the latter, al-Zarqāllu borrowed the table of shadows, but there is no trace of the new eastern acceptance of tangents and other trigonometric functions. As usual, the construction of a sine table was considered carefully as the mathematical basis for the rest of the work. Here we find a new method of constructing the first several sine values which was to retain some currency for several hundred years.[199]

One of the sine tables extant in a large portion of the Latin manuscript tradition of the *Toledan Tables* uses the radius $R = 150$ minutes (the value likely in al-Khwārizmī's *zīj*, originally from Indian sources). Al-Zarqāllu begins by considering sines in multiples of 15°, the so-called *kardajas*.[200] The

[199] The trigonometric canons to the Toledan tables are available in Latin in [Curtze 1900, 337–347], but this publication is rendered obsolete by the recent publication of an edition and translation of the entire Latin version of the Toledan tables in the four-volume set [F. Pedersen 2002].

[200] The term *kardaja*, often thought to be an Arabic version of the Sanskrit *kramajyā*, is in fact a Persian word for "section" [Richard of Wallingford (North) 1976, 46–47]. It referred to the Indian division of a right angle into 24 segments of $3\frac{3}{4}°$, but by the time of the Toledan tables it signified a division into six parts of 15°.

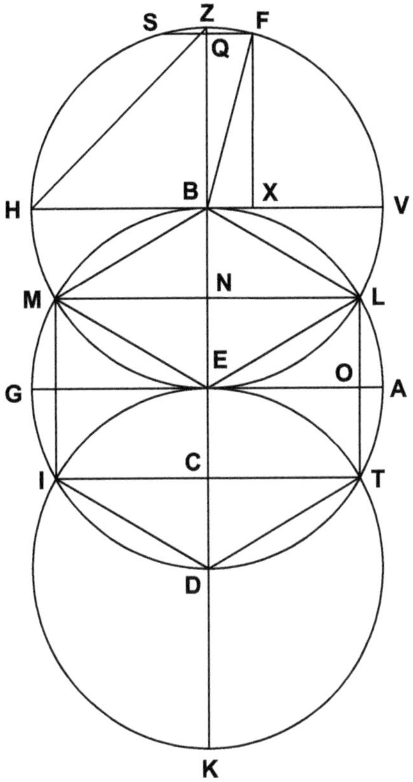

Figure 4.42
Calculating the Sines of the kardajas in the *Toledan Tables*

calculation of the sines of the six *kardajas* from 0° to 90° proceeds as follows. The circles in figure 4.42 each have radius $R = 150$; most of the constructions should be obvious at a glance, except that $\widehat{FZS}$ is asserted to be equal to 30°.[201] Since the segments that form the regular hexagon inscribed in the middle circle are all equal to its radius, we know that Crd 60° = 150, so $GM = \text{Sin } 30° = 75$; thus we may find $MN = \text{Sin } 60°$ easily (using the Pythagorean Theorem) as well as $ML = \text{Crd } 120°$.

[201] Following along in the text [Curtze 1900, 344–346]: circle *ABGD* is divided by two perpendicular lines, *AG* and *BD*. Circle *VZH* is drawn with center *B* and radius *BE*; likewise, circle *TKI* is drawn with center *D* and radius *DE*. Circles *ABGD* and *VZH* intersect at two points *L* and *M*, and $\widehat{LBM} = \widehat{MEL}$. Now $\widehat{LB}$, $\widehat{LE}$, $\widehat{BM}$, and $\widehat{ME}$ are equal, so their chords are also equal. *LM* is cut in half by *N*, and so $NB = NE$ is one quarter the diameter of the circle. We can do the same to the lower part of the figure, so that $ID = DT$ is equal to the other four chords, and we have an inscribed hexagon *LBMIDT* with sides equal to *R*.

Now $ZH = \text{Crd}\,90° = \sqrt{150^2 + 150^2} = \sqrt{45{,}000}$, and so $\text{Sin}\,45° = \frac{1}{2}\,\text{Crd}\,90° = \sqrt{11{,}250}$. Then, from ΔALO (not completely drawn on the figure),

$$AL = \text{Crd}\,30° = \sqrt{(\text{Sin}\,30°)^2 + (R - \text{Sin}\,60°)^2}, \tag{4.75}$$

and half of this is Sin 15°.[202] Moving to the top of the circle (presumably for ease of reference rather than mathematical reasons), $QB = \text{Sin}\ 75°$ may be found from $FB = R$ and $QF = \text{Sin}\ 15°$. We now have the sines of all six *kardajas*, summarized in the small table of figure 4.43.[203]

This work is utterly unremarkable mathematically, but that did not stop the procedure from being adopted widely. The diagram and accompanying calculations (although with different values of R) may be found in European texts as late as the fifteenth century, in works by John of Gmunden and Georg Peurbach (see figure 5.11), both major influences on Regiomontanus.

Our second treatise, not long after 1100, was part of an Andalusian tradition of critical evaluation of scientific work, rather than simply accepting ancient authority. Jābir ibn Aflaḥ's *Iṣlāḥ al-Majisṭī* ("Correction of the *Almagest*") is a lengthy collection of arguments against some of Ptolemy's astronomical inferences; so fierce were some of his words that he would later earn from Copernicus the title "egregious calumniator of Ptolemy."[204]

One of Jābir's attacks on Ptolemy sounds familiar: he complains that the Transversal Figure is too cumbersome, difficult to implement and difficult for students to learn. Jābir seeks to find another theorem to replace it, and he comes up with none other than the Rule of Four Quantities, which he demonstrates directly. From there he goes on to establish the three main theorems of his trigonometric stable:[205]

- the Law of Sines, proved first for a right triangle, then in the general case;

[202] This is essentially an application of the half-angle formula

$$2 \cdot \text{Sin}\,\frac{\alpha}{2} = \sqrt{(\text{Sin}\ \alpha)^2 + (\text{Vers}\ \alpha)^2};$$ see (3.19) for an early occurrence.

[203] [Curtze 1900, 339], where the entry for a *kardaja* of 1 is mistakenly recorded as a 0. It has been argued that al-Zarqāllu's Sine table as a whole was a misguided reconstruction of al-Khwārizmī's table, missing in al-Zarqāllu's source [McCarthy/Byrne 2003, 262–265].

[204] [Lorch 1975, 85]; this paper also contains a summary of the contents of the *Iṣlāḥ* on pp. 94–99. Josep Bellver has recently produced an edition and translation of the *Iṣlāḥ* in his 2007 dissertation at the University of Barcelona. Ironically, some of Copernicus's spherical trigonometry appears to derive from the *Iṣlāḥ*.

[205] A full account of Jābir's development of these results is in [Lorch 1995, 3–9].

Kardaja	Sine
1	39
2	75
3	106
4	130
5	145
6	150

Figure 4.43
Sines of the kardajas in the *Toledan Tables*

- where $\angle C$ is right, $\sin B = \cos A/\cos a$ (Geber's Theorem);
- where $\angle C$ is right, $\cos c = \cos a \cos b$.

Jābir's arguments repeat almost precisely late tenth-century motives for reinventing spherical trigonometry, and even parallel the shift in focus from spherical quadrilaterals to triangles—so much that one can hardly avoid the question of transmission. However, there is no need to insist that it occurred. Although Jābir's objections to the Transversal Figure mimic those of Abū Naṣr Manṣūr and Abū'l-Wafā', the latter two produced more sophisticated responses, in Abū'l-Wafā''s case including the tangent function. The tantalizing presence of Ibn Muʿādh's book in Spain only a century before Jābir also fails to support transmission; it shares very little with Jābir's text beyond superficial similarities. Finally, Jābir was aware of Thābit ibn Qurra's work on the Transversal Figure, which contains a proof of the Rule of Four Quantities not much different from what is found here. So, given that the Transversal Figure is genuinely awkward to use, it is entirely plausible that Jābir simply walked a path that, unbeknownst to him, had been traversed before.[206]

Jābir's plane trigonometry has a different feel from that of other Islamic scholars: it uses chords rather than sines, describes procedures rather than proves theorems, and stays very close to the *Almagest*. However, it does present a systematic approach to solving all plane triangles, and in this Jābir's work is unique for its time. It has even been suggested that Jābir is the source for the Western tradition that was to emerge in Europe.[207] Indeed, we shall see that the *Iṣlāḥ* had considerable influence there through the seventeenth century, partly through a translation by the ubiquitous Gerard of Cremona.

[206] This is the conclusion arrived at in [von Braunmühl 1900/1903, vol. 1, 81–82]. Lorch feels that the coincidence of motives is too strong to be ignored, although he is careful not to hypothesize any specific links [Lorch 1995, 9].
[207] [Lorch 1995, 38].

The transition to European science was a complex and gradual process, but it is signaled by the construction of a new set of astronomical tables organized by King Alfonso X of Castile in the late thirteenth century, partly as an update to the *Toledan Tables*.[208] Alfonso did his best to preserve Arabic knowledge, and this effort included supporting the *Libros del saber de astronomia*, a collection of astronomical works.[209] These treatises became part of the body of Arabic knowledge that was to inform and inspire scientific minds in the West to new heights, just as the light was beginning to fade gradually elsewhere.

[208] See [Chabás/Goldstein 2003].

[209] See [Ausejo 1983] and [Ausejo 1984] for studies of the trigonometry in these works, including the use of Jābir's theorems. Although the trigonometry remained embedded in its astronomical context, Ausejo argues that this is no reason to devalue its contribution [Ausejo 1983, 32–33].

5 The West to 1550

Transmission from the Arab World

The exact sciences were not of much practical use to Europeans in the early medieval period, and little technical content was produced or preserved. Demands on mathematics extended little further than basic arithmetic in trade and agriculture, and even in these areas it would not be until the thirteenth century that decimal place value numeration challenged Roman numerals. As for astronomy, the most important task for mathematics was in determining the date of Easter in the Christian calendar, attempted by the early seventh-century monk the Venerable Bede and others. Finding Easter's date was no easy computational feat, but in this and other astronomical activities of the time, trigonometry had no place.

Although mathematics and astronomy enjoyed renewed interest through the revival of education in Charlemagne's court in the ninth century on the basis of the quadrivium (which included geometry and astronomy), serious attention to quantitative astronomy began in earnest only with various influences from Islam. Learning in general found its way to Europe through several conduits, including Byzantium and Sicily, but the mathematics associated with astronomy filtered in primarily by way of Spain, across the Pyrenees mountains. As early as the tenth century, a set of scientific treatises from Andalusian Spain, composed at the Benedictine monastery of Santa Maria di Ripoll in Catalonia, was circulating in Europe at least as far as Austria.[1] Through the eleventh century pockets of Arabic astronomy could also be found in Germany; and the astrolabe and more traveled as far as England.[2]

The earliest significant presence of quantitative astronomy in Europe was the eleventh-century *Toledan Tables*, which we discussed in chapter 4. Designed by al-Zarqāllu and based on al-Khwārizmī's and al-Battānī's astronomical works, the tables traveled to the rest of Europe in a variety of forms, including a translation by Gerard of Cremona. The tables were to be a dominant force in the astronomical community, such as it was, until the *Alfonsine Tables* replaced them in the late thirteenth and early fourteenth centuries. Although the tabulated functions were ultimately trigonometric, readers did not need to know any trigonometry to use them. But the twelfth century saw a real European astronomical revolution through the translation

[1] [North 1994, 204–205].

[2] See [Kren 1983, 233] for a summary of early Arabic influences in England. Even some Indian astronomy found its way to England; see [Neugebauer/Schmidt 1952].

of many Greek and Islamic astronomical works, and with these, trigonometric methods entered Europe.

Al-Khwārizmī's *zīj* was among the earliest books to be translated; Adelard of Bath produced an edition around 1130, and Robert of Chester adapted it to the meridian of London. It is in Robert's revision that we find the first occurrence of a form of our word "sine"; the Arabic word for sine (*jaib*, which as we have seen is itself a transliteration from the Sanskrit) happens to coincide with the Arabic word for "a fold or inlet," which in Latin is *sinus*.[3] Plato of Tivoli contributed Latin editions of Theodosius's *Spherics*, and al-Battānī's *Zīj* (*De motu stellarum* in Latin) which also contained trigonometric content. But by far the greatest translator in mathematical astronomy was Gerard of Cremona, whose efforts included (among many others) the *Spherics* of Autolycus, Theodosius, and Menelaus, Jābir ibn Aflaḥ's *Iṣlaḥ al-Majisṭī* (in the West *Elementa astronomica*), and most crucially, Ptolemy's *Almagest*.

An Example of Transmission: Practical Geometry

Geometrical knowledge did not disappear completely in early medieval Europe, but lack of direct access to Euclid's *Elements* and the like meant that the flavor of Greek geometry subsided. The geometry that we find from this time does not pay respect to axioms and proofs; we also witness a freer intermingling of geometry with arithmetic and fairly trivial rudimentary mensuration.[4] Indeed, three separate geometrical traditions have been identified, each with its own standards and foci: theoretical, practical, and constructive geometry.

Theoretical geometry, the tradition with which Euclid would have been most comfortable, returned to prominence mostly after the re-introduction of the *Elements* into Europe during the "renaissance" of the twelfth through the fourteenth centuries via translation from Arabic sources. Constructive geometry, hardly a discipline in a formal sense and with little written tradition, was employed by craftsmen, especially masons.[5] Situated between these two disciplines, with some of the features of each, was practical geometry.

Concerning itself with the uses of geometry in what we might today call engineering, practical geometry has its origin in the practices of Roman surveyors (the *agrimensores*), who in turn received some of their knowledge from Heron of Alexandria's works in mensuration.[6] The scope of practical geometry

[3] Hence, for instance, the term "sinus cavity."

[4] See [Evans 1976/77] for a survey of "sub-Euclidean" geometry in the early medieval period.

[5] [Shelby 1972] is a survey of constructive geometry, particularly as applied by masons.

[6] [Hugh of St. Victor (Homann) 1991, 2]. [Dilke 1971] is a study of the habits and practices of the *agrimensores*.

varied somewhat over the centuries, but roughly speaking it encompassed applications that have to do with measurements of physical bodies. Texts often divide into three parts: altimetry, the measurement of heights and depths; planimetry, on surfaces; and stereometry, on volumes.[7] Objects subjected to such measurements included buildings, cities, war machines, instruments of various sorts (astronomical, musical and scientific), and even celestial bodies.

Early treatises, of which Gerbert d'Aurillac's late tenth-century *Geometria* is the first substantial example,[8] contained little mathematical and no trigonometric knowledge. Thus to find the height of an object Gerbert had to rely mostly on similar triangles, which is at least an improvement on the *agrimensores*, who had used only congruent triangles.[9] Hugh of St. Victor's early twelfth-century *Practica Geometriae*, the first treatise to make an explicit distinction between theoretical and practical geometry, followed Gerbert's example. Situated just prior to an influx of translations of Euclid, Archimedes and Ptolemy beginning in the twelfth century, neither of these works were able to make any substantial connection with Greek or Arabic science and mathematics.

In his discussion of altimetry, Hugh of St. Victor describes how to use a simple astrolabe to determine the height of an object some distance away on the surface of the earth, such as a tree, building or mountain.

Text 5.1
Hugh of St. Victor, Using an Astrolabe to Find the Height of an Object
(from the Practica Geometriae)

[See figure 5.1.] To measure the height of an object in front of you without moving from your place, raise the astrolabe to the object. Adjust the medicline until you can see the top of the object through both apertures. Then compare the medicline degree reading with the whole side of the square. The ratio of the medicline reading to the whole side [i.e., to twelve] is the ratio of height to intervening space with surveyor's height added either proportionally or exactly.[10]

Explanation: The medicline of the astrolabe, otherwise known as the alidade, is the ruler that traverses the back of the astrolabe along the line of

[7] Hugh of St. Victor's third section was on cosmimetry, or celestial measurement; the chapters on volumes are sometimes referred to as crassimetry. The changing boundaries of practical geometry through the medieval period and its relation to the craftsmen's constructive geometry are described in [Shelby 1972].

[8] See [Gerbert (Bubnov) 1899] for a Latin edition, and [Gerbert (Olleris) 1867] for a French translation.

[9] [Hugh of St. Victor (Homann) 1991, 16].

[10] From [Hugh of St. Victor (Homann) 1991, 46].

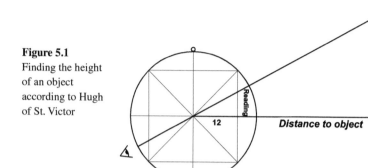

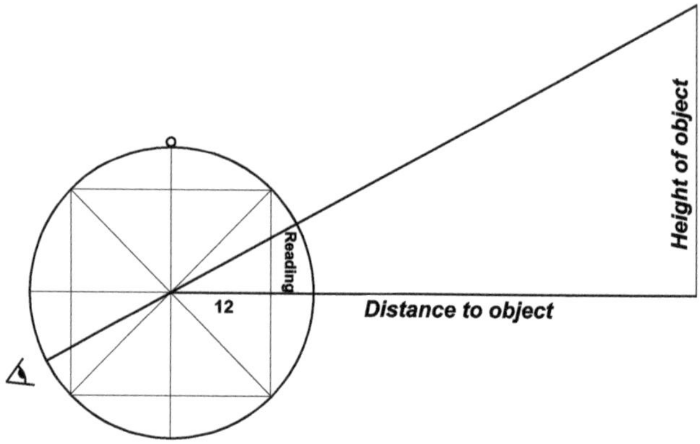

Figure 5.1
Finding the height of an object according to Hugh of St. Victor

sight from the eye to the top of the measured object. The horizontal distance from the center of the astrolabe to the edge of the inscribed square is divided into twelve equal parts. Take a reading of the height of the medicline where it crosses the edge of the square. Then, by similar triangles,

$$\frac{\text{reading}}{12} = \frac{\text{Height}}{\text{Distance}}. \tag{5.1}$$

Of course, since the astrolabe itself is above the ground, the instrument's height must be added to obtain the height of the measured object above ground level.

Our next example of altimetry, from not much later in the twelfth century, illustrates dramatically the impact of exposure to translations of Greek geometrical and astronomical works. The *Artis cuiuslibet consummatio*, an anonymous work adapted to French as the *Pratike de geometrie*, bears little resemblance to its predecessor. Although problems such as measuring the height of a tower using an astrolabe are also to be found (and the method of solution is similarly crude[11]), the section on altimetry is now dominated by astronomy—particularly, problems involving the Sun's altitude. The ability to use both sexagesimal arithmetic and a Sine table is now assumed, although the author still refers the reader to the *Almagest* for proofs.

[11] The use of the astrolabe's shadow square to solve altitude problems seems to have been a staple of practical geometry texts through the fifteenth century; see for instance [Simi/Toti Rigatelli 1993, 458–462].

Text 5.2
Finding the Time of Day from the Altitude of the Sun
(from the Artis cuiuslibet consummatio)

II.11 To Ascertain the Hour of the Day from the Altitude of the Sun.

Let the Sine of the altitude [of the Sun] be multiplied by 60; let the product be divided by the Sine of the meridian altitude of the same day. Let the arc of the Sine which is produced be divided by 15. The quotient will give the hours completed [since sunrise] if the altitude is taken before noon, and those to be completed [before sunset] if after. You will find in the *Almagest* the proofs of this and other things which we do not set forth here lest we make a long work and hoe another man's row.[12]

Explanation: The pattern of calculation in this text derives from figure 5.2. The meridian circle[13] is given a radius of 60; then $AN = \text{Sin } H$, where H is the Sun's altitude at noon. BP is taken to be the Sine of h, the Sun's current altitude. Finally, OP is assumed to be the Sine of time T since sunrise (or, if after noon, until sunset). By the similarity of ΔOAN and, ΔOBP,

$$\frac{\text{Sin}\,T}{60} = \frac{\text{Sin}\,h}{\text{Sin}\,H}, \tag{5.2}$$

from which T may be found. The division by 15 ($=360°/24^h$) the passage converts T from degrees into hours.

Contrary to the author's statement, this approximate formula is not found in the *Almagest*. But we have seen it already in the context of Islamic astronomical timekeeping, where its Indian origin was asserted. Other problems in the *Artis cuiuslibet consummatio* appear to come from India also,[14] perhaps through Islamic intermediaries, so there may be several sources influencing this text.

More rigorous approaches to the use of trigonometry were being developed at the same time in Spain. Abraham bar Ḥiyya (1070–1136), a Jewish official whose job title in the administration of Barcelona gave rise to his common appellation "Savasorda," wrote a treatise on mensuration (*Ḥibbūr ha-meshīḥah we-ha-tishboret*,[15] translated by Plato of Tivoli as *Liber*

[12] [Victor 1979, 243].
[13] The great circle through the north and south points of the horizon, and through the zenith.
[14] Otto Neugebauer, quoted in [Victor 1979, 255–257].
[15] A Hebrew edition is available in [Abraham bar Ḥiyya (Guttmann) 1913]. See also [Lévy 2001] for a survey of the work and a call for a new critical edition.

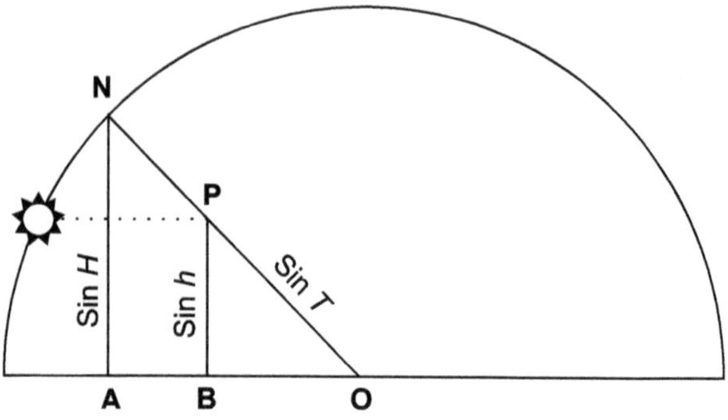

Figure 5.2
Finding the time of day from the Sun's altitude in the *Artis cuiuslibet consummatio*

embadorum[16]) based on the Islamic methods that were available to him. Although its purpose was as practical as the other practical geometries, the *Liber embadorum* was well ahead of its time, bringing to France its first account of Arabic algebra.[17] One of the problems in Abraham's book was to find the area of a segment of a circle, given the length of its chord. To do this he needed a table of arcs as a function of chord lengths. Perhaps the first trigonometric table in Europe, it uses the peculiar radius $R = 14$, chosen so that the largest arc (a semicircle) works out to the whole-numbered length of 44 using the common approximation $\pi \approx 22/7$.[18] Tables of inverse trigonometric functions were rare enough in Islamic trigonometry, let alone European, so the appearance of this one (extracted in figure 5.3) is an extraordinary occasion. In fact, the table is remarkably accurate except for the final few entries, where numerical instability would give any calculator a headache.[19]

Abraham's work was an influential source for one of the most important medieval works of mensuration: the *Practica geometriae* by Leonardo of Pisa, also known as Fibonacci (1220).[20] Although Leonardo was still concerned

[16] Plato of Tivoli's translation of the *Liber embadorum* is available in Latin, and in German translation, in [Abraham bar Ḥiyya (Curtze) 1902].

[17] [Levey 1970, 22].

[18] From [Curtze 1900, 330]. See also [Katz 1998, 292–293]; the arc chord table is again reproduced on p. 294.

[19] The errors are given in figure 5.3 in units of the last sexagesimal place. The correct value for a chord of 25 should be 30;54,53; this entry was either miscalculated or affected by scribal error. For an explanation of the numerical instability see the following section, and figure 5.4.

[20] The standard edition of Leonardo's collected works is [Leonardo of Pisa (Boncampagni) 1857/1862]; the *Practica geometriae* opens the second volume. In recent years, Leonardo's three major works have all been translated into English: the *Book of Squares* [Leonardo of Pisa (Sigler) 1987], the *Liber abaci* [Leonardo of Pisa (Sigler) 2002], and just as this book was being completed, the *Practica geometriae* [Leonardo of Pisa (Hughes) 2008]. See also [Simi 2004] for an account of this work in Italian and a discussion of its roots.

Figure 5.3
Abraham bar Ḥiyya's arc chord table

Chords	Arcs	
1	1;0,2	
2	2;0,8	[−1]
3	3;0,26	[−1]
4	4;0,55	
5	5;1,44	
6	6;2,54	[−4]
7	7;4,42	[+2]
8	8;7,11	[+12]
9	9;9,56	[−3]
10	10;13,42	[−6]
11	11;18,54	[+21]
12	12;24,38	[+15]
13	13;31,9	[−20]
14	14;40,0	
15	15,50,10	[−1]
16	17;2,16	
17	18;16,36	[+2]
18	19;33,27	[−2]
19	20;53,26	[−3]
20	22;17,10	[+1]
21	23;45,6	[−13]
22	25;19,24	[+19]
23	27;0,0	[−1]
24	28;49,56	[+39]
25	31,26,37	[—]
26	33;20,52	[−3]
27	36;27,32	[−116]
28	44;0,0	

with the problems of the surveyor, the great innovation of his work was his inclusion of proofs, moving the subject much closer to theoretical geometry.[21] Indeed, it is a sea change in this tradition to see Leonardo referring extensively to Euclid, Menelaus and Theodosius, and using implicitly the works of Arabic

[21] In fact, title notwithstanding, Victor is inclined to consider the *Practica geometriae* as "a general text-book in geometry instead of a practical geometry" [Victor 1979, 27].

mathematicians including the Banū Mūsā[22]—alongside the usual references to the *agrimensores*.

Much of the *Practica geometriae* is borrowed from the *Liber embadorum*, including the problem of areas of circle segments, but Leonardo's solution differs slightly from Abraham bar Ḥiyya's. Rather than using a table of arc chords, Leonardo refers explicitly to the procedure in Ptolemy's *Almagest*, going so far as to prove Ptolemy's Theorem on a quadrilateral inscribed in a circle.[23] However, Leonardo's table differs from both Ptolemy's and Abraham bar Ḥiyya's in its use of the radius $R = 21$ (again chosen with a view to $\pi = 3^1/_7$, so that the semicircle has the whole-numbered length of 66).

So, the importance of the *Practica geometriae* and similar works was not so much the novelty of their mathematics, but their role as a vehicle for the dissemination of mathematics to Europe. The continuing role of practical geometries in taking mathematics to the people ensured that they were vehicles of transmission at least through the eighteenth century.[24] In what ways their seeds really took root is difficult to say; evidence for the influence of the more sophisticated methods is scanty.[25] However, it is clear that basic trigonometric methods, at least, were now available for use by a more diverse audience than ever before.

Consolidation and the Beginnings of Innovation: The Trigonometry of Levi ben Gerson, Richard of Wallingford, and John of Murs

The trigonometric methods in practical geometries, of course, were not original in any significant way. However, as time progressed and the methods of the ancients and the Arabs (at least from Muslim Spain, not the more powerful trigonometry of eastern Islam) became more well known, European scientists gradually made more contributions within the existing themes. The spread of the *Alfonsine Tables* from Spain through Paris in the 1320s must have provoked some mathematical curiosity. The increasing availability of the *Almagest*, and the disproportionate influence of Seville astronomer Jābir ibn Aflaḥ's *Iṣlaḥ al-Majisṭi*, helped to direct these interests further. This treatise by Geber (as Jābir was known in Europe), via its translation by Gerard of Cremona, shaped subsequent trigonometric work starting as early as the

22 [Folkerts 2004, 98–104].

23 [Leonardo of Pisa (Boncampagni) 1857/1862, 94–96, 104]. See also [Leonardo of Pisa (Hughes) 2008, 344, 354–359].

24 The history of practical geometries has been traced through John Love's 1768 work *Geodesia: Or, The Art of Measuring Land* in [Hugh of St. Victor (Homann) 1991, 20–22].

25 [Shelby 1983, 207].

thirteenth century.[26] Whatever the motivating factors were, it is around this time that we find two new substantial European astronomical works, both with some trigonometric novelty.

Levi ben Gerson (1288–1344), a philosopher and theologian who lived in the south of France, appears to have had access to the knowledge filtering across Europe, although only in the form of Hebrew translations.[27] Nevertheless he seems to have been widely read—his astronomy was influenced by Ptolemy, al-Battānī, Jābir ibn Aflaḥ, and others—and he certainly wrote widely. Although he authored several mathematical works, he is most known for his philosophical treatise *The Wars of the Lord*, a work of such magnitude that the *Astronomy* alone, one section of the fifth of its six parts, occupies over 500 folios.[28] Although the *Astronomy* contains many impressive achievements, including detailed critiques of al-Biṭrūjī's planetary models based on Levi's own observations and a lunar model that improved upon Ptolemy, it is best known for its astronomical tables and for its description of the Jacob's staff, an instrument used for measuring angles.[29]

One portion of the *Astronomy* that circulated on its own, the *De sinibus, chordis et arcubus*,[30] describes how to compute tables of Sines and versed Sines (the *sagitta*)[31] and is clearly inspired by Ptolemy's description of the construction of a table of chords in the *Almagest.* Levi deals only with chords, sines, and versed sines; although the latter can serve the role that cosines do for us, Levi has no equivalent to the tangent. Beginning with the geometrically-determined sines and versed sines of 90°, 30° and 18°, Levi works his way down to the sine and versed sine of $\frac{3}{4}^\circ$ in the conventional manner, using basic identities.

From this point, Levi finds himself in a situation similar to that of Ptolemy and the Islamic table makers: he knows the sines of all multiples of $\frac{3}{4}^\circ$, but not his desired $\frac{1}{4}^\circ$. To reach this goal he begins by applying his half-angle procedure repeatedly to $8\frac{1}{4}^\circ$ and $3\frac{3}{4}^\circ$ (both multiples of $\frac{3}{4}^\circ$), which

[26] Jābir ibn Aflaḥ's presence in Europe is an example of happenstance: the work was in the right place at the right time and rose to undue prominence. See [Lorch 1971] for a full treatment of his influence in Europe, and [Lorch 1975] for a survey of his astronomical contributions.

[27] [Samsó 1973, 279].

[28] The first 20 of 136 chapters of the astronomy appear in Hebrew edition and English translation in [Levi ben Gerson (Goldstein) 1985].

[29] What little trigonometry one needs to work with this device allows us to conclude that Levi was unfamiliar with the tangent, since when a simple tangent will do Levi resorts instead to a more complex procedure involving a sine—similarly to what we have seen in Ptolemy's astronomical calculations.

[30] Part of *De sinibus, chordis et arcubus* appears in English translation in [Espenshade 1967] and [Levi ben Gerson (Goldstein) 1985, 31–47], and in Latin in [Curtze 1898].

[31] Levi ben Gerson's astronomical tables, including his table of sines, have been published in [Goldstein 1974].

brings him down to the sines of $\frac{1}{4}^\circ + \frac{1}{128}^\circ$ and $\frac{1}{4}^\circ - \frac{1}{64}^\circ$ respectively. Here he notices that

$$\frac{\mathrm{Sin}\left(\frac{1}{4}^\circ + \frac{1}{128}^\circ\right)}{\mathrm{Sin}\left(\frac{1}{4}^\circ - \frac{1}{64}^\circ\right)} = \frac{\frac{1}{4}^\circ + \frac{1}{128}^\circ}{\frac{1}{4}^\circ - \frac{1}{64}^\circ}, \tag{5.3}$$

"so far that a difference in fourths of minutes of a fraction of the ratio is not apparent. It is allowable that a little may appear in the fifth place"[32]—i.e., the ratios begin to differ in the fifth sexagesimal place. Thus

$$\frac{\mathrm{Sin}\left(\frac{1}{4}^\circ + \frac{1}{128}^\circ\right)}{\mathrm{Sin}\left(\frac{1}{4}^\circ\right)} \approx \frac{\frac{1}{4}^\circ + \frac{1}{128}^\circ}{\frac{1}{4}^\circ}; \tag{5.4}$$

and solving for $\mathrm{Sin}\frac{1}{4}^\circ$, Levi gets 0;15,42,28,32,27—which is out by 3 in the fourth sexagesimal place. Levi's method is effectively the same as Ptolemy's, although he computes only one bound rather than Ptolemy's two; both assume that the Sine is a linear function for small arcs.[33] This process is an example of a larger class of methods that Levi calls "heuristic reasoning" (in this instance related to the rule of double false position), which is used repeatedly in Levi's *Astronomy* to solve problems for which a demonstrative solution is not possible.[34]

Although most of Levi's Islamic cousins wrote little about heuristic methods, falling as they did outside of an analytic written tradition, Levi spends some time discussing them. For instance, in the context of an astronomical problem, he says:

> It is appropriate to know that it is not possible to do in a quick way a demonstrative research in order to show how our model must be constructed, so that from it will follow what we see by experience for any position [of the planet] . . . , because the computations for our models are deep and difficult. . . . Therefore, the investigations which lead us to the truth necessarily are of the kind of heuristic reasoning, which are made from division and experience, which approach step by step to the truth until it is reached. These types of reasoning belong to the category of conditional reasoning, and there

[32] [Espenshade 1967, 634].

[33] Although Levi ben Gerson's method has been called "linear interpolation," we have seen in chapter 4 that this term is only partly appropriate.

[34] [Mancha 1998] presents a number of astronomical examples of Levi's heuristic reasoning.

are two classes of them. One of which is taken from an excess and a defect; the second one is taken from two experiences in excess or from two experiences in defect.[35]

The methods to which Levi refers here are strongly reminiscent of Indian iterative schemes. Levi argues for their use in the same way that modern numerical analysts do: namely, that analytic techniques are either genuinely or practically impossible to solve the problem at hand. The case of (5.3) is an example of the first of Levi's two classes, an excess in the numerator and a defect in the denominator.

Levi's interest in numerical methods goes into mostly untrodden territory in the next passage from *De sinibus*, where he explains his reason for adopting a step size of $\frac{1}{4}^\circ$ for his trigonometric tables rather than the traditional 1°.

Text 5.3
Levi ben Gerson, The Best Step Size for a Sine Table
(from the *Astronomy*)

We decided to determine these values at intervals of $\frac{1}{4}^\circ$, not being satisfied with finding them at intervals of 1°, as is the case with many such tables, because we noticed that in some places the error reaches about 15 minutes of arc when we seek to find the arc corresponding to a [given] sine, especially when the arc is a little more or less than 90°. To illustrate this, let the sine be 59;59,52. According to the tables arranged at degree intervals, the corresponding arc is 89;45,27° if it is less than 90°, or 90;14,33° if it is greater than 90°. It is clear from the preceding that if this were correct, 0;0,8 would be the versed sine of 0;14,33°. But it follows from the preceding that when the amount of the versed sine is 0;0,8, the square of the sine of that arc is 0;15,59,58,56 for that is the result of multiplying 0;0,8 times the remainder of the diameter. Therefore, the amount of the sine of that arc, 0;14,33°, is 0;30,59,0,53. Therefore, it is clear that the arc corresponding to the aforementioned sine is greater or less than 90° by about 0;29,35°. Thus, the error introduced by the other method is greater than 0;15°, and for this reason we decided to arrange these tables at intervals of $\frac{1}{4}^\circ$ so that no perceptible error arises from linear interpolation.[36]

Explanation: Levi knows precisely where to find the worst numerical instability: by trying to determine the arc Sine of a length nearly equal to the radius of the circle. As indicated in figure 5.4 the Sine curve reaches its maximum at

[35] Quoted from [Mancha 1998, 15–16].
[36] From [Levi ben Gerson (Goldstein) 1985, 41–42].

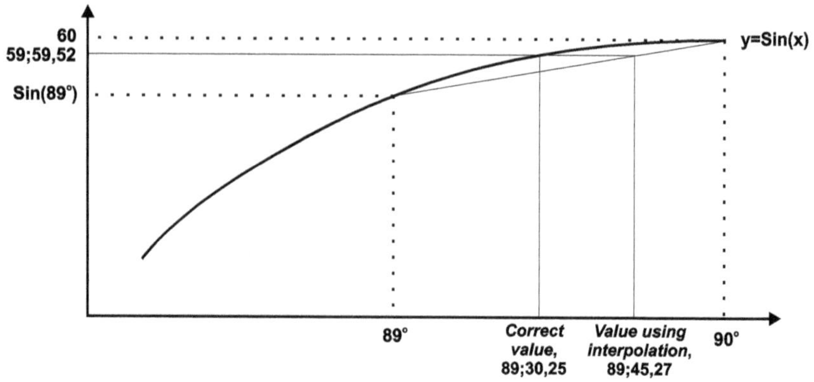

Figure 5.4 Numerical instability in finding arc Sin(59;59,52)

90°; inverse linear interpolation for a Sine value near the peak (like 59;59,52) results in a large error, since the horizontal distance from Sine curve to the line for linear interpolation between 89° and 90° is so large. This phenomenon had already caused problems for Muslim astronomers, although it often remained unnoticed.[37]

Using inverse linear interpolation Levi gets arc Sin(59;59,52) = 89;45,27°, already an error of more than $\frac{1}{4}^\circ$ (the correct value is 89;30,25°). But perhaps because Levi does not trust even his improved tables or because he wishes to make the error appear dramatic, he converts his values to represent Sines of small arcs, as follows. A moment's reflection confirms that if Sin (89;45,27°) = 59;59,52, then Vers (0;14,33°) = 0;0,8. Now from *Elements* III.35 (figure 5.5)[38] we know that

$$\text{Sin}^2\theta = \text{Vers}\,\theta \cdot (2R - \text{Vers}\,\theta) \qquad (5.5)$$

and from this Levi gets Sin(0;14,33°) = 0;30,59,0,33. But, using his own Sine table to compute Sin(0;14, 33°)—with small arguments like this, interpolation within a Sine table is reliable—Levi arrives at the correct value arc Sin(0;30,59,0,33) = 0;29,35°, more than twice the expected value of 0;14,33°!

[37] [Berggren/Van Brummelen 2003] describes the case of al-Samaw'al. In his attempt to correct an error in the work of his predecessor al-Kūhī on finding the dip angle to the horizon, he actually commits an error 50 times the magnitude of the error he is trying to correct, mostly because of this numerical instability.

[38] "If in a circle two straight lines cut one another, the rectangle contained by the segments of the one is equal to the rectangle contained by the segments of the other" [Euclid (Heath) 1925, vol. 2, 71]. In this case the two straight lines are the diameter and the vertical line, the latter extended downward to the lower half of the circle (not drawn).

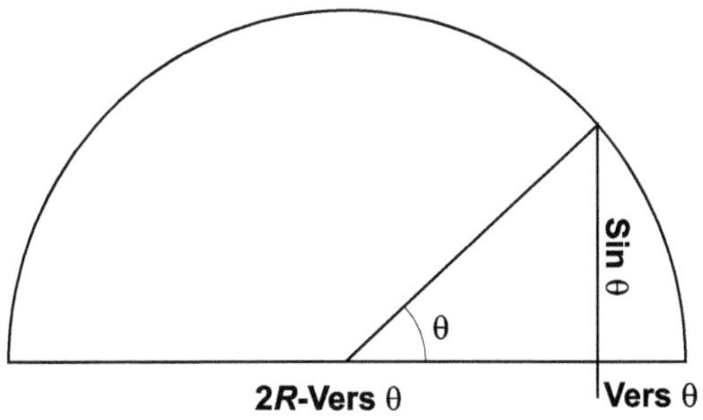

Figure 5.5
A trigonometric lemma used in Levi ben Gerson's *Astronomy*

What is remarkable about this discussion is not only that Levi has spotted the problem of error magnification in interpolation (at its most sensitive spot, indicating some deeper awareness of the issue), but that he writes about it at all. Numerical problems such as this one were simply not discussed in scientific works of the time, even if trace evidences reveal that they were considered behind the scenes.

Levi follows this passage with a collection of prescriptions for solving plane triangles. The presentation, divided into four cases, is clear and concise. One of these cases contains an explicit statement of the Law of Sines, its first appearance in a European work; it is unclear whether Levi discovered this theorem himself, or learned it from an Arabic source.[39] Levi does not enter into spherical trigonometry; astronomical problems involving spherical geometry are reduced to plane problems.[40]

Roughly contemporaneous with Levi but much further north, Richard of Wallingford was abbot of St. Albans in England in the early fourteenth century.[41] He is best remembered for a large astronomical clock that he built at the abbey, but he also wrote a number of astronomical works. He possessed a genius for inventing novel astronomical instruments, including the *rectangulus* (which did the same job as the armillary sphere) and the *Albion* ("all by one"), an unusually flexible equatorium that enjoyed popularity for centuries.

Although instruments were Richard's first love he did not shy away from mathematics; indeed, his instruments required considerable mathematical ingenuity to construct. But Richard also wrote on mathematics proper.

[39] [Bond 1921, 319]; [Samsó 1973, 280].

[40] For instance in [Levi ben Gerson (Goldstein) 1985, 58–66 and 151–154].

[41] Richard of Wallingford's works have been edited and translated to English in [Richard of Wallingford (North) 1976]. His position at the monastery means that much information on his life was recorded; [North, in Richard of Wallingford (North) 1976, vol. II, 1–16] is a thorough and reliable biography.

The work that most concerns us, his *Quadripartitum*,[42] takes readers fairly rapidly from the beginnings of trigonometry to its applications in spherical astronomy. Divided into four parts (hence the name), the *Quadripartitum* begins with a set of theorems that leads to a method for the construction of a table of sines. The work continues in Books II and III with coverage of the eighteen modes of Menelaus's Theorem in the style that we saw in chapter 4 with Thābit ibn Qurra several centuries earlier. Finally, it concludes with a series of applications of Menelaus in spherical astronomy, in the spirit of similar problems in Books I and II of the *Almagest*. Although the argument can be made that the *Quadripartitum* merely compiles mathematical knowledge from the *Almagest* and a couple of other sources, it breaks new ground by dealing with trigonometry more or less apart from its applications, and thus has some claim to the title of first trigonometric work in Christian Europe.[43]

The first part of the *Quadripartitum*, on calculating sines, reads much like the corresponding passage on the table of chords in the *Almagest* and clearly owes a debt of inspiration to it. Much of what we find in Richard's work is expected: various identities are proved involving Sines, chords, and versed Sines[44] (again, we find no tangents), many of which may be found in the *Almagest*, or Islamic and Indian sources. The identities include procedures for finding the Sine of half an angle, the use of the Pythagorean Theorem to find the Sine of the complement of an arc whose Sine is known, and procedures for dealing with sums and differences of arcs. A corollary to 1.7 solves a triangle using our often-occurring acquaintance, the plane version of the Law of Sines.

Book I culminates with several methods for deriving Sin 1°. Richard arrives at this value in the usual way, by applying his identities to known Sines and chords and noticing that the only accessible Sines of whole-numbered arcs are multiples of 3°. Richard gives three solutions. The first is simply a recounting of Ptolemy's approximation to Crd 1° (see chapter 2). The second comes down to the following equation:

[42] The *Quadripartitum* was made available in Latin in [Bond 1923a] and translated into English in [Bond 1923b] (reprinted with minor alterations in [Grant 1974, 188–198]); see also the accompanying article [Bond 1921]. However, this work was rendered obsolete when the entire *Quadripartitum* was published as part of Richard's collected works in [Richard of Wallingford (North) 1976]. We shall rely on the latter.

[43] The occasionally repeated claim that Richard's predecessor and colleague John Maudith was the first European to write on trigonometry is debunked by John North in [Richard of Wallingford (North) 1976, vol. I, 6–7]; North attributes the canons to Maudith's astronomical tables (interesting for their use of tangents—the shadow function—among other reasons) to Richard himself.

[44] Richard uses the convention of $R = 150$ from the Toledan Tables.

$$\text{Sin}1^\circ \approx \text{Sin}\tfrac{15}{16}{}^\circ + \tfrac{1}{3}\left(\text{Sin}\tfrac{18}{16}{}^\circ - \text{Sin}\tfrac{15}{16}{}^\circ\right), \tag{5.6}$$

which, although Richard gives no reasoning, is the same idea (applying an equivalent to linear interpolation) that we saw in the works of Abū'l-Wafā' and al-Kāshī in chapter 4.

Richard's third method, an elaboration of the second, is unique (so far as I know). It extends the differences technique of (5.6) to smaller and smaller increments of arcs, thereby showing how one may find Sin(1°) to arbitrary accuracy.

Text 5.4
Richard of Wallingford, Finding Sin(1°) with Arbitrary Accuracy
(*from the Quadripartitum*)

Thirdly, and most accurately, and without error such as would be detected by the mind's eye, you will proceed as follows for all [the calculations instanced in] this [treatise].[45] Having earlier found $\text{Sin}\,\tfrac{3}{16}{}^\circ$ by 1.3,[46] you will find $\text{Sin}\,\tfrac{3}{32}{}^\circ$, the sinus of half the former arc. Having found this, again by 1.3, you will seek $\text{Sin}\,\tfrac{3}{64}{}^\circ$, the sinus of half the arc. Now find $\text{Sin}\left(\tfrac{3}{64}+\tfrac{15}{16}\right)^\circ$ by 1.7,[47] which—as I am quite sure—is to say find $\text{Sin}\,\tfrac{63}{64}{}^\circ$. Lastly, halve the arc of $\tfrac{3}{64}{}^\circ$ giving $\tfrac{3}{128}{}^\circ$, whose sinus you will find by 1.3. Next you will halve the arc of this. By 1.3, seek $\text{Sin}\,\tfrac{3}{256}{}^\circ$. Find therefore $\text{Sin}\left(\tfrac{63}{64}+\tfrac{3}{256}\right)^\circ$, by 1.7, that is to say, you will find $\text{Sin}\left(1-\tfrac{1}{256}\right)^\circ$. Again resume the work, take $\text{Sin}\,\tfrac{3}{256}{}^\circ$, and by 1.3 seek $\text{Sin}\,\tfrac{3}{512}{}^\circ$. Likewise by 1.3 seek the sinus of half this arc, that is, of $\tfrac{3}{1024}{}^\circ$. By 1.7 you will find $\text{Sin}\left(\tfrac{255}{256}+\tfrac{3}{1024}\right)^\circ$, or $\text{Sin}\left(1-\tfrac{1}{1024}\right)^\circ$. And proceed in this way even to the 9000th part of a degree, or even to the infinitely small, if by working minutely you wish to do so. This method of working is the most precise, but the first is the more to be recommended.

[45] From [Richard of Wallingford (North) 1976, 50–53].
[46] A sine half-angle formula.
[47] Equivalent to a sine summation formula.

Explanation: Richard does not spell out the details, but he is obviously suggesting the following: from $\text{Sin}\frac{3}{16}^\circ$ apply the half-angle formula twice to arrive at $\text{Sin}\frac{3}{64}^\circ$. Combine it with $\text{Sin}\frac{15}{16}^\circ$ using the summation formula and we get $\text{Sin}\frac{63}{64}^\circ$. Presumably, from the latter we can estimate Sin 1° with

$$\text{Sin}1^\circ \approx \text{Sin}\frac{63}{64}^\circ + \frac{1}{3}\left(\text{Sin}\frac{66}{64}^\circ - \text{Sin}\frac{63}{64}^\circ\right). \tag{5.7}$$

We may repeat the process and get

$$\text{Sin}1^\circ \approx \text{Sin}\frac{255}{256}^\circ + \frac{1}{3}\left(\text{Sin}\frac{258}{256}^\circ - \text{Sin}\frac{255}{256}^\circ\right), \tag{5.8}$$

or even

$$\text{Sin}1^\circ \approx \text{Sin}\frac{1023}{1024}^\circ + \frac{1}{3}\left(\text{Sin}\frac{1026}{1024}^\circ - \text{Sin}\frac{1023}{1024}^\circ\right). \tag{5.9}$$

In each iteration the numerator of the angle in the base Sine $\left(\frac{63}{64}, \frac{255}{256}, \frac{1023}{1024}\right)$ must be a multiple of 3° to make the Sine geometrically accessible. This situation occurs only for every other angle bisection, which explains why Richard performs two bisections in each iteration.[48]

The reference to 9000ths of a degree is unclear; it might mean three further bisections to 8192nds of a degree, but this step is not one of the stopping points in the iteration. In any case, Richard could not have stated more clearly his understanding that the process may continue as far as desired—although he concludes the passage by discouraging its use.

As noted earlier, the first three books of the *Quadripartitum* appear to serve the spherical astronomical applications of Book IV: rising times, declinations, diurnal arcs of planets, and so forth. The spirit is that of the *Almagest*, relying heavily on Menelaus's Theorem and often following the *Almagest* solutions—hence quite different from the Arabic approach typified by Geber, which used the Rule of Four Quantities and other relations involving four

[48] Richard could have used the intermediate bisections to generate approximations in a similar fashion, e.g., $\text{Sin}\,1^\circ \approx \text{Sin}\,\frac{33}{32}^\circ - \frac{1}{3}\left(\text{Sin}\,\frac{33}{32}^\circ - \text{Sin}\,\frac{30}{32}^\circ\right)$, but this expression has a slightly different form than (5.6)–(5.9). There is really no need for Richard to formulate it in any case, since he can always proceed to the next iterate and return to the already established form.

trigonometric quantities.[49] Since there is not much new in Book IV we shall not spend time on it—except to note that, at least in one place, Richard was not above the sorts of approximations that we find in Indian spherical astronomy. In IV.15 the goal is to find the diurnal arc of a planet; the first step is to determine the planet's declination (δ in figure 5.6). To get δ Richard instructs the reader to add the planet's latitude (β) to the declination of its place on the ecliptic (δ_E). Now if β is small (as is usually the case) the error incurred is not very large; nevertheless, this method is somewhat crude.

We should note, in fairness, that this example is an exception to the rule. Although Richard's spherical astronomy has a few logical shortcomings, the *Quadripartitum* is for the most part a thorough, rigorous work for its time, and it deserved the influence it had in the following centuries.

Although the works of Levi ben Gerson and Richard of Wallingford stand out, they were hardly the only trigonometric works of the fourteenth century. Interest in the revival and extension of mathematical astronomy continued in both England and France. Simon Bredon of Winchecombe (1300–1372) of Merton College, Oxford,[50] wrote a commentary on the first three books of the *Almagest*, as well as a work on chords which lamentably has been lost.[51] John of Lignères (fl. 1320–1335) and John of Murs (fl. 1317–1345) were both active astronomers (especially in planetary theory) and played important roles in spreading the *Alfonsine Tables*. Clearly their colleagues were still having issues dealing with the sexagesimal arithmetic used in astronomy, since they both wrote on it. Their trigonometric efforts were, as usual, pitched as a precursor to the study of astronomy. John of Lignères's canons to his astronomical tables, the *Primum mobile*, contain some trigonometry and basic spherical astronomy.[52] John of Murs also wrote a small treatise, *Figura inveniendi sinus kardagarum*, on the favorite topic of sine table construction.[53]

Of greater interest is John of Murs's *De arte mensurandi*.[54] Many of its 13 chapters deal with topics related to practical geometry, but this treatise contains much more. The first four chapters and part of the fifth, which contain

[49] In a later, much less successful edition of the *Quadripartitum*, Richard incorporated some of Geber's treatise, especially the proofs [Richard of Wallingford (North) 1976, 23].

[50] Merton College was a center for scientific research; Bredon may have been a younger colleague of Thomas Bradwardine there.

[51] [Talbot 1962–63] is a summary of Bredon's works and includes an abstract of his commentary on the *Almagest*.

[52] The trigonometric part of John of Lignères's canons was published in [Curtze 1900, 390–413]; his sine table is reproduced in facsimile and recomputed in [Glowatzki/Göttsche 1990, 73–79].

[53] This was published in [Curtze 1900, 413–416].

[54] The complete *De arte mensurandi* was published in a Latin edition in [John of Murs (Busard) 1998], superceding [Busard 1974] which contained a Latin edition of a segment of the work.

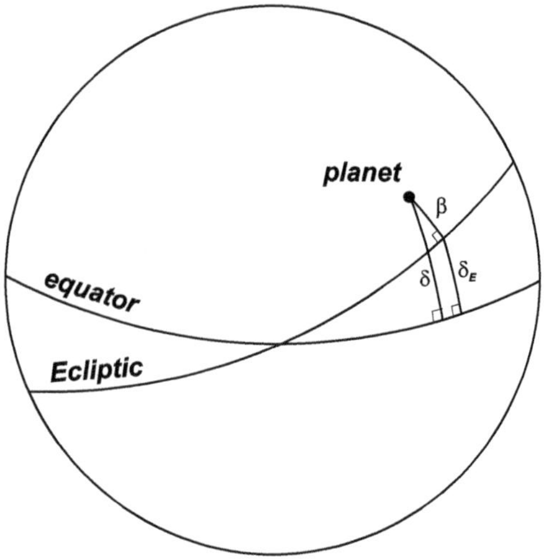

Figure 5.6
An approximation in Richard of Wallingford's spherical astronomy

the text of most interest to us, actually were written by another author.[55] As John says, "[t]his author, whose name I do not know but who was a subtle geometer, certainly covered the art of the measurement of figures in a general way." Without doubt practical geometry, concerned as it was with quantifying geometric quantities, had a natural affinity to trigonometry; it is not surprising that our unknown author put it into practice here. Chapter 2 is devoted entirely to the familiar business of constructing a table of chords, and does so very much in the Ptolemaic style. Of more interest is chapter 3, which contains a reasonably systematic treatment of solving plane triangles.

The heart of the chapter is theorems III.7 through III.9, in which the author tackles various triangles. III.7 determines the angles of a triangle when all three sides are known. His argument may be paraphrased as follows. In figure 5.7, drop perpendicular AD onto BC. Then, by Euclid II.13,

$$AB^2 + 2BC \cdot DC = BC^2 + AC^2. \tag{5.10}$$

Since the right side of this equation involves known quantities and AB is also given, we find

$$BC \cdot DC = \frac{1}{2}(BC^2 + AC^2 - AB^2).$$

And since BC is known we have DC, and thence BD $(= BC - DC)$. But, applying Pythagoras to the triangle on the left,

[55] Marshall Clagett came to this conclusion in [Clagett 1978, 11–18], retracting his previous opinion in [Clagett 1952].

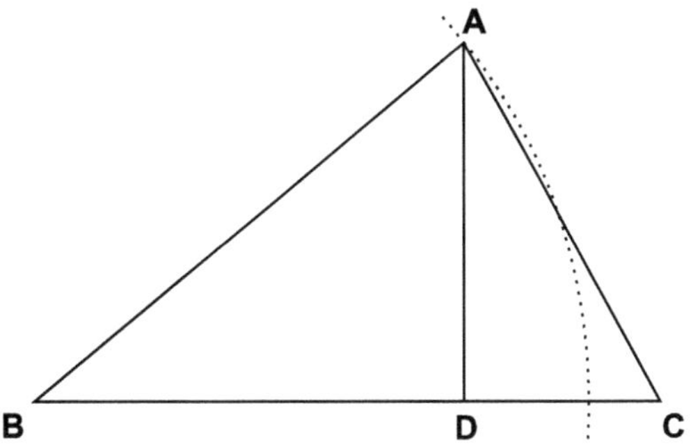

Figure 5.7
Applying the Law of Cosines in the *De arte mensurandi*

$$AB^2 - BD^2 = AD^2,$$

which gives AD. So in fact all the line segments in the diagram are known. Now AD is the Sine of the angle at B in a circle with radius $R = BA$ (in dotted lines in figure 5.7). The author tacitly assumes that we know what to do from here: namely, convert AD to the Sine of an angle with R equal to that used by an available Sine table[56] and use the table to find the arc Sine, which will be the angle at B. A similar process can be used to find the angle at C, and since the angles in a triangle sum to two right angles, we have the angle at A as well. The triangle has been solved.

So our anonymous author has used *Elements* II.13, which we defended against charges of trigonometry early in chapter 1, as the foundation for an application of the Law of Cosines to ΔABC. (Of course Euclid would not have conceived his theorem in this way.) In any case, Theorem III.8 of *De arte mensurandi* uses the Law of Sines to solve triangles where two angles and one side are known; III.9 solves triangles where two sides and an angle are known by reducing to III.7.[57]

At this point the reader may be wondering what has become of the trigonometric functions other than the sine (*sinus rectus*), versed sine (*sinus versus* or *sagitta* = "arrow"), chord (*sinus duplatus* or *corda*), and cosine (*complementi sinus* = "complement of the sine")—in particular the tangent (*umbra versa*),[58] which as we have seen was well known and studied in eastern Islam. Indeed, we have seen repeatedly that Ptolemy would have benefited from bringing the tangent out of the shadows (gnomon lengths), so to

[56] This is an easy application of the Rule of Three: as AD is to AB, so is AD (in the new units) to the radius R of the circle used in the Sine table.

[57] The case where three angles are known is not discussed; but the lengths of the sides in such a triangle may be determined only in ratio to one another. So, one could assign one side a given reference length and apply III.8 to solve the triangle.

[58] The cotangent, when defined, was the "umbra recta."

speak, and into his planetary astronomy. The tangent was hardly ignored at least within astronomical contexts in Europe at this time;[59] for instance, John of Lignères had explicitly defined both tangent and cotangent in the context of altitude problems in the canons to his astronomical tables.[60] We also find the umbra recta and umbra versa on various shadow scales on the backs of medieval European astrolabes, borrowed from Islamic counterparts.[61] But these limited appearances were not enough to promote the tangent to a higher status until Regiomontanus.

Interlude: The Marteloio in Navigation

It was around the fourteenth century that trigonometry found a new and powerful application in seafaring, one that was to keep it in business for many centuries. There was great commercial interest in traversing the seas more efficiently than simply by hugging the coast, and a number of techniques were devised to achieve this from the thirteenth century onwards. These methods seem to have been learned quickly; technological developments that enhance enterprise become ripe for adoption once merchants realize that competitors have a leg up.

Seafaring merchants were decidedly not connected to the academic social groups that would have been the primary milieu for discussion of mathematics. However, they had the advantage of traveling widely, and many of these travelers would have encountered some of the leading centers of knowledge in Spain and the Middle East. Leonardo of Pisa learned mathematics while traveling with his father, a merchants' representative. His *Liber abaci*[62] (1202) introduced decimal numeration for the benefit of commerce and made possible advances in several other areas—including, eventually, navigation.

Two of the new navigational methods that surfaced were the marine chart and the portolan.[63] The former, revolutionary in its accurate portrayal of coastlines, required the use of a few basic facts from Euclidean geometry. The latter was a written compilation of distances and directions between destinations as well as local landmarks, and was not a mathematical document. Most

[59] The sixteenth-century astrologer Lucas Gauricus attributed to the late thirteenth-century astronomer Campanus of Novara a table of tangents [Bond 1921, 323]. This claim has been repeated since then (see for instance [von Braunmühl 1900/1903, vol. 1, 101] or [Hughes in Regiomontanus (Hughes) 1967, 7]), but its legitimacy has been questioned [Benjamin/Toomer in Campanus of Novara (Benjamin/Toomer) 1971, 24] and the table has never been found ([von Braunmühl 1900/1903, vol. 1, 101] and [Zinner 1990, 93]).

[60] [Curtze 1900, 399–402].

[61] [King 2005, 247–252].

[62] For a recent translation, see [Leonardo of Pisa (Sigler) 2002].

[63] [Taylor 1960, 5–11].

techniques did not need to be sophisticated mathematically; indeed, it was an advantage if they were not, as most merchants were not mathematicians. However, one method, the ***marteloio***,[64] was essentially a trigonometric tool.

Developed in the fourteenth century and with its origins possibly as early as the thirteenth,[65] the marteloio was part of a class of techniques known as "dead reckoning," in which the ship's current position is determined solely from knowledge of earlier positions and directions, speeds, and times of travel.[66] This approach is opposed to, say, nautical astronomy, which was to make its appearance in the fifteenth century but would have had the disadvantage of being a fair-weather friend. How the marteloio came about is a matter for conjecture; the wide travels of seamen admit several possibilities—perhaps contact with scholars in Spain, Islam, or even within Europe itself.[67] Navigators of the time kept personal notebooks containing technical knowledge, mostly mathematical, required for their trade. The content in these books was probably copied from one navigator to another, which makes tracking the origins of the marteloio a murky business.

One of the best surviving copies of these notebooks was owned by Michael of Rhodes, a Venetian military and merchant navigator who traveled extensively around the Mediterranean in the early fifteenth century. His manuscript contains a typical discourse on the marteloio and we shall follow excerpts of it here.[68]

Compass directions were measured using the ***windrose***, a division of the directional circle into 32 parts; each quarter circle thus was divided into eight directions, called "quarters," corresponding to $11\frac{1}{4}^\circ$ each. These directions and the distances traveled were crucial to record, because a ship could not sail directly into the wind. Instead it had to start its approach to a windward destination by traveling at an angle almost 90° removed from the desired direction (the *alargar*; see figure 5.8), typically advancing only slightly toward the goal (the *avanzar*). Next, the ship would change its direction to bring the wind to the other side of the ship, returning it to its course somewhat closer to its destination.

[64] The origin of the term *marteloio* is controversial; it might mean "hammering," probably an idiom referring to the change of watch on a ship. See [Kelley 2000, 40, 48–50].

[65] [Richey 1992, 271]. A passage in Ramón Lull's *Arbre de Sciencia* is the source of this claim; see [Taylor 1957, 117–119], [Cortesão 1969, vol. I, 206–207], and [Campbell 1987, 441–442]. [Kelley 2000, 144–146] disagrees, arguing that Venice was the origin, or "at least the center of use," of the marteloio.

[66] See [Taylor 1950] for a summary of methods of dead reckoning.

[67] It has been speculated that the marteloio originated with Leonardo of Pisa's school [Motzo 1947, LI–LII], although evidence is lacking for any theory.

[68] The manuscript is being made available in [Long/McGee/Stahl, to appear]. See also the associated web site http://brunelleschi.imss.fi.it/michaelofrhodes/.

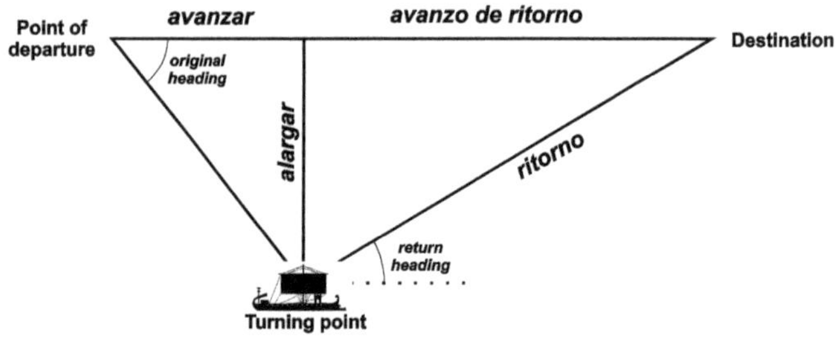

Figure 5.8
Definitions of terms in the *marteloio*

To make the proper course adjustments a small set of tables (figure 5.9), the *toleta de marteloio*, was used. It gives four quantities as functions of the windrose quarters: the *alargar* (distance off course), the *avanzar* (advance), the *ritorno* (return), and *avanzo de ritorno* (advance on return). These tables, which varied little from one to the other,[69] correspond to the modern sine, cosine, cosecant,[70] and cotangent functions respectively—but, as we shall see, we must be careful to read them appropriately.

Figure 5.8 portrays a typical situation where a ship needs to travel upwind. In the first leg of the voyage, the ship travels in a direction as close to the wind as it is able; eventually the winds shift and the ship may return to its desired course. Assuming that the first leg is 100 miles, the alargar gives the number of miles that the ship is off course, while the avanzar gives the distance that the ship has advanced when it is finally able to turn toward its destination. For a ship that is off course (alargar) by ten miles, the ritorno gives the distance along the second leg required to reach the destination, while the avanzo de ritorno gives the distance advanced toward the destination while traveling on the second leg, projected onto the original course.

Text 5.5
Michael of Rhodes, a Navigational Problem from His *Manual*

And for an example of this rule, let's say that a land is 100 miles to the east of you; how much do I want to go east southeast, which is 2 quarters, so that the land will be north by east of me, which is 7 quarters, and how far would I be from that place? And below is written the way according to this technique.

Take the distance off course that you want to remain, which is 7 quarters, which is $\frac{98}{10}$; then we will add the 2 quarters that you go and the 7 that you seek, which will be 9. So here is the return of 9 quarters, which is $10\frac{1}{5}$.

[69] See [Masiero 1984, 403], which reproduces five marteloio tables differing in only a few entries.
[70] [Kelley 2000, 42–43] argues that this may be the first practical application of the cosecant function.

Quarter	Alargar (distance off course)	Avanzar (advance)	Ritorn (return)	Avanzar de ritorno (advance on return)
1	20	98	51	50
2	38	92	26	24
3	55	83	18	15
4	71	71	14	10
5	83	55	12	6½
6	92	38	11	4
7	98	20	10½	2½
8	100	0	10	0
	For every 100 miles		For every 10 miles alargar	

Figure 5.9 Michael of Rhodes's *toleta de marteloio*

> We multiply and we say in this way $\frac{98}{10}\,\frac{51}{5}$, and multiplied together, multiplied and divided, is $99\frac{48}{50}$ miles.[71]

Explanation: In figure 5.10 we wish to travel 100 miles east from A to B, but for some reason (the prevailing winds, or perhaps an intervening obstacle) we must travel in the direction $\alpha = 2$ quarters until at C we are able to change course. We then travel in direction $\beta = 7$ quarters and reach our destination. We deal only with Michael's first problem, which is how far we must travel in the first leg; i.e., the length of AC. The pattern of calculation in Michael's text, which claims to determine AC, is

$$\frac{\text{alargar }(\beta)}{10} \cdot \text{ritorno }(\alpha + \beta); \tag{5.11}$$

in our case,

$$AC = \frac{\text{alargar }(7)}{10} \cdot \text{ritorno }(7+2) = \frac{98}{10} \cdot 10\tfrac{1}{5} = 99\tfrac{48}{50} \text{ miles.} \tag{5.12}$$

If we convert the alargar and ritorno to their modern equivalents (100 sin β and (α + β respectively), (5.11) simplifies to

[71] From the Michael of Rhodes web site (http://brunelleschi.imss.fi.it/michaelofrhodes/).

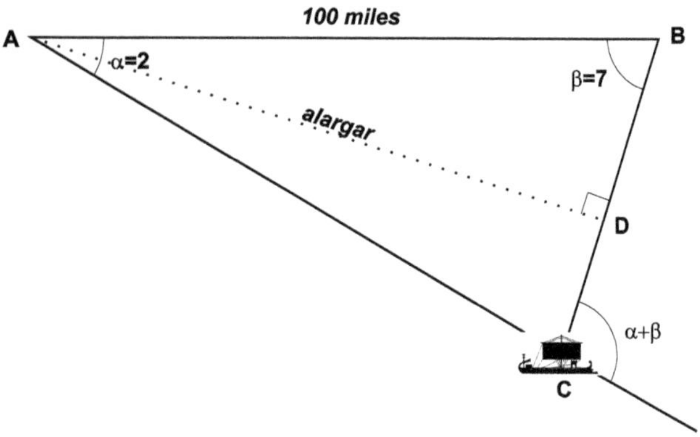

Figure 5.10
A typical *marteloio* problem

$$AC = 100\,\mathrm{Sin}\,\beta \cdot \csc(\alpha + \beta) = \frac{100 \sin \beta}{\sin(\alpha + \beta)} \tag{5.13}$$

Since the angle at C is $180° - (\alpha + \beta)$, and the sine of this quantity is the same as sin $(\alpha + \beta)$, Michael's rule is an application of the Law of Sines to the large triangle in figure 5.10.

Or is it? Although we can only speculate concerning how Michael (or the inventor of the method) thought about the marteloio, the pattern of calculation in the text makes it likely that he reasoned as follows. The length of the perpendicular from A onto BC is the alargar of β, since the hypotenuse of ΔABD is 100 miles. Now, looking at ΔACD, the ritorno of $\angle C$[72] gives the length AC, assuming that the alargar AD is 10 miles. So by the Rule of Three

$$\frac{\text{alargar}}{AC} = \frac{10}{\text{ritorno}},$$ [73]

and solving for AC gives our result.

Had the inventor of the marteloio clearly formulated this rule without the intervening calculations (that is, without breaking down the triangle into two parts) and applied it elsewhere in this streamlined form, we could be confident in asserting that he was aware of the Law of Sines, and we could turn to the question of whether he derived the Law himself or learned it from another source. However, applying the Law of Sines directly to ΔABC would not require determining AD, and in similar situations Michael always goes through this intermediate step. We are inclined to say that the author has not

[72] Whether this means $\angle BCA$ or its supplement is immaterial, since the ritorno is the same.
[73] Both sides of the equality are the ratio of the dotted line to AC, only in different units.

sufficiently codified and applied the Law of Sines to be able to assert that he knew and used it.[74]

One wonders whether or not the marteloio could have been used in practice by seamen. Navigators would have had to know only the principles of decimal arithmetic, which had been introduced to Europe in the early thirteenth century by Leonardo of Pisa in the *Liber abaci*. Perhaps some would have been up to the task; the rest of Michael of Rhodes' book makes it clear that he, at least, was capable in these matters.[75] But not every navigator was Michael of Rhodes. There were alternatives to the marteloio for seamen not as arithmetically inclined, for instance the "circle and square," a graphical instrument that served a similar navigational purpose.[76]

Nautical astronomy proper began to emerge in the fifteenth century, and began to play a significant role in the history of trigonometry late in the sixteenth century. We shall revisit this topic early in the sequel to this volume.

From Ptolemy to Triangles: John of Gmunden, Peurbach, Regiomontanus

Europe was hardly a single place in these years; in the thirteenth and fourteenth centuries knowledge did not travel well. The English and French works that we have seen would not necessarily have been on bookshelves in countries to the east and south, even as late as the fifteenth century. In this constrained environment, the building of university communities could make as much difference to the intellectual climate as any great discovery did. Three successive astronomers, John of Gmunden, George Peurbach, and his student Regiomontanus, contributed in this way to the University of Vienna (which had already started well with an infusion of mathematics from Paris), turning it into a haven of astronomical learning.[77] This is not to say that the three

[74] I am indebted to Joel Silverberg for a correspondence on this topic that led me to this view. [Kelley 2000, 43–45] assumes that the Law of Sines is present, and that its application using the alargar and ritorno (sine multiplied by cosecant, rather than dividing one alargar/sine by the other) is due to the fact that this allows the navigator to avoid division. Obviously, we disagree.

[75] In a related manuscript by Pietro de Versi (now known to be written also by Michael of Rhodes), he says: "This is the *raxion* called *del marteloyo* for navigating mentally. . . . This is done by an artful method for the man who may have the intellect to learn and who might take pleasure in learning the theory." [Kelley 2000, 150] takes this as an indication that the marteloio was more "intellectual challenge than a practical tool."

[76] [Taylor 1960, 13–14]; see especially [Kelley 2000, 57–72] for a more detailed description of the instrument.

[77] The local activities of our three astronomers at the University of Vienna have been studied recently in various articles. The teaching careers of all three are profiled in [Grössing 1983, pp. 67–141].

were not great scholars in their own right. Their roles in mathematical astronomy, culminating with Regiomontanus's systematic *De triangulis omnimodis*, took trigonometry beyond its confines as a precursor to astronomy to stand as an autonomous discipline.

The first of our trio, John of Gmunden (1380–1442),[78] was a student at Vienna and became a lecturer there in 1406. As time went on he focused more and more on mathematics and astronomy, eventually teaching these subjects exclusively. His astronomical work was devoted especially to tables and instruments and was not particularly novel in itself, but he was well acquainted with predecessors such as Levi ben Gerson, Richard of Wallingford, John of Lignères, John of Murs, and Campanus of Novara.[79] By bringing these luminaries into his work, John of Gmunden also brought their intellectual presence to the Viennese astronomical community. Relevant to us in his astronomical writings are a couple of purported applications of the spherical Law of Cosines to solve certain spherical astronomical problems, but this claim is subject to dispute.[80] As we shall see, it is likely that Peurbach deserves the credit for this result.

John of Gmunden's trigonometric treatise, *Tractatus de sinibus, chordis et arcubus*,[81] was important less for its contents (which relied heavily on predecessors) than for the calculation patterns that it set in motion. Its instructions for computing tables of sines begins by following the *Toledan Tables* in finding the sines of the *kardajas* (multiples of 15°). John continues essentially by following Ptolemy: the construction of the regular pentagon gives chords of multiples of 36°; appropriate combinations of chord addition/subtraction and half-angle formulas give chords of multiples of $1\frac{1}{2}°$; and finally Ptolemy's approximation procedure for the value of Crd 1° allows the rest of the table to be filled in.[82]

On university life in Vienna see [Uiblein 1980]; on the curriculum see [Gall 1980]; and on Regiomontanus's activities in Vienna see [Hamann 1980b]. [Shank 1997] discusses interactions between the university and the court on matters of astrology, including the involvements of John of Gmunden and Peurbach.

[78] For comprehensive descriptions of John of Gmunden's life and work see [Klug 1943], the shorter [Mundy 1943], and especially [Vogel 1973]. [Firneis 1988] (a summary of his astronomy) and [Kaiser 1988] (a summary of his mathematics) are part of the proceedings of a conference on the life, work, and culture of John of Gmunden.

[79] Indeed, John's astronomical work *Compositio et usus intrumenti quod Magister Campanus in theorica sua docuit fabricare ad inveniendum vera loca planetarum* (available in a Latin edition in [Benjamin 1954]) is mostly excerpted from Campanus's *Theorica planetarum*.

[80] The first of these instances is a formula used in determining the Sun's altitude; [Klug 1943, 50–51] says that the Law is there, but [Vogel 1973, 118] disagrees. In another place it is asserted that John was able to apply at least a special case of the Law [Firneis 1988, 72].

[81] A Latin edition of John of Gmunden's *Tractatus* is available in [Busard 1971].

[82] Around this time Nicholas of Cusa anticipated a method of computing chords, using a sort of limiting process that eventually equates small arcs with their corresponding chords; it represented

More interesting is John's idea to use new radii for the base circle. In the *Tractatus* he actually gives two sine tables both using established R values, one the Toledan $R = 150$ and the other the Ptolemaic $R = 60$.[83] However, he has also been credited with the idea of using the radius $R = 600{,}000$, to be found later in Georg Peurbach's tables.[84] This combination of decimal and sexagesimal numeration, possibly considered first by John of Lignères,[85] began a journey that would lead eventually to the decimalization of trigonometric tables in the work of Regiomontanus. The choice of a large radius, which allowed the reader to obtain accurate results without using decimal fractions, emphasizes that the modern familiarity with decimal arithmetic was not yet present.

George Peurbach (1421–1461)[86] arrived as a student at the University of Vienna not long after John of Gmunden's death in 1442. We know very little of his experiences there, but it is tempting to think that John's legacy in books and instruments would have benefited Peurbach enormously. Later, when Peurbach returned to Vienna around 1453, his astronomical reputation was well established. A series of lectures he gave the following year (which included a young Regiomontanus in the audience) eventually became his *Theoricae novae planetarum*, a popular introduction to the Ptolemaic cosmos.[87]

Peurbach was a prodigious calculator, as evidenced by his *Tabulae eclipsium* which was used to predict eclipses for over a century. His *Tractatus super propositiones Ptolemaei de sinubus et chordis*, eventually published in 1541 with Regiomontanus's sine tables, is simply an abridgment of John of Gmunden's *Tractatus* (see figure 5.11). But elsewhere, in manuscript, we have from Peurbach a monumental table of sines with $R = 600{,}000$ with arguments

a significant improvement on Ptolemaic methods, but it was not picked up until discovered two centuries later by Snell. See [Nicholas of Cusa (Hofmann) 1980, 160–177], and commentary in [von Braunmühl 1900, vol. 1, 112–113].

[83] These tables are reproduced in facsimile and recomputed in [Glowatzki/Göttsche 1990, 81–93].

[84] John often wrote each value of his calculations as a whole-numbered multiple of the last decimal place in the given quantity; this may have led to the idea.

[85] See [Bond 1921, 320] and [Firneis 1988, 81].

[86] Peurbach's precise date of birth is not certain. See [Goldmann 1923] for a discussion and justification of this date, and [Hellman/Swerdlow 1978] and [Zinner 1990, 17–30] for summaries of Peurbach's life and work.

[87] The *Theoricae novae planetarum*, an improvement on the well-known *Theorica planetarum* probably by Gerard of Cremona, was translated into English in [Aiton 1987]. Its debt to Islamic astronomy is obvious and substantial; it includes a number of Muslim innovations, including the theory of trepidation of the equinoxes. It has even been suggested that Peurbach was aware of the alternate planetary models of the Marāgha astronomers and formulated new models inspired by them [Dobrzycki/Kremer 1996]. See also [Langermann 1998] for the *Theoricae*'s influence on the Hebrew tradition in the sixteenth century.

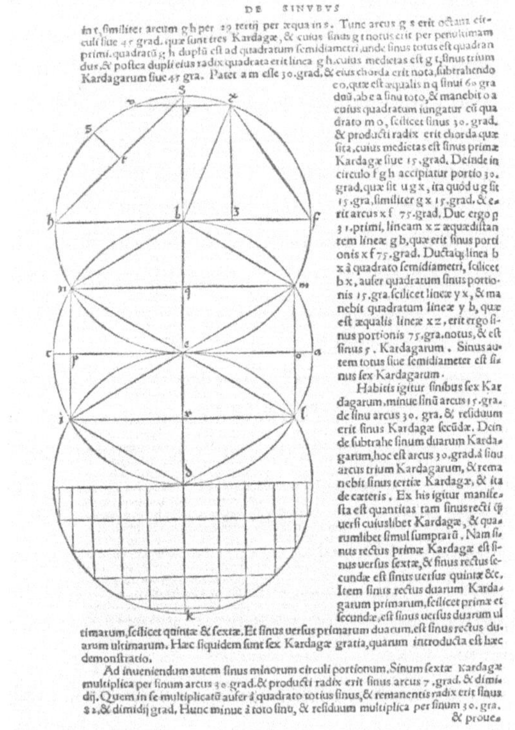

Figure 5.11
A page from Georg Peurbach's *Tractatus* showing the calculation of the *kardajas* (courtesy of the Burndy Library)

separated by only 10′ of arc.[88] Although Peurbach's table does not compare well to Ulugh Beg's, which was composed during Peurbach's lifetime (five sexagesimal places for every minute of arc; Peurbach's table corresponds to a bit more than three places), this was a major step forward for Europe. It set in motion a trend toward greater and greater accuracy that would culminate with the great tables of Rheticus and Pitiscus. Peurbach also went on to construct the first European table of tangents in 1455, which he used in conjunction with a geometric instrument to determine angular altitudes.[89] Evidently, Peurbach and friends had mastered decimal arithmetic to the point that substantial feats of calculation were now within their reach.

Peurbach's greatest achievement was the *Epitome of the Almagest*, an abridgment of Ptolemy's classic. More than a summary of its contents, the

[88] A table of sines by Peurbach with $R = 60{,}000$ is reproduced in facsimile and recomputed in [Glowatzki/Göttsche 1990, 115–125].

[89] [Zinner 1990, 26–27].

Figure 5.12
Peurbach's and Regiomontanus's *Epitome of the Almagest* (courtesy of the Burndy Library)

Epitome enhanced and critiqued the *Almagest* and so played a large role in familiarizing European scientists with the *Almagest*'s strengths and weaknesses. Indeed, the *Epitome* was to be a factor in what led Copernicus eventually to overthrow Ptolemaic astronomy. Peurbach's work on the *Epitome* was cut short by his death; his student Regiomontanus delivered on his promise to bring it to fruition, finishing it in Rome in 1462 and dedicating it to Cardinal Bessarion in 1463. It was eventually published in 1496 (see figure 5.12).

Peurbach's too-short career was extended, academically at least, by Regiomontanus's (1436–1476) equally short career.[90] Johann Müller (his actual name—Regiomontanus is derived from a Latin version of the name

[90] Among the small industry of works about Regiomontanus, [Zinner 1990] (now available in English translation) stands out as the central source. Regiomontanus's writings are available in facsimile in [Regiomontanus 1972]; a number of papers surrounding his life and work are collected in [Hamann 1980a]. One easily-overlooked source is an abridged translation of Gassendi's biography of Regiomontanus in [Regiomontanus (Hughes) 1967, 11–18].

of his home town of Königsberg) was attracted to Vienna by its reputation in astronomy, and he began teaching there as early as 1457. The completion of the *Epitome* was his first major accomplishment, although it was not published until two decades after his death. Peurbach had conceived of a book on trigonometry while streamlining the mathematics of the *Almagest* for the *Epitome*; Regiomontanus realized this goal with the *De triangulis omnimodis* shortly after the *Epitome* was completed. But Regiomontanus was firstly an astronomer; in addition to his work in theoretical astronomy he was a creative instrument, table, and calendar builder. It is impressive that, in such a short life, his interests went beyond astronomy to encompass all areas of mathematics of his time; among his other achievements, he rediscovered part of Diophantus's *Arithmetica* and was well acquainted with the latest developments in arithmetic and algebra.[91]

Regiomontanus's place on our stage is dominated by the *De triangulis omnimodis* ("On triangles of every kind"; see figure 5.13). This comprehensive treatment of triangles and their solutions, both plane and spherical, is far and away the most careful and systematic European work on triangles of its time. Although *De triangulis* is without doubt an independent work on trigonometry, Regiomontanus clearly wished to situate it within the astronomical tradition. At the outset he places *De triangulis* as a necessary antecedent to the *Epitome*:

> You, who wish to study great and wonderful things, who wonder about the movement of the stars, must read these theorems about triangles. Knowing these ideas will open the door to all of astronomy and to certain geometric problems. For although certain figures must be transformed into triangles to be solved, the remaining questions of astronomy require these books.[92]

Regiomontanus goes on to refer to his book as "the foot of the ladder to the stars."[93] Nevertheless the style has changed dramatically: when he refers back to the classics, it is not so much to the *Almagest* that he turns, as to Euclid's *Elements*. The accounts of the construction of sine and chord tables are gone, as are any specific references to astronomical problems. Instead

[91] Several surveys of Regiomontanus's mathematics exist, including [Rose 1975, 90–117], [Kaunzner 1980], [Folkerts 1977], and an update on recent research in [Folkerts 1990]. See also [Folkerts 1996] for his interaction with the Greek mathematical tradition (although not particularly with trigonometry), and [Folkerts 2002] for his familiarity with Italian mathematics. Finally, [de Siebenthal 1993, 268–352], the bulk of a chapter on medieval trigonometry, is a mathematical survey of *De triangulis*, as is [Delambre 1819, 292–323].

[92] [Regiomontanus (Hughes) 1967, 27].

[93] [Regiomontanus (Hughes) 1967, 29].

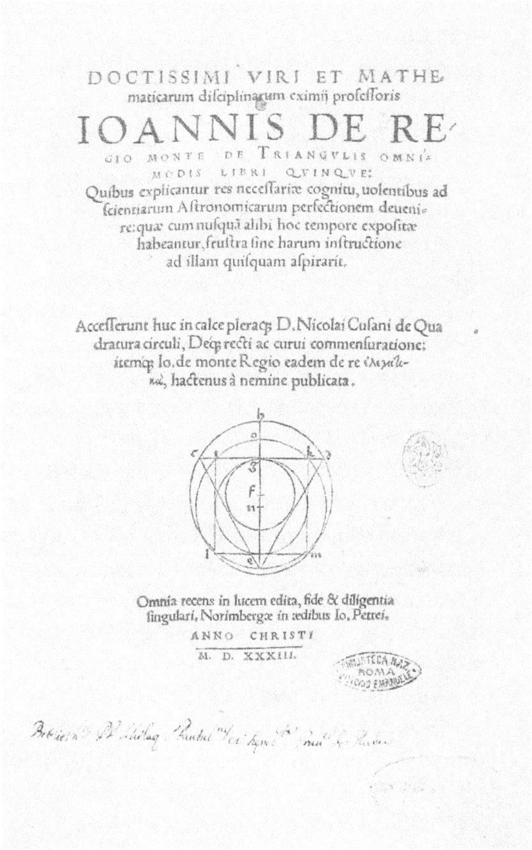

DOCTISSIMI VIRI ET MATHE-
maticarum diſciplinarum eximij profeſſoris
IOANNIS DE RE-
GIO MONTE DE TRIANGVLIS OMNI-
MODIS LIBRI QVINQVE:
Quibus explicantur res neceſſariæ cognitu, uolentibus ad ſcientiarum Aſtronomicarum perfectionem deueni-re: quæ cum nuſquã alibi hoc tempore expoſitæ habeantur, fruſtra ſine harum inſtructione ad illam quiſquam aſpirarit.

Acceſſerunt huc in calce pleraq; D. Nicolai Cuſani de Quadratura circuli, Deq; recti ac curui commenſuratione: itemq; Io. de monte Regio eadem de re ἐλεγκτικά, hactenus à nemine publicata.

Omnia recens in lucem edita, fide & diligentia ſingulari, Norimbergæ in ædibus Io. Petrei.
ANNO CHRISTI
M. D. XXXIII.

Figure 5.13
Petreius's 1533 edition of Regiomontanus's *De triangulis* (courtesy of the Burndy Library)

we find a mathematical structure familiar to us: a first chapter full of definitions and axioms, and a careful build-up of theorems depending one on another which ends up covering the solutions of all possible triangle configurations.

After the definitions and axioms, Book I begins with nineteen theorems dealing with basic calculations of quantities in the style of Euclid's *Data*: "if such-and-such is given, then such-and-such is known."[94] Often ignored because of their seeming triviality when seen algebraically, these theorems nevertheless exemplify the tension between geometry and computation that still endured in the fifteenth century. We excerpt one theorem to give the flavor.

[94] The *Data* is available in three recent translations; see [Euclid (Ito) 1980/1998], [Euclid (McDowell/Sokolik) 1993], and [Euclid (Taisbak) 2003]; the latter contains extensive commentary.

Text 5.6
Regiomontanus, Finding the Side of a Rectangle from Its Area and Another Side
(from *De triangulis,* I.17)

From any given side of a rectangular parallelogram [of] known [area], one can determine the other side.

If [the area of] rectangular parallelogram *ABCD* is known [and] any one side is also known—for example, *AB*—then its other side *AD* can be found.

Extend line *AD* to *Q* and line *CB* to *P*, until each of the lines *AQ* and *BP* is equal to the given line *AB*. Complete the square *QB* by constructing line *PQ*. Now by [Euclid] VI.1 the ratio of square *QB* to parallelogram *AC* will be as that of line *QA* to line *AD*. Therefore the ratio of square *QB* to parallelogram *AC* is given by definition, since each of these areas *QB* and *AC* is given; for indeed, by Theorem 1 above, *QB* is the square of the given line *AB*, and the hypothesis has given parallelogram *AC*. Therefore, the ratio of line *QA* to line *AD* becomes known. But *QA* and *AB* are the equal sides of square *QB*, and hence the ratio of *AB* to *AD* is necessarily known. Therefore since one of the lines—namely *AB*—is given, by Theorem 6 above the other line—namely *AD*—will be known, and thus the remaining side of the parallelogram has been determined. Q.E.D.

[Another proof follows.]

The mechanics in brief. Divide the area of the parallelogram by the given side to find the unknown side. For example, if the area of parallelogram *AC* is given as 36 square feet with side *AB* as 4 linear feet, divide 36 by 4 to get 9. The unknown side is therefore 9 feet long.[95]

Explanation: (See figure 5.14.) The argument itself requires no debriefing, except to note that each step may be justified by a proposition in the *Data*. The implication of the argument is simply that the quotient of two given quantities is known, as the last paragraph illustrates. However, since we are dealing with geometrical magnitudes, a *Data*-style argument is required to justify each apparently trivial arithmetic operation.

In I.20 Regiomontanus defines the sine function. From there he goes through a careful coverage of right-angled triangles, dealing first with situations that can be handled without sines. After these preliminaries, we reach the key theorems in I.27–30:

> **I.27:** When two sides of a right triangle are known, all the sides can be found.

[95] [Regiomontanus (Hughes) 1967, 53–55].

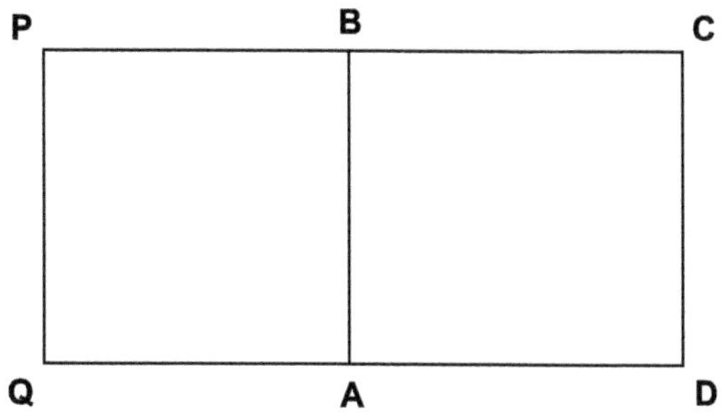

Figure 5.14
Regiomontanus's *De triangulis* I.17

I.28: When the ratio of two sides of a right triangle is given, its sides can be found (if a reference length is specified).
I.29: When one of the two acute angles and one side of a right triangle are known, all the angles and sides can be found.

Text 5.7
Regiomontanus, the Angle-angle-angle Case of Solving Right Triangles (from *De triangulis,* I.30)

If one of the acute angles of a right triangle is given, the ratios of the sides can be found although the [lengths of the] sides themselves are not known.

If right ΔABC has C as the right angle and acute $\angle B$ is given, then the ratios of any two of the sides can be found.

$\angle A$ will be found by Theorem 22 above. Therefore by the table of sines and the directions in Theorem 20 above, each of the sides AC and BC will be known as long as AB is the whole sine. Thus three known numerical values may be assigned to the three sides. Then by the definition of a known ratio, the theorem has indeed been proven.

The work is simple: Find the sine of the arc of the given angle and that of its complement, and you will have numerical values for the two sides that include the right angle. For the third side, find the sine of the quadrant. When this is done, you know that the ratio of any two sides is as the ratio of their corresponding [sine] values. For example, if $\angle B$ is 36[°], find the sine of 36°, namely 35267, and the sine of its complement, 54°, namely 48541. Furthermore, take 60000 for side AB. Therefore the ratio of AB to AC is as 60000 to 35267. And so on.[96]

Explanation: Theorem 22 asserts that knowledge of one acute angle of a right triangle implies knowledge of the other. Theorem 20 simply identifies

[96] [Regiomontanus (Hughes) 1967, 71].

the sine. The calculations rely on the sine table accompanying the *De triangulis*, which uses R = 60,000.

From I.31 to I.41 Regiomontanus deals with isosceles triangles, reducing each case to the right triangle propositions I.27–I.30. Finally he deals with arbitrary triangles in I.42–I.57, handling all the various cases of knowns and unknowns in I.47–I.53:

I.47: side-side-side given;
I.48: two angles given (only the ratios of sides are found);
I.49: side-angle-side given;
I.50: side-side-obtuse angle given;
I.51: side-side-acute angle given;
I.52: angle-side-angle given; and
I.53: angle-angle-side given.

All possible cases and ambiguities are handled. Again, Regiomontanus reduces each problem to a right triangle and appeals to one of I.27–I.30. For instance, in one of the cases of I.53 (figure 5.15) he proceeds as follows: suppose AB, $\angle B$, and $\angle C$ are known. Drop perpendicular AD; then $AD = AB \sin\angle B$ and so

$$AC = \frac{AD}{\sin\angle C} = \frac{AB \sin\angle B}{\sin\angle C}. \tag{5.14}$$

We have again the Law of Sines. But since the perpendicular AD is calculated as an intermediate stage, we are in the same position as we were in the marteloio calculations, not fully able to attribute to this theorem "Law of Sines" status.

Fortunately Regiomontanus remedies this situation at the outset of Book II with a clear statement and proof of the Law of Sines.[97] He goes on to revisit some of the problems in Book I with his new tool. However, most of Book II consists of solutions to a variety of triangles with certain quantities known: the usual angles and sides, but also lengths of perpendiculars, sums or differences of sides, and so forth. Several theorems are worth noting specially:

II.26: If the area of a triangle is given together with the rectangular product of the two sides, then either the angle opposite the base

[97] Since Levi ben Gerson's *De sinibus, chordis et arcubus* was found in Regiomontanus's papers upon his death it was speculated for a time that this may have been his source for the Law of Sines. However, it seems clear that Regiomontanus did not have Levi's work in hand when he was writing the *De triangulis omnimodis* [Zinner 1990, 57].

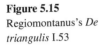

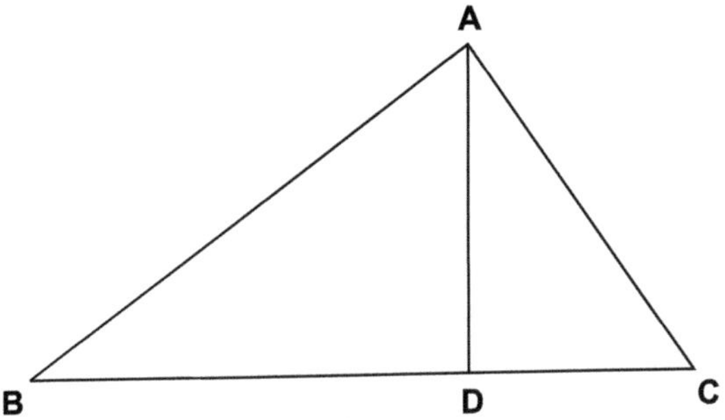

Figure 5.15
Regiomontanus's *De triangulis* I.53

> becomes known or [that angle] together with [its] known [exterior] angle equals two right angles.[98]

Although the area of the triangle is given, Regiomontanus uses the formula Area $= \frac{1}{2} AB \cdot AC \cdot \sin A$ to solve for A; this is the first appearance of the trigonometric area formula for the triangle. This and a couple of other references to the area of a triangle, inconsequential for Ptolemaic astronomy, illustrate again that trigonometry had taken a step away from astronomy into a mathematical subject on its own.

In a few cases (for example II.12, II.13, and II.23) Regiomontanus is compelled to use algebra when geometric methods fail him.[99] For example:

> **II.12:** If the perpendicular is given and the base and the ratio of the sides are known, each side can be found.

In figure 5.16 let $AB / AC = 3/5$, $AD = 5$, and $BC = 20$. If x is the distance between D and the midpoint of BC, then $BD = 10 - x$ and $DC = 10 + x$. Now $AB^2 = BD^2 + AD^2 = x^2 - 20x + 125$, while $AC^2 = DC^2 + AD^2 = x^2 + 20x + 125$. But $AB^2 / AC^2 = 9/25$. So

$$\frac{x^2 - 20x + 125}{x^2 + 20x + 125} = \frac{9}{25}.$$

Cross-multiplication yields $16x^2 + 2000 = 680x$, from which we may solve for x. Then the two segments of the base are known, and the Pythagorean Theorem gives the other two sides of the triangle.

Book III, a transition from plane to spherical geometry, builds the geometrical infrastructure needed before one can approach spherical trigonometry. The first 34 propositions deal with the geometry of great circles on spheres, while the last 22 prove various preliminary propositions about

98 [Regiomontanus (Hughes) 1967, 133].
99 The arithmetic procedures and algebra in *De triangulis* are studied in [Kaunzner 1990].

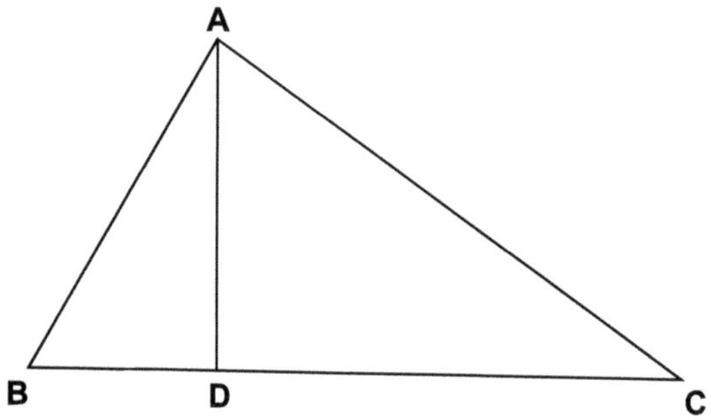

Figure 5.16
Regiomontanus's *De triangulis* II.12; also serves as the diagram for Rheticus's *De triquetris* Praeceptum Nonum

spherical triangles. It has been noted that parts of this Book resemble passages in Menelaus's *Spherics* and Jābir ibn Aflaḥ's *Iṣlāḥ al-Majisṭī*, but as with the rest of the *De triangulis*, Regiomontanus does not disclose his sources.

Roughly speaking, Books IV and V parallel for spherical triangles what Books I and II do for plane triangles. In Book IV, after some initial inequalities are proved, Regiomontanus establishes the fundamental theorems. From there he solves right spherical triangles, and concludes the book with all cases of the general spherical triangle. The unfinished Book V (only 15 propositions)[100] begins with an alternative fundamental theorem (the spherical Law of Cosines) and uses it to solve some triangles; the remaining theorems solve triangles with other sorts of given quantities (differences of sides, differences of arcs).

The real work in Book IV begins as follows:

> **IV.15:** If, in a sphere, two great circles are inclined toward each other, and if two points are marked on the circumference of one of them or one point [is marked] on the circumference of each, and if a perpendicular arc is drawn from either one of the points to the circumference of the other circle, then the ratio of the sine of the arc that is between one of those points and the point of intersection of the circles to the sine of the perpendicular arc extended from that [same marked point] to the other circle is as the ratio of the sine of the arc found between the other [marked] point and the point of intersection to the sine of the arc drawn from that [second marked] point.[101]

If we wade carefully through this verbiage (typical of statements of spherical trigonometric theorems), we find none other than the Rule of Four

[100] [Zinner 1990, 55].
[101] [Regiomontanus (Hughes) 1967, 219–221].

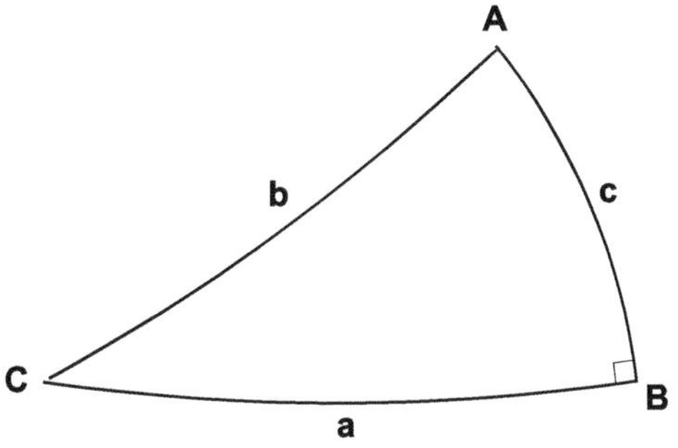

Figure 5.17
Regiomontanus's *De triangulis* IV.18

Quantities.[102] Regiomontanus respects this key result as much as the Arabs had:

> I pray that the present verbose and intricate theorem should not frighten you away at first sight, for in things mathematical you will scarcely make the language be clear enough, much less graceful. Truly you will pluck the sweetest fruit from this tree, however unyielding; and when you have savored this [fruit], you will understand almost the entire present book.[103]

This praise is somewhat overstated since Regiomontanus uses it to prove only two further results, fundamental though they are: in a triangle with right angle at B (figure 5.17),

IV.18: $\sin A \cos c = \cos C$ (Geber's Theorem); and
IV.19: $\cos b = \cos a \cos c$.

The two theorems we skipped over, IV.16 and IV.17, are the spherical Law of Sines, first for a right-angled triangle and then the general case. The Law of Sines is really Regiomontanus's foundation for his solutions of right-angled triangles in Book IV, and these solutions are in turn the bases for his solutions of general triangles. The Law of Sines would not have been new to Regiomontanus; indeed, it is found in Jābir ibn Aflaḥ's book and in the part of the *Epitome* written by Peurbach,[104] and was known in Austria as early as 1431.[105]

Our final dalliance with the *De triangulis* is V.2: in an arbitrary spherical triangle,

[102] To be precise, the case where the angle at the base is right.
[103] [Regiomontanus (Hughes) 1967, 221].
[104] The part of the *Epitome* on spherical astronomy is quite different from the corresponding chapters in the *Almagest*. It is possible that Regiomontanus played a role in the composition of this section [Zinner 1990, 53].
[105] [Zinner 1990, 55, 47].

$$\frac{\operatorname{vers} A}{\operatorname{vers} a - \operatorname{vers}(b - c)} = \frac{1}{\sin b \sin c}. \tag{5.15}$$

Replacing the versed sines with $(1 - \cos)$'s and applying the cosine difference law, we arrive at the spherical Law of Cosines

$$\cos A = \frac{\cos a - \cos b \cos c}{\sin b \sin c}, \tag{5.16}$$ [106]

the workhorse of modern spherical trigonometry. We have here the first appearance of the Law in Europe on its own: although Peurbach had given a solution to a problem involving the Sun's altitude which was effectively equivalent, he never posed the Law independently.[107] It has been speculated that Regiomontanus came up with the Law not from Peurbach, but from al-Battānī's *De motu stellarum*, which also solves an astronomical problem in a manner identical to an application of the Law of Cosines.[108]

Whether the *De triangulis* is something substantially new has been debated. Many of the theorems on plane triangles were already known, and as we have seen, authors such as Levi ben Gerson and John of Murs had approached plane triangles systematically using the sine function.[109] In fact it has been said recently that "Regiomontanus appears on the scene as a very natural and not especially remarkable successor" to many of the scholars already mentioned in this chapter.[110] Certainly there is no reason to pull *De triangulis* outside of its tradition and claim undue originality for it. But what separates the *De triangulis* from its predecessors is—as its title says—its universal coverage of all cases of triangles, plane or spherical, and its demonstrations from first principles of the most important theorems. It is remarkable in the way that Euclid's *Elements* is: not because its results were

[106] See for instance [Zeller 1944, 30–31] for a derivation.

[107] See [Klug 1943, 50] for a claim that the solution had already been known to John of Gmunden, and [Vogel 1973, 118] for a reply, firmly attributing the result to Peurbach. [Vogel 1973, 55–65] gives a Latin edition and German commentary on Peurbach's work on the solar altitude. [Zinner 1990, 23] also gives a brief account and quotation from Peurbach's discovery.

[108] See [von Braunmühl 1900/1903, vol. 1, 130–131] and [Zinner 1990, 57–58], although neither von Braunmühl [von Braunmühl 1900/1903, vol. 1, 513–54] nor Nallino, in [al-Battānī (Nallino) 1899/1903/1907, vol. 1, 185], believe that the law was really on al-Battānī's mind.

[109] [Levi ben Gerson (Goldstein) 1985, 43–47] and [John of Murs (Busard) 1998, 20–22, 128–142]; recall however that this part of Murs's treatise was composed by an anonymous earlier author. There is no direct connection between John of Murs and Regiomontanus; the proofs of the theorems are different.

[110] [North, in Richard of Wallingford (North) 1976, vol. 2, 30–31], an opinion earlier stated by [Delambre 1819, 292]. Indeed, Gerolamo Cardano accused Regiomontanus of borrowing heavily from Geber, especially the spherical trigonometry (quoted in [Zinner 1990, 190]). The extent to which this borrowing actually occurred is outlined carefully in [Lorch 1995, 10–11].

new, but its structure codified the subject for the future. Although not published until 1533, the *De triangulis* was to be the foundation of trigonometric work for centuries, and was a source of inspiration for Copernicus, Rheticus, and Brahe, among many others.[111]

There are several possible Arabic sources that Regiomontanus might have used for the *De triangulis*. It is natural to attempt to connect *De triangulis* with Naṣīr al-Dīn al-Ṭūsī's *Treatise on the Quadrilateral*, a book that systematized the study of plane and spherical triangles for eastern Islam as much as *De triangulis* did for Europe. Alas, some of the more important theorems in al-Ṭūsī's work do not appear in *De triangulis*, and the proofs of the theorems that do appear in both often differ, so if there is a link, it is not direct.[112] Rather, as the absence of the tangent function in the *De triangulis* suggests, Regiomontanus's Arabic debt seems to lie mostly in the tradition of the *Toledan Tables* and Jābir ibn Aflaḥ,[113] whose writings were still being published after Regiomontanus's death. Several Arabic antecedents have been suggested for particular theorems in *De triangulis*, but the smoking gun of transmission awaits discovery.[114]

The most remarkable omission from the *De triangulis* is the tangent. Regiomontanus began to use this function around this time, but it was not until the *Tabula directionum* (written in 1467, published 1490)[115] that it entered his scientific work. The use of the tangent in the *Tabula directionum* reveals that he was now aware of its advantages in astronomy, as opposed to its limited use in altimetry in Peurbach's writings:

> It is not in vain that this table [of tangents] is called 'fruitful', for, by resembling a tree, it is accustomed to produce great and admirable effects.[116]

As a feat of calculation, the "tabula fecunda" (figure 5.18) does not match Regiomontanus's best efforts—it gives tangents for intervals of 1°, and uses a "radius" of 100,000 (that is, the function tabulated is 100,000 tan θ). But the table is after all a first attempt, and where entries are vulnerable to numerical instability for arguments near 90°, Regiomontanus's values could be

[111] See for instance [Hughes, in Regiomontanus (Hughes) 1967, 3–4, 8–9].
[112] See the extended study [von Braunmühl 1897b]. [Folkerts 1977, 237] states that the link has not yet been decided upon, and that the issue should be revisited.
[113] [Lorch 1971, 62–69].
[114] See [Hairetdinova 1970] and [Hairetdinova 1986].
[115] [Rosińska 1981] argues that this work depends on Giovanni Bianchini's astronomical tables. See also [Rosińska 1980] for a more general survey of the spread of trigonometric tables in the fifteenth century.
[116] *Tabula directionum*, quote translated in [Zeller 1944, 34].

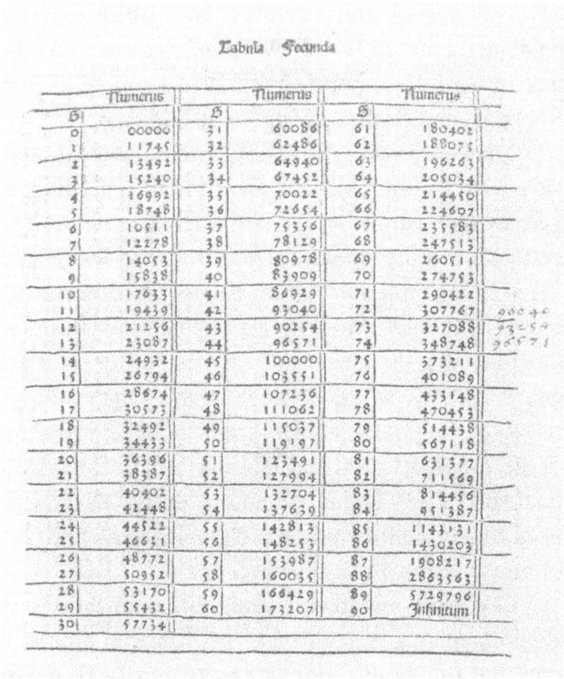

Tabula Secunda

G	Numerus	G	Numerus	G	Numerus
0	00000	31	60086	61	180402
1	11745	32	62486	62	188075
2	13492	33	64940	63	196263
3	15240	34	67452	64	205034
4	16992	35	70022	65	214450
5	18748	36	72654	66	224607
6	10511	37	75356	67	235583
7	12278	38	78129	68	247513
8	14053	39	80978	69	260511
9	15838	40	83909	70	274753
10	17633	41	86929	71	290422
11	19439	42	93040	72	307767
12	21256	43	90254	73	327088
13	23087	44	96571	74	348748
14	24932	45	100000	75	373211
15	26794	46	103551	76	401089
16	28674	47	107236	77	433148
17	30573	48	111062	78	470453
18	32492	49	115037	79	514438
19	34433	50	119197	80	567118
20	36396	51	123491	81	631377
21	38387	52	127994	82	711569
22	40402	53	132704	83	814456
23	42448	54	137639	84	951387
24	44522	55	142813	85	1143131
25	46631	56	148253	86	1430203
26	48772	57	153987	87	1908217
27	50952	58	160035	88	2863563
28	53170	59	166429	89	5729796
29	55432	60	173207	90	Infinitum
30	57734				

Figure 5.18
The first printed tangent table in Europe, in Regiomontanus's *Tabula directionum* (courtesy of the Burndy Library)

worse.[117] Strictly speaking this was not the first European tangent table—Giovanni Bianchini's *Tabulae magistrales* contained one for $R = 1000$ as early as 1463[118]—but Regiomontanus's was the first to appear in print.

Regiomontanus's choice of radius here is noteworthy, for it signals the completion of the shift from sexagesimal to decimal radii in trigonometric tables. The sine table in the *Tabula directionum*, the first one printed in Europe,[119] uses the same value $R = 60{,}000$ that he had adopted in the *De triangulis* (presumably influenced by Peurbach's earlier choice of $R = 600{,}000$). However, Regiomontanus suggests changing R to 100,000, due to the

117 A few sample entries: Tan 20° = 36396 [−1]; Tan 40° = 83909[−1]; Tan 60° = 173207 [+2]; and at the trouble spot near the end of the table, Tan 88° = 2863***563*** (correct is 2863625), and Tan 89° = 572***9796*** (correct is 5728996). See [Glowatzki/Göttsche 1990, 183] for a complete recomputation.

118 See [Rosińska 1981b] for a description of Bianchini's tables, which include a cosecant table for $R = 10{,}000$, and sine, cosine and cotangent tables for $R = 6000$ and $R = 60{,}000$. They are reproduced in facsimile and recomputed in [Glowatzki/Göttsche 1990, 92–115].

119 The occasionally repeated claim that Peter Apian's 1533 sine tables were the first to be printed in Europe (for instance [Koyré 1964]) might be a misinterpretation of statements that they were the first printed tables with a decimal radius (e.g., [Kish 1970, 179]). The first decimal tables were Bianchini's [Rosińska 1981b], but they were not printed. The tables in the *Tabula directionum* were published as early as 1490.

obvious computational advantage of a purely decimal scheme. He may have gained inspiration for this from Bianchini (with whom he corresponded[120]), whose earlier tangent and cosecant tables also adopt a decimal base.[121] Shortly thereafter Regiomontanus computed sine tables with the staggering values of $R = 6{,}000{,}000$ and $R = 10{,}000{,}000$ for every minute of arc; these were eventually published in 1541, along with Peurbach's *Tractatus*, in the *Compositio tabularum sinuum rectorum.*[122]

The *Tabula directionum* in particular must have played a role in the spread of the tangent function and of trigonometric tables in general, for its popularity rivaled that of *De triangulis*. Eleven editions were published, and copies found their way into the grateful hands of Copernicus and Kepler, among many others.[123] Especially through these two works, Regiomontanus directly influenced the works of the best scientists through the sixteenth and well into the seventeenth century.

We would be remiss if we moved on from Regiomontanus without describing, however briefly, his 1463 *Tabula primi mobilis*. This large double-argument table of the function arc sin ($\sin x \sin y$) was eventually used to solve 63 different, mostly spherical problems. It is thus an heir to the Arabic tradition of auxiliary functions. Indeed, the tabulated function is identical to the approximate solar timekeeping formula we saw in the previous chapter—although there is no evidence of transmission of the notion of auxiliary function to the West. European auxiliary tables were constructed through the early seventeenth century, for instance by Magini;[124] and Regiomontanus's table in various forms continued to be used as late as the nineteenth century.[125]

[120] The surviving Regiomontanus-Bianchini correspondence, which contains many specific trigonometric and astronomical calculations, may be found in [Curtze 1902, 185–291], and recently in [Gerl 1989].

[121] [Rosińska 1987]. On pp. 419–420 Rosińska infers that Bianchini must have had sine and cosine tables with a decimal base, but these tables have never been found.

[122] Regiomontanus gives a method of computing Sin 1° for a radius of $R = 600{,}000{,}000$, using an equivalent to Ptolemy's method for his upper bound $\left(\frac{4}{3}\sin 45'\right)$ and an Arabic method for his lower bound $\left(\frac{1}{3}\sin 1.5° + \frac{2}{3}\sin 45'\right)$, arriving, at Sin 1° = 10,471,*468* (as opposed to the correct value of 10,471,*444*). See [Glowatzki/Göttsche 1990, 27] (noting that the inequalities are mistakenly reversed). The entries in Regiomontanus's sine table for $R = 6{,}000{,}000$ rarely exceed two units in the last place [Glowatzki/Göttsche 1990, 47].

[123] [Zinner 1990, 93].

[124] See [King 2004, 188–190] for more on Regiomontanus's and Magini's tables (including similarities between the tables of Regiomontanus and Ḥabash al-Ḥāsib), as well as a description of a similar table by Adolf Heegeman dated as late as 1849.

[125] On the problems solved by the *Tabula primi mobilis* see [von Braunmühl 1900, vol. 1, 122–123]; on other auxiliary tables see [Glowatski/Göttsche 1990, 200-206].

Successors to Regiomontanus: Werner and Copernicus

Regiomontanus's most prominent successor in spherical trigonometry was the Nuremberg astronomer and geographer Johann Werner (1468–1522). His *De triangulis sphaericis*, not published until 1907,[126] nevertheless made its mark as one of the best trigonometric works of its time; it has been compared favorably to Regiomontanus's *De triangulis omnimodis*.[127] The general structure of the book, in fact, mimics the spherical parts of *De triangulis omnimodis*. The first of the four books deals with the geometrical basics of triangles on the sphere. Book II begins with a buildup to Menelaus's Theorem (II.8 and II.9), but the key theorems come next. In particular, the Rule of Four Quantities for right-angled triangles (II.11) is Werner's favorite device for solving right triangles in the second half of Book II. When Werner approaches general spherical triangles in Book III he usually relies either on the Rule of Four Quantities or on some previous triangle solution—often, the case of a right-angled triangle with two known sides and the third to be found (II.20). This case, in turn, is solved using the relation $\cos a \cos b = \cos c$ (where the right angle is at C).

Of most interest in Werner's book is Book IV. As a practicing scientist he would have been excruciatingly aware of the tediousness of multiplication, especially of figures with five or even more decimal places such as sine and cosine values. A clever way around this is suggested by the triangle solutions in Book IV, which rely on equivalents of one of the so-called ***product formulas***:

$$\sin\alpha\sin\beta = \tfrac{1}{2}\left[\cos(\alpha-\beta) - \cos(\alpha+\beta)\right]. \tag{5.17}$$

Products like the term on the left side of this equation occur all the time, especially in spherical astronomy. Replacing the product on the left with the difference on the right was to become a useful labor-saving tool, known as ***prosthaphaeresis***. It was used for several decades until, and even after, the more powerful logarithms came along and achieved a similar end.

Since Werner's work was not published the propagation of prosthaphaeresis was not immediate; in fact, the story of its development is downright murky. It first appeared decades later in 1588, in Nicolai Ursus's *Fundamen-*

[126] [Werner (Bjornbo) 1907] is a Latin edition; the companion work *De meteoroscopis* is in [Werner (Bjornbo) 1913].

[127] See for instance [Folkerts 1976, 274]. Also see [Cantor 1907/ 1913/ 1901/ 1908, vol. 2, 452–459] for a survey of Werner's mathematics, especially pp. 454–455 for topics of interest to us.

tum astronomicum, along with the corresponding rule for the product of two cosines,

$$\cos\alpha\cos\beta = \tfrac{1}{2}\left[\cos(\alpha-\beta) - \cos(\alpha+\beta)\right].^{128} \qquad (5.18)$$

However, it is clear that Tycho Brahe and his assistants were already using the method by this time. For instance, in their trigonometric manual *Triangulorum planorum et sphaericorum praxis arithmetica*,[129] the spherical Law of Cosines

$$\cos a = \cos b \cos c + \sin b \sin c \cos A \qquad (5.19)$$

is used to solve a side-angle-side triangle. But it is modified using the product formulas to

$$\begin{aligned}\cos a &= \tfrac{1}{2}[\cos(b-c) + \cos(b+c)] \\ &\quad + \tfrac{1}{2}[\cos(b-c) - \cos(b+c)]\cos A,\end{aligned} \qquad (5.20)$$

thereby removing two of the three products.[130] The possibility that their inspiration was Johann Werner himself (through Tycho's junior colleague Paul Wittich), once rejected by historians, has recently arisen again.[131] In any case, prosthaphaeresis became public around 1588 and the trials of computing products were somewhat relieved, at least in trigonometry.[132]

Our next astronomer Nicholas Copernicus (1473–1543), only five years younger than Werner, is of course famous for introducing the heliocentric model of the solar system. He was fortunate to begin his studies at the University of Cracow, then a leading academic center. He had the opportunity there to study the ancient classics such as Euclid and Ptolemy, but also newer

[128] It is sometimes stated in the secondary literature that this formula had appeared as early as the tenth century with Ibn Yūnus (originally [Delambre 1819, 112, 164]; see also [von Braunmühl 1896, 105] and [Juschkewitsch 1964, 300], among others), but this is incorrect (see [King 1972, 7, 149] or [King 1973b, 360]). The third of the product formulas is $\sin\alpha \cos\beta = \tfrac{1}{2}\left[\sin(\alpha+\beta) + \sin(\alpha-\beta)\right]$.

[129] This manual was published in facsimile in [Brahe (Studnička) 1886].

[130] Joost Bürgi devised a method to remove the third product as follows: let $\tfrac{1}{2}\left[\cos(b-c) - \cos(b+c)\right] = \cos x$ and apply the cosine product formula, arriving at $\cos a = \tfrac{1}{2}\left[\cos(b-c) + \cos(b+c)\right] + \tfrac{1}{2}\left[\cos(x-A) + \cos(x+A)\right]$. See [Dreyer 1916], especially pp. 128–129.

[131] See [Thoren 1988], which also contains a summary of the debate surrounding the beginnings of prosthaphaeresis that occurred around the end of the nineteenth century.

[132] See [von Braunmühl 1899] on the origin and especially the development of prosthaphaeresis.

European works such as the Alfonsine Tables and Sacrobosco's *De sphaera*. Although Copernicus traveled occasionally his academic life was somewhat quiet; until the very end of his life he published hardly anything. The arrival of his young colleague Georg Rheticus (1514–1574) in 1539 changed this completely: within a year Rheticus published the *Narratio prima*, an enthusiastic account of Copernicus's heliocentric theory. With Rheticus's encouragement and assistance, the *De revolutionibus orbium coelestium* was printed in the year of Copernicus's death, 1543.[133]

The story of Copernicus and his heliocentric system has been told countless times and we shall not repeat it here, preferring to concentrate on the mathematics. Rheticus had arranged for the trigonometric portion of *De revolutionibus* to be published as a separate work, the *De lateribus et angulis triangulorum*, a year earlier in 1542. The only difference between the trigonometry of the two works is that the sine table in *De lateribus* is computed to two extra digits ($R = 10{,}000{,}000$ rather than $R = 100{,}000$), and with intervals of $1'$ rather than $10'$.[134] Copernicus begins, as customary, with a description of the methods used to construct a table of sines. He follows Ptolemy almost to the letter—even referring exclusively to chords rather than sines. Indeed, although the entries in the table are clearly sines, he and Rheticus studiously avoid this modern terminology, referring to sines instead as "half-chords subtending double arcs."[135] This conservatism, ironic in a work as revolutionary as *De revolutionibus*, was not to last much longer.

Whether or not Copernicus calculated the table in *De lateribus* remains unclear; it has been suggested that Rheticus corrected some of the entries using Regiomontanus's table as a guide.[136] This reliance on existing tables seems to have been fairly common; Regiomontanus's own tables in the *Tabula directionum* seem to have borrowed Bianchini's tables in a similar manner.[137]

As one might expect, after the tables are done Copernicus deals with plane and spherical triangles. This work is also not very original, partly inspired by the *Almagest*, partly by colleagues. Whether or not Copernicus relied specifically on Regiomontanus's *De triangulis omnimodis* has been a subject of some debate.[138] Copernicus owned a copy of the *Tabula direc-*

[133] The two most recent translations of *De revolutionibus* are [Copernicus (Duncan) 1976] and [Copernicus (Rosen) 1992]; the place to go for a technical understanding of the text is [Swerdlow/Neugebauer 1984].

[134] See the detailed comparison in [Zeller 1944, 44–46].

[135] The sine table with $R = 100{,}000$ may be found in [Copernicus (Rosen) 1992, 32–39].

[136] [Rosińska 1983]. Swerdlow and Neugebauer's comparison of the table in *De revolutionibus* with Regiomontanus [Swerdlow/Neugebauer 1984, part I, 101] is inconclusive.

[137] [Rosińska 1981].

[138] For a summary of some of the early debates on this point see [Zeller 1944, 53–56].

tionum, but he seems to have worked on trigonometry independently of Regiomontanus until Rheticus brought him a copy of the 1533 edition of *De triangulis omnimodis*.[139] This may have prompted Copernicus to make certain changes to the section on spherical trigonometry,[140] but whether or not he did so, the revised work still pales in comparison to *De triangulis omnimodis*. At least one can say that Copernicus was modern enough to consider the triangle as the primitive trigonometric entity, rather than chords in a circle or Menelaus configurations.

Text 5.8
Copernicus, the Angle-angle-angle Case of Solving Triangles
(from *De revolutionibus,* I.13, Theorem 1)

If the angles of a triangle are given, the sides are given.

I say, let there be a triangle *ABC*. Circumscribe a circle around it, in accordance with Euclid, Book IV, Problem 5. Then the arcs *AB*, *BC*, and *CA* will likewise be given, according to the system in which 360° are equal to two right angles. But when the arcs are given, the sides of the triangle inscribed in the circle are also given as chords, in the Table set forth above, in units whereof the diameter is assumed to have 200,000.[141]

Explanation: (See figure 5.19.) Of course the triangle cannot be determined unless some length is chosen in advance; Copernicus sets the radius of the circumscribed circle equal to 100,000. He then uses the same technique that Ptolemy had applied when measuring arcs and corresponding angles; for instance, since $\overset{\frown}{AB} = 2\angle C$ $\overset{\frown}{AB}$ will have the same value in demidegrees that $\angle C$ has in degrees. From this, chord *AB* may be found using the table.

By now it is probably no surprise that a familiar result lurks behind this. Applying this theorem twice and combining the results in ratio, we cancel the arbitrary $R = 100{,}000$ and arrive at an equivalent to the ubiquitous Law of Sines ($AB / BC = \text{Crd}\,(2\angle C)/\text{Crd}\,(2\angle A)$).

[139] See [Zinner 1990, 183–185], also for a more comprehensive discussion of Regiomontanus's impact on Copernicus.

[140] [Rosen, in Copernicus (Rosen) 1992, 367]. This speculation, specifically that section I.13, theorems 13–15 in *De revolutionibus* are the result of his study of Regiomontanus's *De triangulis*, is questioned in [Swerdlow/Neugebauer 1984, part I, 103–104] on the grounds that the source material for these results was readily available elsewhere, and that Copernicus's account of these theorems is inferior to Regiomontanus's.

[141] [Copernicus (Rosen) 1992, 40].

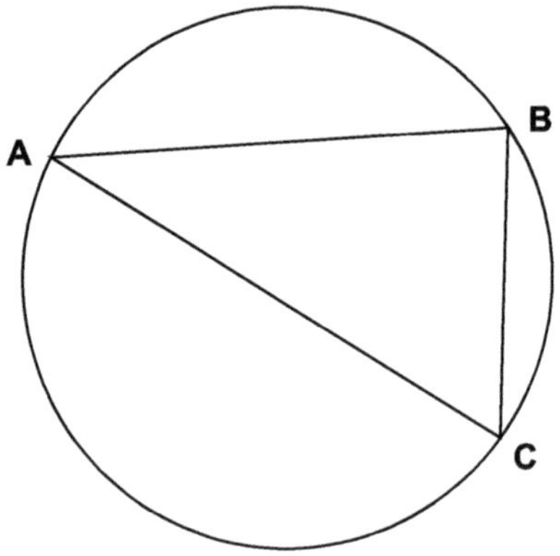

Figure 5.19
Copernicus's *De revolutionibus* I.13, Theorem I

Copernicus goes on to solve seven types of plane triangles, including the side-side-side case which he handles using Euclid's *Elements* II.13 (the Law of Cosines), although he omits the special cases that arise when two angles and a side are known.[142] Turning to spherical triangles (I.14), his fundamental result is:

> **Theorem 3:** In right spherical triangles, the ratio of the chord subtending twice the side opposite the right angle to the chord subtending twice either one of the sides including the right angle is equal to the ratio of the diameter of the sphere to the chord subtending twice the angle included, on a great circle of the sphere, between the remaining side and the hypotenuse.[143]

This is the spherical Law of Sines applied to a right triangle; it is never generalized to arbitrary triangles. Thus when Copernicus solves spherical triangles in the following propositions, he consistently drops a perpendicular and considers the resulting pair of right triangles.[144] Often in his astronomy, he is dealing with right-angled triangles in any case. For instance, when find-

[142] [Rossi 1973] gives an account of the plane trigonometry in *De revolutionibus*, including the theorems leading to the construction of a sine table.

[143] [Copernicus (Rosen) 1992, 43]; see an exposition of the proof in [Zeller 1944, 48–49].

[144] One of his results is equivalent to Geber's Theorem for right-angled triangles. Copernicus received a copy of Geber's work from Rheticus in 1539 and uses Geber's Theorem in the *De revolutionibus* to find the angle of intersection of the ecliptic with the meridian (see [Swerdlow/Neugebauer 1984, part I, 102]). However, the theorem was also available in Regiomontanus's *Epitome* and *De triangulis*, so we cannot conclude that Geber was Copernicus's source.

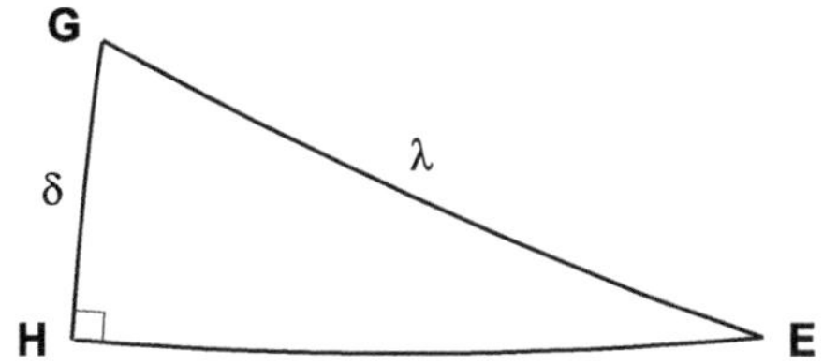

Figure 5.20
Finding the declination in the *De revolutionibus*

ing the declination δ of a given arc of the ecliptic λ (figure 5.20), he simply applies Theorem 3 and gets

$$\frac{\text{Crd}\,2\widehat{EG}}{\text{Crd}\,2\widehat{GH}} = \frac{2R}{\text{Crd}\,2\varepsilon}. \tag{5.21}$$

Since $\widehat{EG} = \lambda$ and $\widehat{GH} = \delta$, this is equivalent to the standard formula $\sin\delta = \sin\lambda \sin\varepsilon$.

As a systematic mathematical work Copernicus's spherical trigonometry is somewhat deficient; it was pointed out very early (for instance, by Pedro Nuñes and Christopher Clavius[145]) that several of his statements are in error, omitting certain possible triangle configurations. These situations were not corrected by Rheticus when editing the *De lateribus*. Since Copernicus's purpose in presenting the mathematics was primarily to serve his astronomy and the errors did not lead to adverse effects in this regard, we are inclined not to judge him too harshly or dwell on the matter any further.

For an example of Copernicus's use of the new trigonometry, we turn to his determination of the solar eccentricity from three observations of the Sun's position—the same problem we saw solved by Hipparchus and Ptolemy in chapter 2. The solution is a curious mix of the old and new, perhaps what one should expect in a work steeped in tradition yet breaking new ground. The Sun is now at the center and the Earth in orbit around it, but the methods remind us sharply of the *Almagest* solution of the same problem.

In figure 5.21 *ADBC* is the Earth's circular orbit around the Sun. From observations of the time it takes to travel from autumnal equinox *B* to 45° further on, and then on to the vernal equinox *A*, Copernicus has already found $\widehat{BC} = 44;37°$ and $\widehat{BCA} = 176;19°$. We pick up the argument from here:

[145] Some of Nuñes's and Clavius's criticisms, including discussions of an ambiguity in a plane triangle case, are translated in Rosen's notes to his translation of *De revolutionibus*, [Copernicus (Rosen) 1992, 365–367].

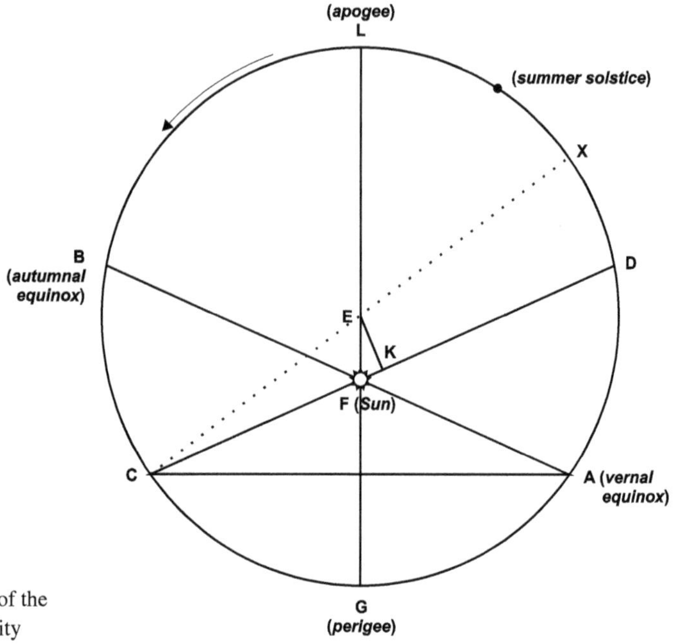

Figure 5.21
Copernicus's determination of the solar eccentricity

Text 5.9
Copernicus, Determining the Solar Eccentricity
(from *De revolutionibus,* III.16)

Therefore $\angle BAC$ is given in terms of $360^{\circ\circ} = 2$ right angles. BFC, the angle of the apparent motion, is 45° in terms of $360^\circ = 4$ right angles; but on the basis of $360^{\circ\circ} = 2$ right angles, $\angle BFC = 90^\circ$. Hence, the remainder, $\angle ACD[= \angle BFC - \angle BAC]$, which intercepts $\overset{\frown}{AD}$, is $45;23^{\circ\circ}[= 90^{\circ\circ} - 44;37^{\circ\circ}]$. But the entire portion $\overset{\frown}{ACB} = 176;19^\circ$. When $\overset{\frown}{BC}$ is subtracted [from $\overset{\frown}{ACB}$], the remainder $\overset{\frown}{AC} = 131;42^\circ[= 176;19^\circ - 44;37^\circ]$. When this figure is added to $\overset{\frown}{AD}[= 45;23^\circ]$, the sum, $\overset{\frown}{CAD}, = 177;5\frac{1}{2}{}^\circ$. Therefore, since each segment $\overset{\frown}{ACB}[= 176;19^\circ]$ and $\overset{\frown}{CAD}$ is less than a semicircle, the center is clearly contained in $\overset{\frown}{BD}$, the rest of the circle. Let the center be E, and through F draw the diameter $LEFG$. Let L be the apogee, and G the perigee. Drop EK perpendicular to CFD. Now the chords subtending the given arcs are derived from the Table: $AC = 182{,}494$, and $CFD = 199{,}934$ units, of which the diameter $= 200{,}000$. Then the angles of ΔACF are given. According to I on Plane Triangles [I,13], the ratio of the sides will also be given: $CF = 97{,}967$ of the units of which $AC = 182,\ 494$. Therefore $FD[= CFD - CF = 199{,}934 - 97{,}967 = 101{,}967]$

exceeds half [of $CFD = 199{,}934 \div 2$ or 99,967], the excess being $FK = 2{,}000$ of the same units [101,967 − 99,967]. The segment $\widehat{CAD}$ [$\cong 177;6°$] is less than a semicircle by 2;54°. Half of the chord subtending this arc is equal to EK and has 2534 units. Therefore, in ΔEFK the two sides FK and KE which form the right angle are given. Of the given sides and angles, EF will have 323 units, of which EL has 10,000; and angle EFK has $51\frac{2}{3}°$, when 360° = 4 right angles. Therefore, the whole angle AFL [$= \angle EFK + (\angle AFD = \angle BFC = 45°)$] has $96\frac{2}{3}°$[$= 51\frac{2}{3}° + 45°$], and the remainder, $\angle BFL$[$= 180° - \angle AFL$] has $83\frac{1}{3}°$. If EL has 60 units, EF will be approximately 1;56 units. This was the sun's distance from the center of the circle, having now become barely $\frac{1}{31}$, whereas to Ptolemy it seemed to be $\frac{1}{24}$. Furthermore, the apogee, which then preceded the summer solstice by $24\frac{1}{2}°$, now follows it by $6\frac{2}{3}°$.[146]

Explanation: Again we witness an application of Ptolemy's demi-degree technique: since $\angle BAC = \frac{1}{2}\widehat{BC}, \angle BAC = 44;37^{\circ\circ}$ where $1^{\circ\circ} = \frac{1}{2}°$. Likewise, once $\angle ACD$ is determined to be 45;23°°, we know that $\widehat{AD} = 45;23°$, and from this addition and subtraction of arcs leaves us with $\widehat{CAD} = 177;5\frac{1}{2}°$.[147]

Our next step is to apply Copernicus's table of sines (see figure 5.22) to find AC and CFD—although in this case a chord table would have been much easier. To find AC we divide $\widehat{AC} = 131;42°$ by 2, look up the result (65;51°) in the table, and multiply the result 91,247 (obtained by interpolation) by two to get 182,494; similarly $CFD = 199{,}934$. All of this, of course, assumes that $R = 100{,}000$.

We turn to ΔACF; since arcs $\widehat{CB}$ and $\widehat{AC}$ are known we may apply section I.13, Theorem 1 above to arrive at the ratio CF/AC; but since $AC = 182{,}494$, we have $CF = 97{,}967$. From here it is an easy matter to determine $FK = \frac{1}{2}CFD - CF = 2000$.

The remaining sides of ΔEFK are easily found. Copernicus's claim that $EK = \frac{1}{2}\text{Crd}\left(180° - \widehat{CAD}\right)$ may be verified by extending CE to a new point X on the orbit circle. Then $DX = \text{Crd}\left(180° - \widehat{CAD}\right)$; and since CEX is a diameter, $\angle D$ is right. Therefore $\Delta CEK \sim \Delta CXD$ and is exactly half its size.

146 [Copernicus (Rosen) 1992, 159].

147 Copernicus is obviously working with numbers taken to slightly more precision than is shown in the text; thus the intermediate calculations occasionally do not appear to work out perfectly.

Figure 5.22
Extracts from the sine table in Copernicus's *De revolutionibus* (interpolation column omitted)

Arcs	Half-chords subtending double arcs
0;10	291
0;20	582
0;30	873
⋮	⋮
1;0	1745
⋮	⋮
29;50	49748
30;0	50000
30;10	50252
⋮	⋮
45;0	70711
⋮	⋮
60;0	86602
⋮	⋮
65;40	91116
65;50	91235
66;0	91354
⋮	⋮
89;40	99998
89;50	99999
90;0	100000

From here e = EF is found using Pythagoras.[148] Converting to $R = 60$ Copernicus finds $e = 1;56 \approx \frac{1}{31}$, as opposed to Ptolemy's value $e = 2;29,30 \approx \frac{1}{24}$.

Working as it does with a sine table but relying on Ptolemy's techniques, Copernicus's work here occasionally appears cumbersome to modern eyes. The increasing adoption of the sine function and associated methods would remove these sorts of artifacts in the following century.

[148] The conversion to $R = 10{,}000$ to determine the position of the apogee may be a remnant of the use of an earlier sine table.

Breaking the Circle: Rheticus, Otho, Pitiscus and the *Opus Palatinum*

The name of Georg Rheticus (1514–1574) will always be associated with Copernicus, and rightly so, for it is anyone's guess how Copernicus's work would have propagated if not for the efforts of his junior colleague. But Rheticus is quite an intriguing character in his own right,[149] and he initiated major changes in how trigonometry was approached. His name refers to the region of Rhaetia, overlapping the modern nations Austria, Switzerland, and Germany; he was forced to abandon his family name of Iserin when his father, a physician, was executed for financial crimes against his patients.[150] Rheticus was coming into his own as a scholar just as scientific and mathematical books were reaching the printed press in healthy numbers. In his regular visits to eminent scientists he would present recently-printed copies of important works (including the auspicious book presentations to Copernicus); he must have seemed like a sixteenth-century technology expert, distributing his new wares to the older generation.

After Copernicus, Rheticus devoted much of his astronomical life to trigonometry. He suffered a mental illness in 1547, perhaps provoked by an unsuccessful scholarly trip to Italy, but published the pamphlet *Canon doctrinae triangulorum* just a few years later in 1551.[151] The tables in this work were revolutionary for two reasons: they related trigonometric functions directly to angles in the triangle as we do today, rather than to circular arcs; and they contained all six of the modern trigonometric functions (although under different names), given to seven decimal places. In these respects the *Canon* foreshadowed his later masterwork, the *Opus palatinum* (figure 5.23); we shall dwell on these achievements shortly.[152]

After an incident involving a purported homosexual episode with a student, Rheticus was forced to flee his academic position at Leipzig. He spent most of the rest of his life practicing medicine in Cracow, but continued working part-time on astronomical matters, especially the construction of a

[149] The standard reference work on Rheticus is [Burmeister 1967–1968]; see also the new biography [Danielson 2006].

[150] It is often said that he was executed for sorcery, but according to [Danielson 2006, 15–17] the truth is more mundane.

[151] The *Canon doctrinae triangulorum*, like any document quickly superseded by another, is exceedingly rare. But Rheticus's works are rare in any case, since they were placed on the *Index expurgatorius* [De Morgan 1845, 517]; [Archibald 1949b, 554]. See [Archibald 1953, 131] for a description of the *Canon*, and [DeMorgan 1845] for a discussion of its near disappearance.

[152] The tables were eventually extended by François Viète; see [Hunrath 1899] for a detailed comparison.

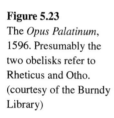

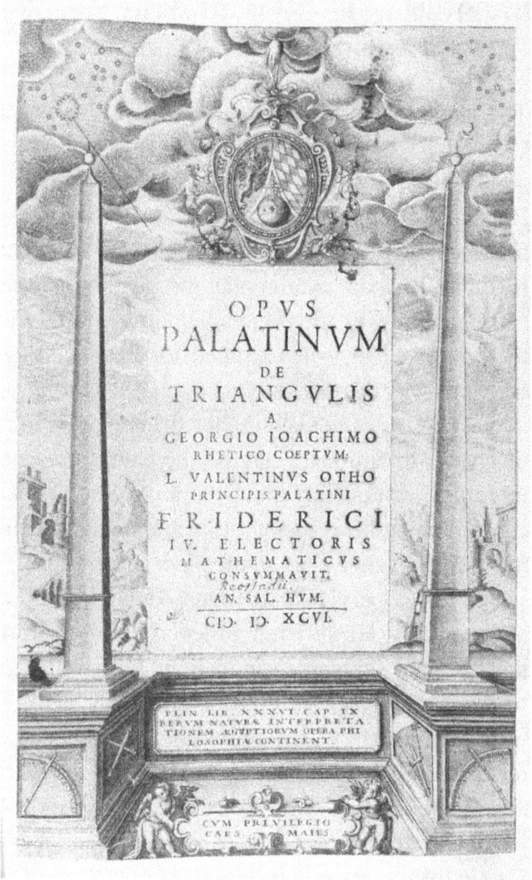

Figure 5.23
The *Opus Palatinum*, 1596. Presumably the two obelisks refer to Rheticus and Otho. (courtesy of the Burndy Library)

new, massive set of trigonometric tables. In 1574, the year of his death, an enthusiastic young student, Lucius Valentin Otho,[153] arrived on his doorstep. Recognizing the parallel to his time with Copernicus, Rheticus took Otho under his wing. Upon Rheticus's death, Otho took on the project of completing Rheticus's magnum opus—but it was not until 1596, under the sponsorship of the palatinate of Frederick IV, that the *Opus palatinum* finally saw the light of day.

No trifling work, the *Opus palatinum* (see figure 5.23) runs to over 1,400 folio-sized pages and splits roughly evenly into text and tables. The text, a comprehensive approach to every corner of trigonometry, is divided quite conventionally into four parts:

[153] On Otho (eventually a professor at the University of Heidelberg) and the *Opus palatinum*, see [Hilfstein 1986/87]. It has been suggested [Danielson 2006, 190–191] that Otho's appearance might not have been an accident.

- *De fabrica canonis doctrinae triangulorum*, a detailed description of the methods used to construct the tables;
- *De triquetris rectarum linearum in planitie*, a short section on plane triangles;
- *De triangulis globi cum angulo recto*, on right-angled spherical triangles; and
- *De triangulis globi sine angulo recto*, a 340-page tome on general spherical triangles, written by Otho.[154]

Perhaps not surprisingly, the text of *Opus palatinum* seems to have proved too overwhelming for colleagues and historians to approach; very little has been published on it.[155] On the other hand the tables were to be the foundation of the most intense calculations for centuries in many fields, such as astronomy, geography, and surveying.

We begin with the introduction to the *De triquetris*, since this is where the six trigonometric functions are defined. Rheticus divides right triangles into three "species," as follows (figure 5.24):

a) if the circle's radius (R) is the hypotenuse, then the perpendicular and the base are the sine and cosine;
b) if the circle's radius is the base, then the hypotenuse and the perpendicular are the secant and tangent; and
c) if the circle's radius is the perpendicular, then the base and the hypotenuse are the cotangent and cosecant.

But Rheticus, following Copernicus, rejects these modern names for the trigonometric functions, and instead refers to them simply as the hypotenuse, base, and perpendicular of triangles of the three species (see the sample page of tables in figure 5.25), reminiscent of the names we saw in Islamic gnomonics and astronomy in chapter 4. (Thus, although Rheticus introduces and tabulates the cosine here, the actual term "cosine" did not arise until twenty-four years later, with E. Günther.[156]) This elegant configuration is what led Rheticus to be the first European to publish a unified set of tables of all six functions. Appearances to the contrary, the circles actually play little role in his trigonometry: the arguments of the six functions are the angles in the triangles, not the circular arcs. This novelty is illustrated by the

[154] Many copies, but not the one to which I had access, contain an additional section by Otho entitled *Meteoroscopium numerorum*.

[155] Almost the only descriptions of the mathematics of the *Opus palatinum* in the secondary literature are a survey in [Delambre 1821, 1–16], and a few extracts in [Zeller 1944, 56–63].

[156] Before Günther, Pitiscus had used the term *sinus complementi*, the sine of the complement [J. Miller]. Likewise, although Rheticus tabulates the tangent and the secant, the terms themselves are due to Thomas Fincke [Archibald 1949a, 397].

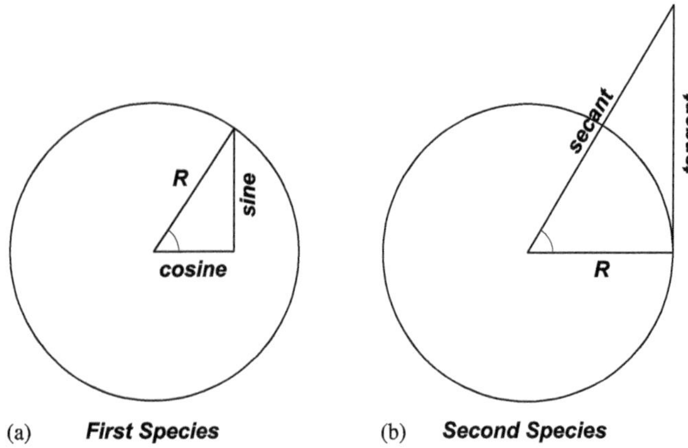

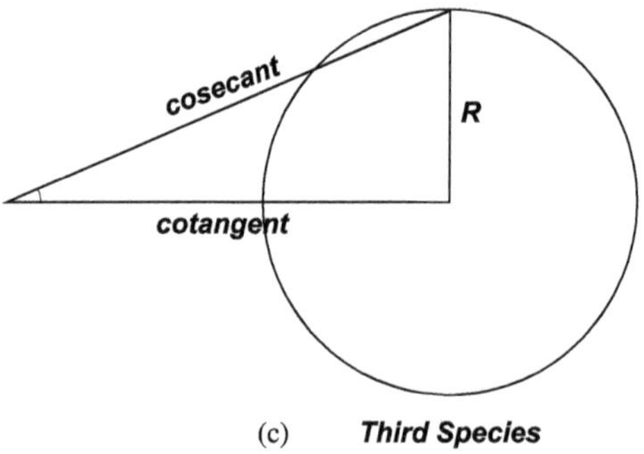

Figure 5.24
Rheticus's definitions of the six trigonometric functions. Rheticus does not use any of the modern terms written here; rather, he refers simply to the base, perpendicular, and hypotenuse of each of the three species.

lack of reference to any circle in this account of a typical result from *De triquetris:*

> **"Praeceptum Nonum,"** first case (figure 5.16): In ΔABC with obtuse angle at A: given AB, BC and $\angle B$, find the remaining side and angles.

Drop perpendicular AD onto BC. Then ΔABD and ΔADC will both be right; but in ΔABD, $\angle B$ is given. Thus all the angles in ΔABD are known, so its sides are given in ratio to each other (set some side length to $R = 10{,}000{,}000{,}000$, use the table to find the other side lengths, and set the sides in ratio to each other). Since AB is given, AD and BD may be found. But $BC = BD + DC$, so DC is given. Thus in right ΔADC two sides are given. Consider ΔADC as a triangle of the third species and set $AD = R = 10{,}000{,}000{,}000$; then we can look up the adjacent side DC in the table and read off the corresponding side AC along the same row of the table in units

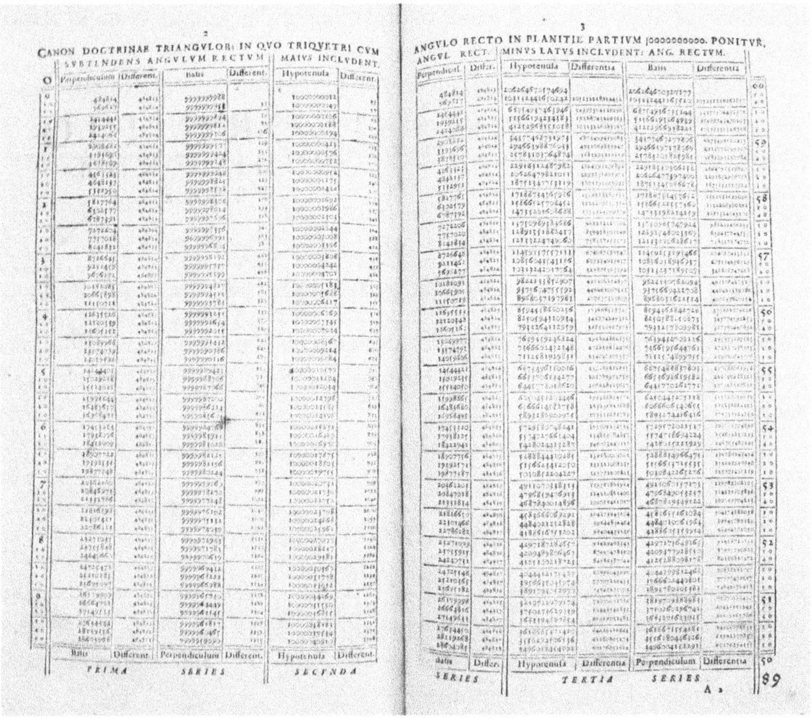

Figure 5.25 The first two of 541 pages of the trigonometric tables in Rheticus's *Opus palatinum*. If we ignore the difference columns, successive columns from left to right represent the sine, cosine, secant, tangent, cosecant, and cotangent of angles from 0 to 10′ on these pages, increasing at 10" intervals. The flawed tangent and secant entries that Romanus complained about may be read at the top right, since cotangents/cosecants of small angles correspond to tangents/secants of large angles. (courtesy of the Burndy Library)

where $AD = R$, as well as the value of $\angle C$. Since two angles in the original ΔABC are now known, the third (obtuse) $\angle BAC$ is known. Finally, since AD's actual length is known, it is a simple matter to compute the length of AC.

The *De fabrica canonis doctrinae triangulorum*, which occupies the first eighty pages of the *Opus palatinum*, describes the evaluation of all six trigonometric functions but concentrates on the sine. The initial lemmas and theorems cover the usual topics: the addition and subtraction theorems, the half-angle formula, and so on. Using the conventional Ptolemaic geometric methods but calculating with a staggering $R = 1{,}000{,}000{,}000{,}000{,}000$, Rheticus builds a table of sines and cosines with argument increments of 45′ for angles from 0° to 90°. But now he is in the usual bind: to build a table in multiples of 1° or some nice fraction of it, like the 10″ of arc that will be the increment of these tables, he needs to break out of the geometric mold somehow. He decides to calculate Sin 30″.

Rheticus begins with a gigantic project to find successively smaller sines and cosines by applying the half-angle formula repeatedly to sin (0;45°), generating the sine and cosine of 0;22,30°, then 0;11,15°, and so on for 43 iterations, until he finally reaches

$$\begin{aligned}&\sin(0;0,0,0,0,0,0,0,14,19,16,33,45,8,27,54,11,9,23,\\&\quad 49,6,5,37,30) = 0.000000000000001.\end{aligned} \quad (5.22)$$

This is the moment when his sine values reach one unit in the last (fifteenth) decimal place (see figure 5.26);[157] beyond this point all entries would read zero.

Unfortunately, the table does not get Rheticus very close to sin (30″). The seventh line gives

$$\sin(0;0,42,11,15°) = 0.000204530770292068, \quad (5.23)$$

while the eighth line dips below 30″ to 0;0,21,5,37,30°. However, Rheticus somehow notices that he can go back to an entry in the original 45′-increment sine table for aid. He takes the entry for $55\frac{1}{2}°$ and applies the half-angle formula 14 times. This results in

$$\sin(0;0,12,11,41,22,1,52,30°) = 0.000059122176177431, \quad (5.24)$$

which just happens to give the sine of an arc almost exactly 30″ less than we found in (5.23). So, we apply the sine subtraction formula to (5.23) and (5.24):

$$\sin(0;0,29,59,33,37,58,7,30°) = 0.000145408594115. \quad (5.25)$$

Rheticus's final step, transforming this value to one for sin (0;0,30°)—the heart of the entire operation, and of the tables themselves—is simply an application of a procedure that goes back at least to Levi ben Gerson, and through him, back to Ptolemy. In modern language, since the sine function is nearly linear for small angles, the ratio of two sines will be very close to the ratio of their corresponding angles:

$$\frac{\sin(0;0,29,59,33,37,58,7,30)}{\sin(0;0,30)} \approx \frac{0;0,29,59,33,37,58,7,30}{0;0,30}. \quad (5.26)$$

[157] We use the modern sine function, since the use of $R = 10^{10}$ makes the modern function hardly different from the historical one. Throughout the *De fabrica* Rheticus calculates cosines alongside the sines; we suppress references to the cosines, to save space. We have also removed some of Rheticus's checking procedures, to get to the heart of the process.

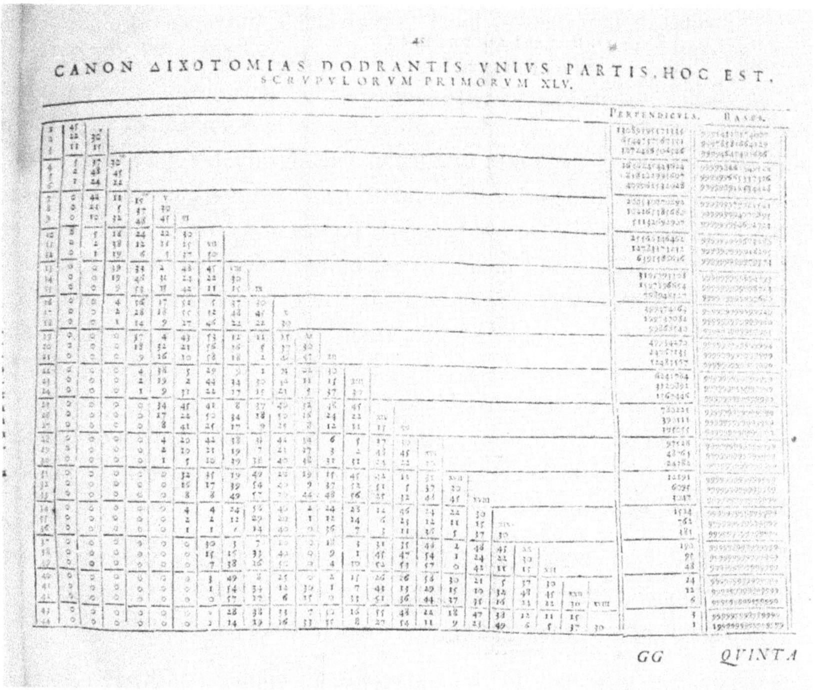

Figure 5.26 Rheticus's successive application of the sine half-angle formula, beginning with 45′ (courtesy of the Burndy Library)

This result allows Rheticus finally to reach his goal:

$$\sin(0;0,30) \approx 0.000145444102929.^{158} \qquad (5.27)$$

Now the work is downhill, if intensive. Rheticus is still in the same bind he started with—his smallest sine is of 30″, three times the smallest sine he wants (10″)—but he knows how to deal with this problem now. He applies a similar process to (5.23) through (5.27) and arrives at an estimate of sin (0;0,5°), from which the sines of 10″, 20″, 30″, 40″, and 50″ are found.

[158] The method is extremely accurate for such small arcs. Had Rheticus entered into (5.26) with correct sine values, he would have found sin (0;0,30) = 0.0001454441038203, out by only 2 in the sixteenth decimal place. However, the correct value for the sine in (5.25) is 0.000145408595***006***, and this error transfers to sin 30″, which is actually 0.0001454441***038201***. (The error appears to arise in the consecutive halving of sin $\left(55\frac{1}{2}°\right)$, leading to the value in (5.24), which is wrong in its last five digits.) Rheticus only needs 10 decimal places for his sine tables, so the error is not immediately a problem. But, as we shall see, other errors were to propagate to his tangent tables in damaging ways.

The sines of multiples of the $1'$ are available to us now, and Rheticus gives a couple of examples of how to use the sines of $10''$, $20''$, $30''$, $40''$, and $50''$ to fill in the gaps between the $1'$ increments. It is now a simple matter, requiring five (human) computers a mere twelve years of effort,[159] to fill in all 32,400 entries of the entire sine table. The entire table, including all six functions and their first-order differences for use in interpolation, comprises 388,800 entries and fills the last half (700 pages) of the *Opus palatinum*.

Rheticus's table is an amazing feat of computation, but on a mathematical level it is no more sophisticated than the calculation of the table of chords in the *Almagest*; one wonders whether Rheticus thought he might do better. Rheticus did pay a visit to the great algebraist Gerolamo Cardano in 1545, the very year that Cardano's *Ars Magna* was published containing solutions to the cubic and quartic equations. Rheticus was hopeful that he would come away from his trip with something useful, but he left empty-handed. In fact, as late as 1554 he wrote that

> from Italy I am receiving neither books nor letters. I really would have loved to see Cardano's completed work, for I was hoping it would be of some use to me in grappling with the science of triangles.[160]

It is tempting to infer that Rheticus was after the solution of the cubic equation whose root is sin 1° (see (4.11)); this would have transformed utterly the *De fabrica*. As is clear from our account, if Rheticus ever did gain anything from Cardano he did not exploit it, in the *Opus palatinum* or anywhere else.

The *Opus palatinum* finally came out around the time that several other mathematicians had been working on similar projects, so when it was finally published it received careful and detailed scrutiny. Unfortunately, the *Opus* did not measure up. The same numerical instability that had plagued other scholars' attempts to compute tangent and secant tables, namely dividing by a small cosine value, afflicted Rheticus as well. The last few entries in his tangent table (see figure 5.25), compromised by having to divide by the cosine of an angle close to 90°, are in error by as many as nine of the fifteen significant figures.

Adrianus Romanus (1561-1615), the first to notice the flaw, had suspended work on his own trigonometric tables when he had heard of the impending publication of the *Opus palatinum*.[161] One of his rather clever methods of checking the reliability of the *Opus* tables relies on the identity

[159] [Glaisher et al. 1873, 45].
[160] [Danielson 2006, 121].
[161] Our account of Romanus's critique is based on [Bockstaele 1992], which contains a Latin edition of the relevant passage.

$$\sec\alpha + \tan\alpha = \tan\left[\alpha + \tfrac{1}{2}\left(90^\circ - \alpha\right)\right]. \tag{5.28}$$

This may be verified by Romanus's figure (5.27). ΔABC is a triangle of the second species, so $BC = \tan\ \alpha$ and $AC = \sec\ \alpha$. If $CD = AC$ then $\angle CAD = \frac{1}{2}\left(90^\circ - \alpha\right)$,[162] and so $BD = \tan\left[\alpha + \frac{1}{2}\left(90^\circ - \alpha\right)\right]$.

With (5.28) in hand Romanus sums several of Rheticus's entries for sec α and tan α and compares the result to entries for the corresponding tangent; for instance,

sec 80°	56712818196
tan 80°	57587704831
Sum	1143005230***27***
tan 85°	1143005230***91***

This isn't bad, but as α gets closer to 90° matters get progressively worse, until finally

sec 89; 59, 40°	103132441650242
tan 89;59,40°	103132441165520
Sum	206264***882815762***
tan 89;59,50°	206264***670327177***,

an "inexcusable error."[163]

Romanus's disappointment with the quality of this part of the *Opus* seems to have led him to a sort of error analysis. He established a set of rules outlining how many decimal places are needed to be kept in sine values in order to compute accurate tables: in tables like in the *Opus*, generally five extra decimal places, but up to ten places in entries affecting the trouble spots. As we have seen, Rheticus had computed five extra places in his 45′-increment table, but never ten.

Rheticus had been long dead when the errors came to light; tragically, Otho was already ailing and unable to effect repairs. The work of correcting the tables fell to Bartholomew Pitiscus, court chaplain of Frederick IV and mathematician in his own right; indeed, he was the originator of the word "trigonometry."[164]

162 $\angle ACB = 90^\circ - \alpha$, so $\angle ACD = 90^\circ + \alpha$; since ΔACD is isosceles, this implies that $\angle CAD = \frac{1}{2}\left(90^\circ - \alpha\right)$.

163 Romanus, in [Bockstaele 1992, 66]. The correct values are tan 85° = 11430052302*8* and tan 89;59;50° = 2062648*06085492*.

164 The word first appears in Pitiscus's 1595 *Trigonometriae*, which we shall examine in the anticipated sequel to this book.

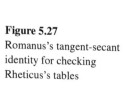

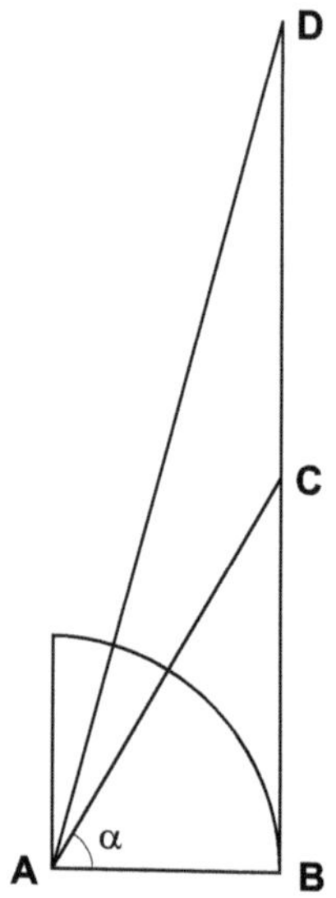

Figure 5.27
Romanus's tangent-secant identity for checking Rheticus's tables

Pitiscus came to suspect that Rheticus had computed yet another sine table to 15 places for every 10″ of arc. Otho confirmed the surmise, but in his weakened mental state he could not locate the table. After Otho's death in 1602 Pitiscus actually found it, in bad condition but still usable.

Taking Romanus's advice and with Rheticus's posthumous tabular help, Pitiscus recomputed the sines of angles up to 7° to 20 decimal places. He used these new values as a firm basis for correcting the *Opus*'s tangent and secant tables, and in 1607 republished Rheticus's tables with eighty-six pages of improved entries in *Magnus canon doctrinae triangulorum.* Six years later he released a further enhancement, *Thesaurus mathematicus.*[165] With these improvements Rheticus's tables became the authority and the computational foundation of many scientific fields for centuries. It took over three centuries before it was eventually replaced less than 100 years ago, by Andoyer's 20-place tables of 1915–18.[166]

We conclude on a curious tangent; strictly speaking it is outside of our chronology, but its connection to topics in this volume make it hard to resist here. Among the most contentious figures in philosophy and theology of his time, Giordano Bruno (1548–1600) was seen as a heretic and eventually

[165] On these two publications see [Delambre 1821, vol. 2, 17–35] and [Archibald 1949a, 394–396]. Pitiscus did not use the decimal point; Magini was the first to use it in his 1592 *De planis triangulis*. He was followed quickly by Clavius in 1593 [Koyré 1964, 48].

[166] [Archibald 1949b, 558]; [Andoyer 1915–18]. In an official British report on mathematical tables as late as 1873 [Glaisher et al. 1873, 45], Rheticus's tables were praised: "to this day [they] remain unsuperceded and the ultimate authorities."

burned at the stake—although (as with the Galileo incident) the full story is more complex. What interests us here is his opposition to trigonometry, expressed in *De triplici minimo et mensura*.[167] This work contains a sustained attack on trigonometry in general, and on tables in particular. Bruno seems to object to trigonometry because of its use of approximate methods. Indeed, as we have seen many times, the key to constructing trigonometric tables is the determination of the chord (or sine) of the smallest arc in the table, which inevitably requires a crucial approximative step.

However, Bruno's complaint has more to do with the proper application of method in what was being called a mathematical science. As opposed to practical sciences, geometry was defined by its axioms and the deductive buildup from them; the approximations involved in constructing tables fall outside of the defining characteristic of mathematics, and hence must lead to error.[168] This attitude distinctly recalls the twelfth-century astronomer al-Samaw'al's rejection of the Ptolemaic approximation of Crd 1° (see chapter 4).[169] One wonders whether Bruno would have approved of al-Samaw'al's solution (the adoption of a 480° circle in order to avoid the approximations), and so approved of trigonometry's resuscitation to the world of mathematics.

[167] This is one of three works reprinted in facsimile in [Bruno (Canone) 2000].

[168] An account of Bruno's views of mathematics may be found in [Gatti 1999, 143–170]; on trigonometry see especially pp. 151–152, 160–161.

[169] Another example of a similar attitude is al-Kūhī's polemic against approximative methods in his correspondence with al-Ṣābī; see [Berggren 1983].

Concluding Remarks

The story of trigonometry is far from over, but the nature of the subject does begin to change along with the rest of mathematics in the late sixteenth century. Through the work of François Viète and many others, symbolic algebra gave mathematicians the power to mechanize algebraic processes that had before been challenging, intensive, and often unattainable. Viète himself derived many trigonometric identities and encoded them in tabular form. And symbolic algebra triggered further transformations: first the coordinatization of geometry, and then the calculus. Oddly enough it would take some time for trigonometric functions to enter the calculus,[1] but the writing was on the wall.

At the same time the Age of Discovery was starting to place greater demands on ocean-going exploration, and navigators increasingly turned to spherical trigonometry to map their trajectories across the featureless seas. This new application of trigonometry, along with astronomy and geography, was enhanced by the invention of logarithms in the early seventeenth century. Logarithms made the calculations needed to pilot a ship or find a planet more efficient, and therefore more reliable.

It was Leonhard Euler who established the language of functions that we use so subconsciously today in applied mathematics. In the 1730s he discovered that trigonometric functions play a fundamental role in the solutions of differential equations representing harmonic oscillations. Further, through this innovation he found a startling unexpected connection between sines/cosines and exponential functions. These discoveries brought trigonometry belatedly into analysis, and also brought the hyperbolic trigonometric functions into the world. Thus trigonometry was drawn into the library of functions; it gradually became the fundamental tool for modeling periodic phenomena in the sciences.

This is not to say that trigonometry has not participated in some of the most important mathematical developments since then. It took on a significant role, for instance, in Jean-Baptiste Fourier's solution of Laplace's differential equation of heat flow in 1807. That solution introduced the notion that any function may be built of infinite sums of sines and cosines in the same way that Taylor series expand functions into infinitely long polynomials. The controversy that arose regarding the validity of Fourier's series forced deep reflection on the nature of series, functions, and continuity, and led to important developments in what are now called real analysis and set theory. Meanwhile, Fourier series became a fundamental tool in the analysis of sound waves.

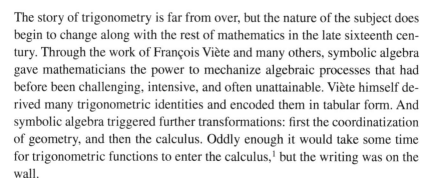

Spherical trigonometry remained a strong component of the school curriculum until the mid-1950s, motivated usually by applications in navigation and astronomy. While it is still a vital tool in these areas and programmed into our GPS systems today, hardly anyone studies it anymore. Its fate in mathematical practice was only slightly less ignominious; it was subsumed into a special case of the trigonometry of non-Euclidean geometries.

All of these episodes, and others besides, will be the subject of a sequel to this volume. In the meantime I beg the reader's patience. The richness of the story to come will make it worth the wait.

Bibliography

Aaboe, Asger. Al-Kāshī's iteration method for the determination of sin 1°, *Scripta Mathematica* **20** (1954), 24–29.

Aaboe, Asger. On the Babylonian origin of some Hipparchian parameters, *Centaurus* **4** (1955), 122–125.

Aaboe, Asger. *Episodes from the Early History of Mathematics*, Random House / The L. W. Singer Company, 1964. Reprinted Washington, DC: Mathematical Association of America, 1998.

Aaboe, Asger. On period relations in Babylonian astronomy, *Centaurus* **10** (1965), 213–231.

Aaboe, Asger. Observation and theory in Babylonian astronomy, *Centaurus* **24** (1980), 14–35.

Abraham bar Ḥiyya. Der "*Liber embadorum*" des Savasorda in der Übersetzung des Plato von Tivoli, Latin edition and German translation by Maximilian Curtze, *Abhandlungen zur Geschichte der mathematischen Wissenschaften* **12** (1902), 1–183.

Abraham bar Ḥiyya. *Chibbur ha-Meschicha weha-Tischboreth. Lehrbuch der Geometrie des Abraham bar Chija*, ed. Michael Guttmann, Berlin: Itzkowski, 1913.

Africa, Thomas W. Copernicus' relation to Aristarchus and Pythagoras, *Isis* **52** (1961), 403–409.

Ahmedov, A. A.; and Rosenfeld, Boris A. Who was the author of the *Treatise on the Determination of the Sine of One Degree*? [in Russian], *Obshchestvennye nauki v Uzbekistane* **10** (1975), 51–53.

Aiton, E. J. Peurbach's *Theoricae novae planetarum*, a translation with commentary, *Osiris* (2) **3** (1987), 5–43.

Andoyer, Henri. *Nouvelles Tables Trigonométriques Fondamentales*, 3 vols., Paris: A. Hermann, 1915–18.

Archibald, Raymond Claire. Bartholomäus Pitiscus (1561–1613), *Mathematical Tables and Other Aids to Computation* **3** (1949), 390–397.

Archibald, Raymond Claire. Rheticus, with special reference to his *Opus palatinum, Mathematical Tables and Other Aids to Computation* **3** (1949), 552–561.

Archibald, Raymond Claire. The *Canon Doctrinae Triangvlorvm* (1551) of Rheticus (1514–1576), *Mathematical Tables and Other Aids to Computation* **7** (1953), 131.

Archimedes. *The Works of Archimedes*, transl. Thomas Little Heath, Cambridge: Cambridge University Press, 1897, reprinted New York: Dover, 1953.

Āryabhaṭa. *Āryabhaṭīya*, part II, ed. K. Sāmbaśiva Śāstrī, Trivandrum: Government Press, 1931.

Āryabhaṭa. *Āryabhaṭīya of Āryabhaṭa*, eds. K. S. Shukla and K. V. Sarma, New Delhi: Indian National Science Academy, 1976.

Aujac, Germaine. Sphérique et sphéropée en Grèce ancienne, *Historia Mathematica* **3** (1976), 441–447.

Ausejo, Elena. Sobre los conocimientos trigonometricos en los *Libros del saber de astronomia* de Alfonso X el Sabio, *LLULL* **6** (1983), 5–36.

Ausejo, Elena. Trigonometría y astronomía en el *Tratado del Cuadrante Sennero* (c. 1280), *Dynamis* **4** (1984), 7–22.

Autolycus. *The Books of Autolykos. On a Moving Sphere and On Risings and Settings*, edited and translated by Frans Bruin and Alexander Vondjidis, Beirut: American University, 1971.

Autolycus. *Autolycus de Pitane: Histoire du Texte, Suivie l'Édition Critique des Traités De la Sphère en Mouvement et Des Levers et Couchers*, edited and translated by J. Mogenet, Louvain: Université de Louvain, 1950.

Ayyangar, A. A. Krishnaswami. The Hindu sine-table, *Journal of the Indian Mathematical Society* **15** (1923), 121–126.

Ayyangar, A. A. Krishnaswami. Remarks on Bhaskara's approximation to the sine of an angle, *Mathematics Student* **18** (1950), 12.

Bag, Amulya Kumar. Trigonometrical series in the *Karaṇapaddhati* and the probable date of the text, *Indian Journal of History of Science* **1** (1966), 98–106.

Bag, Amulya Kumar. Mādhava's sine and cosine series, *Indian Journal of History of Science* **11** (1976), 54–57.

Bag, Amulya Kumar. *Mathematics in Ancient and Medieval India*, Varanasi: Chaukhambha Orientalia, 1979.

Bag, Amulya Kumar. A note on the sine table in ancient India, in eds. G. Kuppuram and K. Kumudamani, *History of Science and Technology in India*, Delhi: Sundeep Prakashan, 1990, vol. 4, 247–252.

Bagheri, Mohammad. *Books I and IV of Kūshyār ibn Labbān's Jāmiʿ Zīj: An Arabic Astronomical Handbook by an Eleventh-Century Iranian Scholar*, doctoral dissertation, University of Utrecht, 2006.

al-Battānī. *Opus Astronomicum*, edition and Latin translation by C. A. Nallino, 3 vols., Milan: Pubblicazioni del Reale Osservatorio di Brera in Milano, vol. 1: 1903, vol. 2: 1907, vol. 3: 1899.

Bellosta, Hélène. Thābit ibn Qurra's treatise on the *Sector-figure, Arabic Sciences and Philosophy* **14** (2004), 145–168.

Benjamin Jr., Francis S. John of Gmunden and Campanus of Novara, *Osiris* **11** (1954), 221–246.

Berggren, J. Lennart. A comparison of four analemmas for determining the azimuth of the qibla, *Journal for the History of Arabic Science* **4** (1980), 69–80.

Berggren, J. Lennart. On al-Bīrūnī's "method of the *zījes*" for the *qibla, Proceedings of the XVIth International Congress for the History of Science* (Bucharest, 1981), 237–245.

Berggren, J. Lennart. al-Bīrūnī on plane maps of the sphere, *Journal for the History of Arabic Science* **6** (1982), 47–112.

Berggren, J. Lennart. The correspondence of Abū Sahl al-Kūhī and Abū Isḥāq al-Ṣābī, *Journal for the History of Arabic Science* **7** (1983), 39–123.

Berggren, J. Lennart. The origins of al-Bīrūnī's "method of the *zījes*" in the theory of sundials, *Centaurus* **28** (1985), 1–16.

Berggren, J. Lennart. *Episodes in the Mathematics of Medieval Islam*, New York: Springer-Verlag, 1986.

Berggren, J. Lennart. Spherical trigonometry in Kūshyār ibn Labbān's *Jāmiʿ Zīj*, in David A. King and George Saliba, eds., *From Deferent to Equant: A Volume of Studies in the History of Science in the Ancient and Medieval Near East in Honor of E. S. Kennedy*, New York: New York Academy of Sciences, 1987, 15–33.

Berggren, J. Lennart. Ptolemy's maps of Earth and the heavens: a new interpretation, *Archive for History of Exact Sciences* **43** (1991), 133–144.

Berggren, J. Lennart. The relation of Greek spherics to early Greek astronomy, in Alan C. Bowen, ed., *Science and Philosophy in Classical Greece*, New York: Garland, 1991, 227–248.

Berggren, J. Lennart. Ḥabash's analemma for representing azimuth circles on the astrolabe, *Zeitschrift für Arabisch-Islamischen Wissenschaften* **7** (1991/92), 23–30.

Berggren, J. Lennart. The transmission of Greek geometry to medieval Islam, *Cubo* **4** (2002), 1–13.

Berggren, J. Lennart; and Sidoli, Nathan. Aristarchus's *On the Sizes and Distances of the Sun and the Moon*: Greek and Arabic texts, *Archive for History of Exact Sciences* **61** (2007), 213–254.

Berggren, J. Lennart; and Thomas, Robert. *Euclid's Phaenomena: A Translation and Study of a Hellenistic Treatise in Spherical Astronomy*, New York/London: Garland, 1996.

Berggren, J. Lennart; and Van Brummelen, Glen. Abū Sahl al-Kūhī on rising times, *SCIAMVS* **2** (2001), 31–46.

Berggren, J. Lennart; and Van Brummelen, Glen. al-Samaw'al versus al-Kūhī on the depression of the horizon, *Centaurus* **45** (2003), 116–129.

Bhāskara I. *Mahābhāskarīya*, ed. and transl. K. S. Shukla, Lucknow: Lucknow University, 1960.

Bhāskara II. *The Siddhānta Śiromaṇi* (with the author's own commentary called *Vāsanābhāṣya*), ed. and transl. Bapu Deva Sastri, Benares (Varanasi): Chowkhamba Sanskrit Series Office, 1929.

Biot, J. B. The Oriental astronomer, etc., quatrième article, *Journal des Savants* **1859**, 401–418.

al-Bīrūnī, Abu al-Rayḥān Muḥammad ibn Aḥmad. *Rasā'ilu 'l-Bīrūnī*, Hyderabad, 1948.

al-Bīrūnī, Abu al-Rayḥān Muḥammad ibn Aḥmad. *al-Qānūn al-Masʿūdī*, Hyderabad, 1954.

al-Bīrūnī, Abu al-Rayḥān Muḥammad ibn Aḥmad. *Taḥdīd Nihāyāt al-Amākin li-Taṣḥiḥ Masāfāt al-Masākin*, edition by P. G. Bulgakov, *Majallat Maʿhad al-Makhṭūṭāt al-Arabiya* **8** (1962).

al-Bīrūnī, Abu al-Rayḥān Muḥammad ibn Aḥmad. The treatise on the determination of chords in a circle by means of a broken line inscribed in it [in Russian], translated by Boris A. Rosenfeld, S. A. Krasnova, and M. M. Rožanskaja, in *From the History of Science and Technology in the East, No. 3* [in Russian], Moscow: Izdat. Vostočn. Lit., (1963), 93–141.

al-Bīrūnī, Abu al-Rayḥān Muḥammad ibn Aḥmad. *The Determination of the Coordinates of Positions for the Correction of Distances Between Cities*, translated by Jamil Ali, Beirut: American University of Beirut, 1967.

al-Bīrūnī, Abu al-Rayḥān Muḥammad ibn Aḥmad. *The Exhaustive Treatise on Shadows*, 2 vols., translated by Edward S. Kennedy, Aleppo: University of Aleppo, 1976.

al-Bīrūnī, Abu al-Rayḥān Muḥammad ibn Aḥmad. *Kitāb Maqālīd ʿilm al-Hay'a. La Trigonométrie Sphérique chez les Arabes de l'Est à la Fin du Xe Siècle*, translated by Marie-Thérèse Debarnot, Damascus: Institut Français de Damas, 1985.

Björnbo, Axel Anthon. Studien über Menelaos' *Sphärik, Abhandlungen zur Geschichte der mathematischen Wissenschaften* **14** (1902), 1–154.

Björnbo, Axel Anthon. al-Chwārizmī's trigonometriske Tavler, in *Festskrift til H. G. Zeuthen*, Copenhagen: Kgl. Hofboghandel Andr. Fred. Høst & Søn, 1909, 1–17.

Björnbo, Axel Anthon. Thabits Werk über den Transversalensatz (liber de figura sectora), *Abhandlungen zue Geschichte der Naturwissenschaften und der Medizin* **7** (1924), 1–91.

Bockstaele, Paul. Adrianus Romanus and the trigonometric tables of Georg Joachim Rheticus, in Sergei S. Demidov, Menso Folkerts, David E. Rowe, and Christoph J. Scriba, eds., *Amphora: Festschrift für Hans Wussing zu seinem 65. Geburtstag*, Basel/Boston/Berlin: Birkhäuser, 1992, 55–66.

Bond, John David. The development of trigonometric methods down to the close of the XVth century, *Isis* **4** (1921), 295–323.

Bond, John David. Quadripartitum Ricardi Walynforde de Sinibus Demonstratus, *Isis* **5** (1923), 99–115.

Bond, John David. Richard Wallingford's *Quadripartitum, Isis* **5** (1923), 339–363.

Brack-Bernsen, Lis. The path of the Moon, the rising points of the Sun, and the oblique great circle on the celestial sphere, *Centaurus* **45** (2003), 16–31.

Brahe, Tycho. *Tychonis Brahe triangulorum planorum et sphaericorum praxis arithmetica. Qua maximus eorum, praesertim in astronomicis usus compendiose explicatur*, ed. F. I. Studnička, Prague: Farský, 1886; reprinted Vaduz, Liechtenstein: Sändig Reprint Verlag, 1984.

Brahmagupta. *The Khaṇḍakhādyaka: An Astronomical Treatise of Brahmagupta*, transl. Prabodh Chandra Sengupta, Calcutta: University of Calcutta, 1934.

Braunmühl, von. *See* von Braunmühl.

Bressoud, David. Was calculus invented in India? *College Mathematics Journal* **33** (2002), 2–13.

Brown, David. *Mesopotamian Planetary Astronomy-Astrology*, Leiden: Brill, 2000.

Bruno, Giordano. *Poemi Filosofici Latini*, ed. Eugenio Canone, La Spezia: Agorà Edizioni, 2000.

Bulmer-Thomas, Ivor. Theodosius of Bithynia, in Charles Coulston Gillispie, ed., *Dictionary of Scientific Biography*, vol. 13, New York: Charles Scribner's Sons, 1976, 319–321.

Burgess, Ebenezer. *Translation of the Sûrya-Siddhânta: A Text-Book of Hindu Astronomy*, originally published by Burgess and W. D. Whitney, *Journal of the American Oriental Society* **6** (1860), 141–498; reprinted by the University of Calcutta, 1935 (with different pagination).

Burmeister, Karl Heinz. *Georg Joachim Rheticus, 1514–1574. Eine Bio-Bibliographie*, 3 vols., Wiesbaden: Pressler-Verlag, 1967–1968.

Busard, H.L.L. Der Traktat *De sinibus, chordis et arcubus* von Johannes von Gmunden, *Österreichische Akademie der Wissenschaften, Mathematisch-Naturwissenschaftliche Klasse, Denkschriften* **116** (1971), 73–113.

Busard, H.L.L. The second part of Chapter 5 of the *De arte mensurandi* by Johannes de Muris, in R. S. Cohen, J. J. Stachel, and M. W. Wartofsky, eds., *For Dirk Struik*, Dordrecht/Boston: Reidel, 1974.

Campanus of Novara. *Campanus of Novara and Medieval Planetary Theory. Theorica planetarum*, Latin edition and English translation by Francis S. Benjamin, Jr. and Gerald J. Toomer, Madison/Milwaukee/London: University of Wisconsin Press, 1971.

Campbell, Tony. Portolan charts from the late thirteenth century to 1500, in J. B. Harley and D. Woodward, eds., *The History of Cartography*, Chicago: University of Chicago Press, 1987, vol. 1, 371–463.

Cantor, Moritz. *Vorlesungen über Geschichte der Mathematik*, 4 vols., Leipzig: Teubner, vol. 1: 1907; vol. 2: 1913; vol. 3: 1901; vol. 4: 1908.

Carandell, Juan. An analemma for the determination of the azimuth of the *qibla* in the *Risāla fīʿilm al-ẓilāl* of Ibn al-Raqqām, *Zeitschrift für Geschichte der Arabisch-Islamischen Wissenschaften* **1** (1984), 61–72.

Carmody, Francis J. *The Astronomical Works of Thabit b. Qurra*, Berkeley / Los Angeles: University of California Press, 1960.

Carra de Vaux, M. le Baron. L'*Almageste* d'Abû'lWéfa Albûzdjâni, *Journal Asiatique* (8) **19** (1892), 408–471.

Chabás, José; and Goldstein, B. R. *The Alfonsine Tables of Toledo*, Dordrecht/Boston/London: Kluwer, 2003.

Chace, Arnold Buffum. *The Rhind Mathematical Papyrus*, Oberlin, OH: Mathematical Association of America, 1927/1929; an abridged version reprinted Reston, VA: National Council of Teachers of Mathematics, 1978.

Chakravarty, A. K. *The Sūryasiddhānta. The Astronomical Principles of the Text*, Kolkata: The Asiatic Society, 2001.

Charette, François. A monumental medieval table for solving the problems of spherical astronomy for all latitudes, *Archives Internationales d'Histoire des Sciences* **48** (1998), 11–64.

Charette, François. *Mathematical Instrumentation in Fourteenth-Century Egypt and Syria: The Illustrated Treatise of Najm al-Dīn al-Misrī*, Leiden: Brill, 2003.

Charette, François; and Schmidl, Petra. al-Khwārizmī and practical astronomy in ninth-century Baghdad. The earliest extant corpus of texts in Arabic on the astrolabe and other portable instruments, *SCIAMVS* **5** (2004), 101–198.

Chatterjee, Bina. Geometrical interpretation of the motion of the Sun, Moon and the five planets as found in the *Mathematical Syntaxis* of Ptolemy and in the Hindu astronomical works, *Journal of the Royal Asiatic Society of Bengal, Science* **15** (1949), 41–89.

Christianides, J.; Dialetis, D.; and Gavroglu, K. Having a knack for the non-intuitive: Aristarchus's heliocentrism through Archimedes's geocentrism, *History of Science* **40** (2002), 146–168.

Clagett, Marshall. The use of the Moerbeke translations of Archimedes in the works of Johannes de Muris, *Isis* **43** (1952), 236–242.

Clagett, Marshall. Archimedes, in Charles Coulston Gillispie, ed., *Dictionary of Scientific Biography*, vol. 1, New York: Charles Scribner's Sons, 1970, 213–231.

Clagett, Marshall. *Archimedes in the Middle Ages*, volume 3: *The Fate of the Medieval Archimedes 1300–1565*, parts I and II, Philadelphia: American Philosophical Society, 1978.

Clagett, Marshall. *Ancient Egyptian Science. Volume 2: Calendars, Clocks, and Astronomy*, Philadelphia: American Philosophical Society, 1995.

Clagett, Marshall. *Ancient Egyptian Science: A Sourcebook. Volume 3: Ancient Egyptian Mathematics*, Philadelphia: American Philosophical Society, 1999.

Clark, Walter E. *The Āryabhaṭīya of Āryabhaṭa*, Chicago: University of Chicago Press, 1930.

Copernicus, Nicholas. *Copernicus: On the Revolutions of the Heavenly Spheres*, transl. A. M. Duncan, Newton Abbot/Devon: David & Charles, 1976.

Copernicus, Nicholas. *On the Revolutions*, transl. Edward Rosen, Warsaw/Cracow: Polish Scientific Publishers, 1978; reprinted Baltimore/London: Johns Hopkins University Press, 1992.

Cortesão, Armando. *History of Portuguese Cartography*, 2 vols., Coimbra: Junta de Investiações do Ultramar-Lisboa, vol. 1: 1969, vol. 2: 1971.

Crozet, Pascal. Thābit ibn Qurra et la composition des rapports, *Arabic Sciences and Philosophy* **14** (2004), 175–211.

Curtze, Maximilian. Die Abhandlungen des Levi ben Gerson über Trigonometrie und den Jacobstab, *Bibliotheca Mathematica* **12** (1898), 97–112.

Curtze, Maximilian. Urkunden zur Geschichte der Trigonometrie im christlichen Mittelalter, *Bibliotheca Mathematica* (3) **1** (1900), 321–416.

Curtze, Maximilian. Der Briefwechsel Regiomontans mit Giovanni Bianchini, Jacob von Speier und Christian Roder, in Maximilian Curtze, *Urkunden zur Geschichte der Mathematik im Mittelalter und der Renaissance*, vol. I, Leipzig: Teubner, 1902, 185–336.

Czwalina, A., transl. *Autolykos. Rotierende Kugel und Aufgang und Untergand der Gestirne. Theodosius von Tripolis. Sphaerik*, Leipzig: Akademische Verlagsgesellschaft, 1931.

Dalen, van. *See* van Dalen.

Dallal, Ahmad S. Ibn al-Haytham's universal solution for finding the direction of the *qibla* by calculation, *Arabic Sciences and Philosophy* **5** (1995), 145–193.

Danielson, Dennis. *The First Copernican: Georg Joachim Rheticus and the Rise of the Copernican Revolution*, New York: Walker Books, 2006.

Datta, Bibhutibhusan; and Singh, Avadhesh Narayan. *History of Hindu Mathematics: A Source Book*, Lahore: Motilal Banarsi Das, vol. 1: 1935, vol. 2: 1938.

Datta, Bibhutibhusan; and Singh, Avadhesh Narayan. Hindu trigonometry, revised by Kripa Shankar Shukla, *Indian Journal of History of Science* **18** (1983), 39–108.

Dauben, Joseph W., ed. *The History of Mathematics from Antiquity to the Present: A Selective Annotated Bibliography*, revised edition on CD-ROM edited by Albert C. Lewis, American Mathematical Society, 2000.

Davis, Samuel. On the astronomical computations of the Hindus, *Asiatic Researches* **2** (1799), 225–287.

Debarnot, Marie-Thérèse. Introduction du triangle polaire par Abū Naṣr b. ʿIrāq, *Journal for History of Arabic Science* **2** (1979), 126–136.

Debarnot, Marie-Thérèse. The *Zīj* of Ḥabash al-Ḥāsib: a survey of MS Istanbul Yeni Cami 784/2, in David A. King and George Saliba, eds., *From Deferent to Equant: A Volume of Studies in the History of Science in the Ancient and Medieval Near East in Honor of E. S. Kennedy*, New York: New York Academy of Sciences, 1987, 35–69.

Debarnot, Marie-Thérèse. Trigonometry, in Roshdi Rashed, ed., *Encyclopedia of the History of Arabic Science*, London/New York: Routledge, 1996, vol. 2, 495–538.

Delambre, J.B.J. *Histoire de l'Astronomie Ancienne*, 2 vols., Paris: Courcier, 1817; reprinted New York/London: Johnson Reprint Corporation, 1965.

Delambre, J.B.J. *Histoire de l'Astronomie du Moyen Âge*, Paris: Courcier, 1819.

Delambre, J.B.J. *Histoire de l'Astronomie Moderne*, 2 vols., Paris: Courcier, 1821.

De Morgan, Augustus. On the almost total disappearance of the earliest trigonometrical canon, *Philosophical Magazine* (3) **26** (1845), 517–526.

Derenzini, G. L'eliocentrismo di Aristarco da Archimede a Copernico, *Physis* **16** (1974), 289–308.

De Siebenthal, Jean. *Les Mathématiques dans l'Occident Médiéval*, Lausanne: Éditions Terre haute, 1993.

Dijksterhuis, E. J. *Archimedes*, Copenhagen: Munksgaard, 1956; reprinted Princeton: Princeton University Press, 1987.

Dilke, O. A. W. *The Roman Land Surveyors: An Introduction to the Agrimensores*, Newton Abbot: David and Charles, 1971.

Diller, Aubrey. Geographical latitudes in Eratosthenes, Hipparchus and Posidonius, *Klio* **27** (1934), 258–269.

Diocles. *On Burning Mirrors*, edited and translated by Gerald J. Toomer, Berlin/Heidelberg/New York: Springer-Verlag, 1976.

Dittrich, Arnošt. Náhrada astronomických tabulek babylonských trigonometrickými vzorci, *Časopis pro Pěstování Matematiky a Fisiky* **63** (1934), 82–96.

Dobrzycki, Jerzy; and Kremer, Richard L. Peurbach and Marāgha astronomy? The ephemerides of Johannes Angelus and their implications, *Journal for the History of Astronomy* **27** (1996), 187–237.

Dold-Samplonius, Yvonne. Archimedes: Einander berührende Kreise, *Sudhoffs Archiv* **57** (1973), 15–40.

Donnay, J.D.H. *Spherical Trigonometry after the Cesàro Method*, New York: Interscience, 1945.

Dreyer, J.L.E. On Tycho Brahe's manual of trigonometry, *The Observatory* **39** (1916), 127–131.

Duke, Dennis. Hipparchus' coordinate system, *Archive for History of Exact Sciences* **56** (2002), 427–433.

Duke, Dennis. Hipparchus' eclipse trios and early trigonometry, *Centaurus* **47** (2005), 163–177.

Duke, Dennis. The equant in India: the mathematical basis of ancient Indian planetary models, *Archive for History of Exact Sciences* **59** (2005), 563–576.

Duke, Dennis. Comment on the origin of the equant papers by Evans, Swerdlow, and Jones, *Journal for the History of Astronomy* **36** (2005), 1–6.

Erhardt, von. *See* von Erhardt.

Espenshade, Pamela H. A text on trigonometry by Levi ben Gerson (1288–1344), *The Mathematics Teacher* **60** (1967), 628–637.

Euclid. *The Thirteen Books of Euclid's Elements*, 2nd ed., transl. Thomas Little Heath, Cambridge: Cambridge University Press, 1925, reprinted New York: Dover, 1956.

Euclid. *The Medieval Latin Translation of the Data of Euclid*, transl. Shuntaro Ito, Tokyo: Tokyo University Press, 1980; Basel: Birkhäuser, 1998.

Euclid. *The Data of Euclid*, transl. George L. MacDowell and Merle A. Sokolik, Baltimore: Union Square Press, 1993.

Euclid. *Euclid's Data: The Importance of Being Given*, transl. Christian Marinus Taisbak, Copenhagen: Museum Tusculanum Press, 2003.

Evans, Gillian R. The "sub-Euclidean" geometry of the early Middle Ages, up to the mid-twelfth century, *Archive for History of Exact Sciences* **16** (1976/77), 105–118.

Evans, James. On the function and the probable origin of Ptolemy's equant, *American Journal of Physics* **52** (1984), 1080–1089.

Evans, James. *The History and Practice of Ancient Astronomy*, New York/Oxford: Oxford University Press, 1998.

Fibonacci. See Leonardo of Pisa.

Firneis, Maria G. Johannes von Gmunden—der Astronom, *Österreichische Akademie der Wissenschaften*, Phil.-Hist. Kl., Sitzungsberichte **497** (1988), 65–84.

Folkerts, Menso. Werner, Johannes, in Charles Coulston Gillispie, ed., *Dictionary of Scientific Biography*, vol. 14, New York: Charles Scribner's Sons, 1976, 272–277.

Folkerts, Menso. Regiomontanus als Mathematiker, *Centaurus* **21** (1977), 214–245.

Folkerts, Menso. New results on the mathematical activity of Regiomontanus, in Ernst Zinner, *Regiomontanus: His Life and Work*, Amsterdam/New York/Oxford/Tokyo: North-Holland, 1990, 363–372.

Folkerts, Menso. Regiomontanus' role in the transmission and transformation of Greek mathematics, in Jamil and Sally Ragep, eds., *Tradition, Transmission,*

Transformation: Proceedings of Two Conferences on Pre-Modern Science Held at the University of Oklahoma, Leiden: Brill, 1996, 89–113.

Folkerts, Menso. Regiomontanus' role in the transmission of mathematical problems, in Yvonne Dold-Samplonius, Joseph W. Dauben, Menso Folkerts, and Benno van Dalen, eds., *From China to Paris: 2000 Years Transmission of Mathematical Ideas*, Stuttgart: Franz Steiner, 2002, 411–428.

Folkerts, Menso. Leonardo Fibonacci's knowledge of Euclid's *Elements* and of other mathematical texts, *Bollettino di Storia delle Scienze Matematiche* **24** (2004), 93–113.

Fowler, David H. *The Mathematics of Plato's Academy: A New Reconstruction*, Oxford: Clarendon Press, 1987, 2nd ed. 1999.

Fowler, David H. Logistic and fractions in early Greek mathematics: a new interpretation, in Paul Benoit, Karine Chemla, and Jim Ritter, eds., *Histoire de fractions, fractions d'histoire*, Basel/Boston: Birkhäuser Verlag, 1992, 133–147.

Fowler, David H.; and Robson, Eleanor. Square root approximations in Old Babylonian mathematics: YBC 7289 in context, *Historia Mathematica* **25** (1998), 366–378.

Friberg, Jöran. Methods and traditions of Babylonian mathematics I: Plimpton 322, Pythagorean triples, and the Babylonian triangle parameter equations, *Historia Mathematica* **8** (1981), 277–318.

Gall, Franz. Die Wiener Universität zur Zeit des Regiomontanus, in Günther Hamann, ed., *Regiomontanus-Studien*, Vienna: Verlag der Österreichischen Akademie der Wissenschaften, 1980, 211–216.

Gandz, Solomon. The origin of angle-geometry, *Isis* **12** (1929), 452–481.

Gatti, Hilary. *Giordano Bruno and Renaissance Science*, Ithaca/London: Cornell University Press, 1999.

Gerbert d'Aurillac. *Oeuvres de Gerbert,* transl. A. Olleris, Paris/Clermont/Ferrand: F. Thibaud, 1867.

Gerbert d'Aurillac. *Gerberti Opera Mathematica*, ed. N. Bubnov, Berlin: Friedländer, 1899.

Gerl, Armin. *Trigonometrisch-astronomisches Rechnen kurz vor Copernicus*, Stuttgart: Steiner Verlag, 1989.

Gillings, Richard J. *Mathematics in the Time of the Pharaohs*, Cambridge, MA: MIT Press, 1972.

Gingerich, Owen. Did Copernicus owe a debt to Aristarchus? *Journal for the History of Astronomy* **16** (1985), 37–42.

Gingerich, Owen. The trouble with Ptolemy, *Isis* **93** (2002), 70–74.

Glaisher, J.W.L.; Cayley, Arthur; Stokes, G. G.; Thomson, W.; and Smith, H.J.S. Report of the committee on mathematical tables, in *Report of the British Association for the Advancement of Science*, London: Taylor and Francis, 1873, 1–175.

Glowatski, Ernst; and Göttsche, Helmut. *Die Sehnentafel des Klaudios Ptolemaios*, Munich: Oldenbourg, 1976.

Glowatski, Ernst; and Göttsche, Helmut. *Die Tafeln des Regiomontanus. Ein Jahrhundertwerk*, München: Institut für Geschichte der Naturwissenschaften, 1990.

Gold, David; and Pingree, David. A hitherto unknown Sanskrit work concerning Mādhava's derivation of the power series for sine and cosine, *Historia Scientiarum* **42** (1991), 49–65.

Goldmann, Arthur. Familienname, Geburtsort und Geburtsdatum des Astronom Georg Peuerbach, *Heimatgaue: Zeitschrift für Oberösterreichische Geschichte, Landes- und Volkeskunde* **4** (1923), 75–77.

Goldstein, Bernard R. The astronomical tables of Levi ben Gerson, *Transactions of the Connecticut Academy of Arts and Sciences* **45** (1974), 1–285.

Goldstein, Bernard R.; and Bowen, Alan C. A new view of early Greek astronomy, *Isis* **74** (1983), 330–340.

Goldstein, Bernard R. The making of astronomy in early Islam, *Nuncius* **1** (1986), 79–92.

Goldstein, Bernard R.; and Bowen, Alan C. The introduction of dated observations and precise measurement in Greek astronomy, *Archive for History of Exact Sciences* **43** (1991), 93–132.

Goldstein, Bernard R. An anonymous *zīj* in Hebrew for 1400 A.D.: a preliminary report, *Archive for History of Exact Sciences* **57** (2003), 151–171.

Grant, Edward, ed. *A Source Book in Medieval Science*, Cambridge, MA: Harvard University Press, 1974.

Grössing, Helmuth. *Humanistische Naturwissenschaft. Zur Geschichte der Wiener mathematischen Schulen des 15. und 16. Jahrhunderts*, Baden-Baden: Valentin Koerner, 1983.

Gupta, Radha Charan. Bhāskara I's approximation to sine, *Indian Journal of History of Science* **2** (1967), 121–136.

Gupta, Radha Charan. Second order interpolation in Indian mathematics up to the fifteenth century, *Indian Journal of History of Science* **4** (1969), 86–98.

Gupta, Radha Charan. Fractional parts of Āryabhaṭa's sines and certain rules found in Govindasvāmi's *Bhāṣya* on the *Mahābhāskarīya, Indian Journal of History of Science* **6** (1971), 51–59.

Gupta, Radha Charan. Early Indians on second order sine differences, *Indian Journal of History of Science* **7** (1972), 81–86.

Gupta, Radha Charan. Indian approximations to sine, cosine and versed sine, *The Mathematics Education* **6** (Section B) (1972), 59–60.

Gupta, Radha Charan. Addition and subtraction theorems for the sine and the cosine in medieval India, *Indian Journal of History of Science* **9** (1974), 164–177.

Gupta, Radha Charan. An Indian form of third order Taylor series approximation of the sine, *Historia Mathematica* **1** (1974), 287–289.

Gupta, Radha Charan. Sines and cosines of multiple arcs as given by Kamalākara, *Indian Journal of History of Science* **9** (1974), 143–150.

Gupta, Radha Charan. Sines of submultiple arcs as found in the *Siddhānta-Tattva-Viveka, Ranchi University Mathematical Journal* **5** (1974), 21–27.

Gupta, Radha Charan. Solution of the astronomical triangle as found in the *Tantrasaṃgraha* (A.D. 1500), *Indian Journal of History of Science* **9** (1974), 86–99.

Gupta, Radha Charan. Mādhava's power series computation of the sine, *Gaṇita* **27** (1976), 19–24.

Gupta, Radha Charan. Sine of eighteen degrees in India up to the eighteenth century, *Indian Journal of History of Science* **11** (1976), 1–10.

Gupta, Radha Charan. Indian values of the *sinus totus, Indian Journal of History of Science* **13** (1978), 125–143.

Gupta, Radha Charan. A mean-value-type formula for inverse interpolation of the sine, *Gaṇita* **30** (1979), 78–82.

Gupta, Radha Charan. Munīśvara's modification of Brahmagupta's rule for second order interpolation, *Indian Journal of History of Science* **14** (1979), 66–72.

Gupta, Radha Charan. The *Marīci* commentary on the *Jyotpatti, Indian Journal of History of Science* **15** (1980), 44–49.

Gupta, Radha Charan. On derivation of Bhāskara I's formula for the Sine, *Gaṇita-Bhāratī* **8** (1986), 39–41.

Gupta, Radha Charan. Mādhava's rule for finding angle between the ecliptic and the horizon and Āryabhaṭa's knowledge of it, in G. Swarup, A. K. Bag, and K. S. Shukla, eds., *History of Oriental Astronomy*, Cambridge, UK: Cambridge University Press, 1987, 197–202.

Gutas, Dimitri. *Greek Thought, Arabic Culture: The Graeco-Arabic Translation Movement in Baghdad and Early ʿAbbāsid Society (2nd–4th/8th–10th Centuries)*, London/New York: Routledge, 1998.

Gutas, Dimitri; Eickelman, D. F.; De Blois, F. C.; Sadgrove, P. C.; Afshar, Iradj; and Balim, Çiğdem. Tardjama, in *Encyclopaedia of Islam*, 2nd edition, vol. X, Leiden: Brill, 1998, 225–231.

Hairetdinova, N. G. On the oriental sources of the Regiomontanus' trigonometrical treatise, *Archives Internationales d'Histoire des Sciences* **23** (1970), 61–66.

Hairetdinova, N. G. On spherical trigonometry in the medieval Near East and in Europe, *Historia Mathematica* **13** (1986), 136–146.
[See also Khaĭretdinova, N. G.]

Hamadanizadeh, Javad. A medieval interpolation scheme for oblique ascensions, *Centaurus* **9** (1963), 257–265. Reprinted in David A. King and Mary Helen Kennedy, eds., *Studies in the Islamic Exact Sciences by E. S. Kennedy, Colleagues and Former Students*, Beirut: American University of Beirut, 1983, 526–534.

Hamadanizadeh, Javad. Interpolation schemes in *Dastūr al-munajjimīn, Centaurus* **22** (1978), 44–52.

Hamadanizadeh, Javad. The trigonometric tables of al-Kashi in his *Zīj-i Khāqānī, Historia Mathematica* **7** (1980), 38–45.

Hamadanizadeh, Javad. A second-order interpolation scheme described in the *Zīj-i Ilkhānī, Historia Mathematica* **12** (1985), 56–59.

Hamadanizadeh, Javad. A survey of medieval Islamic interpolation schemes, in David A. King and George Saliba, eds., *From Deferent to Equant: A Volume of Studies in the History of Science in the Ancient and Medieval Near East in Honor of E. S. Kennedy*, New York: New York Academy of Sciences, 1987, 143–152.

Hamann, Günther. *Regiomontanus-Studien*, Vienna: Verlag der Österreichischen Akademie der Wissenschaften, 1980.

Hamann, Günther. Regiomontanus in Wien, in Günther Hamann, ed., *Regiomontanus-Studien*, Vienna: Verlag der Österreichischen Akademie der Wissenschaften, 1980, 53–74.

Hamilton, N. T.; Swerdlow, Noel M.; and Toomer, Gerald J. The Canobic Inscription: Ptolemy's earliest work, in J. L. Berggren and Bernard Goldstein, eds., *From Ancient Omens to Statistical Mechanics: Essays on the Exact Sciences Presented to Asger Aaboe*, Copenhagen: University Library, 1987, 55–73.

Hayashi, Takao; Kusuba, T.; and Yano, Michio. The correction of the Mādhava series for the circumference of a circle, *Centaurus* **33** (1990), 149–174.

Hayashi, Takao. A note on Bhāskara I's rational approximation to sine, *Historia Scientiarum* **42** (1991), 45–48.

Hayashi, Takao. Āryabhaṭa's rule and table for sine-differences, *Historia Mathematica* **24** (1997), 396–406.

Hayashi, Takao; Kusuba, T.; and Yano, Michio. *Studies in Indian Mathematics: Series, Pi and Trigonometry* [in Japanese], Tokyo: Kōseisha Kōseikaku, 1997.

Heath, Thomas Little. *Aristarchus of Samos. The Ancient Copernicus*, Oxford: Clarendon Press, 1913; reprinted New York: Dover, 1981.

Heath, Thomas Little. *A History of Greek Mathematics*, 2 vols., Oxford: Clarendon Press, 1921; reprinted New York: Dover, 1981.

Heilbron, John. *The Sun in the Church: Cathedrals as Solar Observatories*, Cambridge, MA: Harvard University Press, 1999.

Hellman, C. Doris; and Swerdlow, Noel M. Peurbach, Georg, in Charles Coulston Gillispie, ed., *Dictionary of Scientific Biography*, vol. XV (Supplement 1), New York: Charles Scribner's Sons, 1978, 473–479.

Heron of Alexandria. *Vermessungslehre und Dioptra*, Greek edition and German translation by Hermann Schöne, vol. 3 of *Heronis Alexandrini. Opera quae Supersunt Omnia*, Leipzig: Teubner, 1903.

Herz-Fischler, Roger. *The Shape of the Great Pyramid*, Waterloo, Canada: Wilfrid Laurier University Press, 2000.

Hilfstein, Erna. Was Valentinus Otho a mathematics professor at the University of Heidelberg? *Organon* **22/23** (1986/87), 221–225.

Hogendijk, Jan P. Review of [al-Bīrūnī (Debarnot) 1985], *Mathematical Reviews* **88i:01016** (1988).

Hogendijk, Jan P. al-Khwārizmī's table of the "sine of the hours" and the underlying sine table, *Historia Scientiarum* (2) **42** (1991), 1–12.

Hogendijk, Jan P. The qibla table in the *Ashrafi Zīj*, in Anton von Gotstedter, ed., *AD RADICES. Festband zum fünfzigjährigen Bestehen des Instituts für Geschichte der Naturwissenschaften der Johann Wolfgang Goethe-Universität Frankfurt am Main*, Stuttgart: Steiner, 1994, 81–94.

Hogendijk, Jan P. Traces of the lost *Geometrical Elements* of Menelaus in two texts of al-Sijzī, *Zeitschrift für Geschichte der Arabisch-Islamischen Wissenschaften* **13** (1999–2000), 129–164.

Hogendijk, Jan P. al-Nayrīzī's mysterious determination of the azimuth of the qibla at Baghdād, *SCIAMVS* **1** (2000), 49–70.

Høyrup, Jens. Mathematics, algebra, and geometry, in David Noel Freedman, ed., *Anchor Bible Dictionary*, New York: Doubleday, 1992, vol. IV, 602–612.

Huber, Peter. Ueber den Nullpunkt der Babylonischen Ekliptik, *Centaurus* **5** (1958), 192–208.

Hugh of St. Victor. *Practical Geometry [Practica Geometriae] Attributed to Hugh of St. Victor*, transl. Frederick A. Homann, Milwaukee: Marquette University Press, 1991.

Hughes, Barnabas. The chord tables (?) of Hipparchus of Rhodes, *CSHPM Bulletin* **26** (2000), 4–5, 11.

Hultsch, Friedrich. Λήμματα εἰς τὰ σφαιρικά. Reste einer verloren geglaubten Schrift, *Neue Jahrbücher für Philologie und Paedagogik* **127** (1883), 415–420.

Hultsch, Friedrich. Winkelmessungen durch die Hipparchische Dioptra, *Abhandlung zur Geschichte der Mathematik* **9** (1899), 193–209.

Hultsch, Friedrich. Die Sehnentafeln der griechischen Astronomen, *Das Weltall* **2** (1901/1902), 49–55.

Hunrath, Karl. Des Rheticus *Canon doctrinae triangulorum* und Vieta's *Canon mathematicus, Zeitschrift für Mathematik und Physik* **44** (1899), supplement Festschrift um siebstigen Geburtstage Moritz Cantors, 213–240.

Hypsicles. *Die Aufgangszeiten der Gestirne*, edited and translated by V. De Falco and Max Krause, *Abhandlungen der Akademie der Wissenschaften in Göttingen, Philologisch-Historische Klasse* (3) **62**, 1–85, Göttingen: Vandenhoeck und Ruprecht, 1966.

Ibn al-Muthannā. *Ibn al-Muthannâ's Commentary on the Astronomical Tables of al-Khwârizmî*, Bernard R. Goldstein, transl., New Haven/London: Yale University Press, 1967.

ʿId, Yusif. An analemma construction for right and oblique ascensions, *Mathematics Teacher* **62** (1969), 669–672. Reprinted in David A. King and Mary Helen Kennedy, eds., *Studies in the Islamic Exact Sciences by E. S. Kennedy, Colleagues and Former Students*, Beirut: American University of Beirut, 1983, 495–498.

Imhausen, Annette. *Ägyptische Algorithmen*, Wiesbaden: Harrassowitz, 2003.

Imhausen, Annette. Egyptian mathematical texts and their contexts, *Science in Context* **16** (2003), 367–389.

Inamdar, M. G. A formula of Bhaskara for the chord of a circle leading to a formula for evaluating sin α°, *Mathematics Student* **18** (1950), 9–11.

Irani, Rida A. K. *The "Jadwal al-Taqwīm" of Ḥabash al-Ḥāsib*, master's thesis, American University of Beirut, 1956.

Jensen, Claus. Abū Naṣr Manṣūr's approach to spherical astronomy as developed in his treatise "The Table of Minutes," *Centaurus* **16** (1972), 1–19.

John of Muris. *Johannes de Muris, De arte mensurandi*, edited by H.L.L. Busard, Stuttgart: Franz Steiner Verlag, 1998.

Jones, Alexander. Hipparchus's computations of solar longitudes, *Journal for the History of Astronomy* **22** (1991), 101–125.

Jones, Alexander. The adaptation of Babylonian methods in Greek numerical astronomy, *Isis* **82** (1991), 440–453.

Jones, Alexander. Peripatetic and Euclidean theories of the visual ray, *Physis* **31** (1994), 47–76.

Jones, Alexander. On Babylonian astronomy and its Greek metamorphoses, in Jamil and Sally Ragep, eds., *Tradition, Transmission, Transformation: Proceedings of Two Conferences on Pre-Modern Science Held at the University of Oklahoma*, Leiden: Brill, 1996, 139–155.

Jones, Alexander. Studies in the astronomy of the Roman period. II. Tables for solar longitude, *Centaurus* **39** (1997), 211–229.

Jones, Alexander. *Astronomical Papyri from Oxyrhynchus*, 2 vols., Philadelphia: American Philosophical Society, 1999.

Jones, Alexander. Eratosthenes, Hipparchus, and the obliquity of the ecliptic, *Journal for the History of Astronomy* **33** (2002), 15–19.

Jones, Alexander. A route to the ancient discovery of nonuniform planetary motion, *Journal for the History of Astronomy* **35** (2004), 375–386.

Jones, Alexander. The Keskintos astronomical inscription: text and interpretations, *SCIAMVS* **7** (2006), 3–41.

Juschkewitsch, A. P. *Geschichte der Mathematik im Mittelalter*, transl. V. Ziegler, Basel: Pfalz, 1964.

Kaiser, Hans K. Johannes von Gmunden und seine mathematischen Leistungen, *Österreichische Akademie der Wissenschaften*, Phil.-Hist. Kl., Sitzungsberichte **497** (1988), 85–100.

Kak, Subhash. Birth and early development of Indian astronomy, in Helaine Selin, ed., *Astronomy Across Cultures: The History of Non-Western Astronomy*, Dordrecht/Boston/London: Kluwer, 2000, 303–340.

Karpinski, Louis C. The place of trigonometry in the development of mathematical ideas, *Scripta Mathematica* **11** (1945), 268–272.

Katz, Victor J. The calculus of the trigonometric functions, *Historia Mathematica* **14** (1987), 311–324.

Katz, Victor J. Ideas of calculus in Islam and India, *Mathematics Magazine* **68** (1995), 163–174.

Katz, Victor J. *A History of Mathematics: An Introduction*, 2nd ed., Reading, MA: Addison Wesley Longman, 1998.

Kaunzner, Wolfgang. Über Regiomontanus als Mathematiker, in Günther Hamann, ed., *Regiomontanus-Studien*, Vienna: Verlag der Österreichischen Akademie der Wissenschaften, 1980, 125–145.

Kaunzer, Wolfgang. On the transmission of mathematical knowledge to Europe, *Sudhoffs Archiv* **71** (1987), 129–140.

Kaunzner, Wolfgang. On Regiomontanus's arithmetic and algebra in *De triangulis omnimodis libri quinque*, in Ernst Zinner, *Regiomontanus: His Life and Work*, Amsterdam/New York/Oxford/Tokyo: North-Holland, 1990, 373–386.

Kaye, G. R. Ancient Hindu spherical astronomy, *Journal and Proceedings of the Royal Asiatic Society on Bengal* (2) **15** (1919), 153–189.

Kazim, M. A. al-Bīrūnī and trigonometry, in Victor Courtois, ed., *al-Bīrūnī Commemoration Volume*, Calcutta: Iran Society, 1951, 161–170; reprinted in Fuat Sezgin, ed., *Abū l-Rayḥān al-Bīrūnī: Texts and Studies III*, Frankfurt: Institut für Geschichte der Arabisch-Islamischen Wissenschaften, 1998, 87–96.

Keller, Agathe. *Expounding the Mathematical Seed: A Translation of Bhāskara I on the Mathematical Chapter of the Āryabhaṭīya*, 2 vols., Basel: Birkhäuser, 2006.

Kelley Jr., James E. *Analog and Digital Navigation in the Late Middle Ages*, Melrose Park, PA: Sometime Publishers, 2000.

Kennedy, Edward S. A Survey of Islamic Astronomical Tables, *Transactions of the American Philosophical Society* **46** (1956), 123–177.

Kennedy, Edward S. Parallax theory in Islamic astronomy, *Isis* **47** (1956), 33–53. Reprinted in David A. King and Mary Helen Kennedy, eds., *Studies in the Islamic Exact Sciences by E. S. Kennedy, Colleagues and Former Students*, Beirut: American University of Beirut, 1983, 164–184.

Kennedy, Edward S. Bīrūnī's graphical determination of the local meridian, *Scripta Matḥematica* **24** (1959), 251–255. Reprinted in David A. King and Mary Helen Kennedy, eds., *Studies in the Islamic Exact Sciences by E. S. Kennedy, Colleagues and Former Students*, Beirut: American University of Beirut, 1983, 613–617.

Kennedy, Edward S. A medieval interpolation scheme using second order differences, in Walter B. Henning and Ehsan Yarshater, eds., *A Locust's Leg: Studies in Honour of S. H. Taqizadeh*, London: Lund, Humphries and Co., 1962, 117–120. Reprinted in David A. King and Mary Helen Kennedy, eds., *Studies in the Islamic Exact Sciences by E. S. Kennedy, Colleagues and Former Students*, Beirut: American University of Beirut, 1983, 522–525.

Kennedy, Edward S. al-Bīrūnī on determining the meridian, *Mathematics Teacher* **66** (1963), 635–637. Reprinted in David A. King and Mary Helen Kennedy, eds., *Studies in the Islamic Exact Sciences by E. S. Kennedy, Colleagues and Former Students*, Beirut: American University of Beirut, 1983, 618–620.

Kennedy, Edward S. An early method of successive approximations, *Centaurus* **13** (1969), 248–250. Reprinted in David A. King and Mary Helen Kennedy, eds., *Studies in the Islamic Exact Sciences by E. S. Kennedy, Colleagues and Former Students*, Beirut: American University of Beirut, 1983, 541–543.

Kennedy, Edward S. The history of trigonometry, in *Historical Topics for the Mathematics Classroom*, 31st yearbook, Reston, VA: National Council of Teachers of Mathematics, 1969, 333–359. Reprinted in David A. King and Mary Helen Kennedy, eds., *Studies in the Islamic Exact Sciences by E. S. Kennedy, Colleagues and Former Students*, Beirut: American University of Beirut, 1983, 3–29.

Kennedy, Edward S. al-Bīrūnī's *Maqālīd ʿilm al-hay'a, Journal of Near Eastern Studies* **30** (1971), 308–314. Reprinted in David A. King and Mary Helen Kennedy, eds., *Studies in the Islamic Exact Sciences by E. S. Kennedy,*

Colleagues and Former Students, Beirut: American University of Beirut, 1983, 596–602.

Kennedy, Edward S. al-Bīrūnī's Masudic Canon, *al-Abhath* **24** (1971), 59–81. Reprinted in David A. King and Mary Helen Kennedy, eds., *Studies in the Islamic Exact Sciences by E. S. Kennedy, Colleagues and Former Students*, Beirut: American University of Beirut, 1983, 573–595.

Kennedy, Edward S. *A Commentary upon Bīrūnī's Kitāb Taḥdīd al-Amākin, an 11th Century Treatise in Mathematical Geography*, Beirut: American University of Beirut, 1973.

Kennedy, Edward S. The motivation of al-Bīrūnī's second order interpolation scheme, in A. Y. al-Hassan, G. Karmi, and N. Namnum, eds., *Proceedings of the First International Symposium for the History of Arabic Science*, Aleppo, Institute for the History of Arabic Science, 1978, vol. II, 67–71. Reprinted in David A. King and Mary Helen Kennedy, eds., *Studies in the Islamic Exact Sciences by E. S. Kennedy, Colleagues and Former Students*, Beirut: American University of Beirut, 1983, 630–634.

Kennedy, Edward S. Applied mathematics in the tenth century: Abu'l-Wafā' calculates the distance Baghdad-Mecca, *Historia Mathematica* **11** (1984), 193–206.

Kennedy, Edward S. Spherical astronomy in Kāshī's *Khāqānī Zīj, Zeitschrift für Geschichte der Arabisch-Islamischen Wissenschaften* **2** (1985), 1–46.

Kennedy, Edward S. Ibn al-Haytham's determination of the meridian from one solar altitude, *Zeitschrift für Geschichte der Arabisch-Islamischen Wissenschaften* **5** (1989), 141–144.

Kennedy, Edward S. Mathematical geography, in Roshdi Rashed, ed., *Encyclopedia of the History of Arabic Science*, London/New York: Routledge, 1996, vol. 1, 185–201.

Kennedy, Edward S. *On the Contents and Significance of the Khāqānī Zīj by Jamshīd Ghiyāth al-Dīn al-Kāshī*, Frankfurt: Institute for the History of Arabic-Islamic Science, 1998.

Kennedy, Edward S.; and ʿId, Yusuf. A letter of al-Bīrūnī. Ḥabash al-Ḥāsib's analemma for the qibla, *Historia Mathematica* **1** (1974), 3–11. Reprinted in David A. King and Mary Helen Kennedy, eds., *Studies in the Islamic Exact Sciences by E. S. Kennedy, Colleagues and Former Students*, Beirut: American University of Beirut, 1983, 621–629.

Kennedy, Edward S.; Kunitzsch, Paul; and Lorch, Richard P. *The Melon-Shaped Astrolabe in Arabic Astronomy*, Stuttgart: Franz Steiner Verlag, 1999.

Kennedy, Edward S.; and Muruwwa, Ahmad. Bīrūnī on the solar equation, *Journal of Near Eastern Studies* **17** (1958), 112–121. Reprinted in David A. King and Mary Helen Kennedy, eds., *Studies in the Islamic Exact Sciences by E. S. Kennedy, Colleagues and Former Students*, Beirut: American University of Beirut, 1983, 603–612.

Kennedy, Edward S.; and Sharkas, Haydar. Two medieval methods for determining the obliquity of the ecliptic, *Mathematics Teacher* **55** (1962), 286–290. Reprinted in David A. King and Mary Helen Kennedy, eds., *Studies in the Islamic Exact*

Sciences by E. S. Kennedy, Colleagues and Former Students, Beirut: American University of Beirut, 1983, 517–521.

Kennedy, Edward S.; and Transue, W. R. A medieval iterative algorism, *American Mathematical Monthly* **63** (1956), 80–83. Reprinted in David A. King and Mary Helen Kennedy, eds., *Studies in the Islamic Exact Sciences by E. S. Kennedy, Colleagues and Former Students*, Beirut: American University of Beirut, 1983, 513–516.

Kennedy, Edward S.; and Ukashah, Walid. The chandelier clock of Ibn Yūnis, *Isis* **60** (1969), 543–545. Reprinted in David A. King and Mary Helen Kennedy, eds., *Studies in the Islamic Exact Sciences by E. S. Kennedy, Colleagues and Former Students*, Beirut: American University of Beirut, 1983, 499–501.

Khaĭretdinova, N. G. History of the spherical tangent theorem [in Russian], *Istoriko-matematicheskie Issledovaniya* **29** (1985), 48–54.

Khaĭretdinova, N. G. History of the polar triangle [in Russian], *Istoriko-matematicheskie Issledovaniya* **29** (1985), 154–159.
[See also Hairetdinova, N. G.]

al-Khwārizmī, Muḥammad ibn Mūsā. *Die Astronomischen Tafeln des Muḥammed ibn Mūsā al-Khwārizmī in der Bearbeitung des Maslama ibn Aḥmed al-Madjrīṭī und der Latein. Uebersetzung des Athelard von Bath auf Grund der Vorarbeiten von A. Bjørnbo und R. Besthorn in Kopenhagen*, edited by Heinrich Suter, *Det Kongelige Danske Videnskabernes Selskab Skrifter*, 7. R., *Hist. og Filos. Afd.* **3** (1), Copenhagen, 1914.

al-Khwārizmī, Muḥammad ibn Mūsā. *The Astronomical Tables of al-Khwārizmī*, translated by Otto Neugebauer, *Det Kongelige Danske Videnskabernes Selskab, Historisk-filosofiske Skrifter* **4** (2), Copenhagen: Munksgaard, 1962.

King, David A. *The Astronomical Works of Ibn Yūnus*, doctoral dissertation, Yale University, 1972.

King, David A. al-Khalīlī's auxiliary tables for solving problems of spherical astronomy, *Journal for the History of Astronomy* **4** (1973), 99–110.

King, David A. Ibn Yūnus's *Very Useful Tables* for reckoning time by the Sun, *Archive for History of Exact Sciences* **10** (1973), 342–394.

King, David A. al-Khalīlī's *qibla* table, *Journal of Near Eastern Studies* **34** (1975), 81–122.

King, David A. al-Khwārizmī and new trends in mathematical astronomy in the ninth century, *Occasional Papers on the Near East* **2**, Hagop Kevorkian Center for Near Eastern Studies, New York University, 1983, 43 pp.

King, David A. The sacred direction in Islam: A study of the interaction of religion and science in the Middle Ages, *Interdisciplinary Science Reviews* **10** (1985), 315–328.

King, David A. The earliest Islamic mathematical methods and tables for finding the direction of Mecca, *Zeitschrift für Geschichte der Arabisch-Islamischen Wissenschaften* **3** (1986), 82–149. Reprinted in David A. King, *Astronomy in the Service of Islam*, Aldershot, UK / Brookfield, VT, 1993, article XIV.

King, David A. Some early Islamic tables for determining lunar crescent visibility, in David A. King and George Saliba, eds., *From Deferent to Equant: A Volume of*

Studies in the History of Science in the Ancient and Medieval Near East in Honor of E. S. Kennedy, New York: New York Academy of Sciences, 1987, 185–225.

King, David A. Mīḳāt: astronomical timekeeping, in *Encyclopaedia of Islam*, vol. 7, Leiden: Brill, 1990, 27–32. Reprinted in David A. King, *Astronomy in the Service of Islam*, Aldershot, UK / Brookfield, VT, 1993, article V, 1–20.

King, David A. Ḳibla: sacred direction, in *Encyclopaedia of Islam*, vol. 5, Leiden: Brill, 1991, 83–88. Reprinted in David A. King, *Astronomy in the Service of Islam*, Aldershot, UK / Brookfield, VT, 1993, article IX, 1–18.

King, David A. Rubʿ, in *Encyclopaedia of Islam*, vol. 8, Leiden: Brill, 1995, 574–575.

King, David A. Astronomy and Islamic society. Qibla, gnomonics and timekeeping, in Roshdi Rashed, ed., *Encyclopedia of the History of Arabic Science*, London/New York: Routledge, 1996, vol. 1, 128–184.

King, David A. *World-Maps for Finding the Direction and Distance to Mecca*, London: al-Furqān Islamic Heritage Foundation / Leiden: Brill, 1999.

King, David A. A *vetustissimus* Arabic treatise on the *quadrans vetus, Journal for the History of Astronomy* **33** (2002), 237–255.

King, David A. *In Synchrony with the Heavens. Studies in Astronomical Timekeeping and Instrumentation in Medieval Islamic Civilization. Volume I: The Call of the Muezzin*, Leiden/Boston: Brill, 2004.

King, David A. *In Synchrony with the Heavens. Studies in Astronomical Timekeeping and Instrumentation in Medieval Islamic Civilization. Volume II: Instruments of Mass Calculation*, Leiden/Boston: Brill, 2005.

King, David A.; and Samsó, Julio. Astronomical handbooks and tables from the Islamic world (750–1900): an interim report, *Suhayl* **2** (2001), 9–105.

Kish, George. Apian, Peter, in Charles Coulston Gillispie, ed., *Dictionary of Scientific Biography*, vol. 1, New York: Charles Scribner's Sons, 1970, 178–179.

Klein, Jacob. Die griechische Logistik und die Entstehung der Algebra I, *Quellen und Studien* B **3** (1934), 18–105.

Klein, Jacob. Die griechische Logistik und die Entstehung der Algebra II, *Quellen und Studien* B **3** (1934), 122–235.

Klintberg, Bo. Hipparchus's 3600′ based chord table and its place in the history of ancient Greek and Indian trigonometry, *Indian Journal of History of Science* **40** (2005), 169–203.

Klug, Rudolf. Johannes von Gmunden, der Begründer der Himmelskunde auf deutschem Boden, *Akademie der Wissenschaften Wien*, Phil.-Hist. Kl., Sitzungsberichte **222** (4) (1943), 1–93.

Knorr, Wilbur. Ancient versions of two trigonometric lemmas, *Classical Quarterly* **35** (1985), 362–391.

Knorr, Wilbur. The practical element in ancient exact sciences, *Synthese* **81** (1989), 313–328.

Knorr, Wilbur. New readings in Greek mathematics: sources, problems, publications, *Impact of Science on Society* **159** (1990), 207–218.

Knorr, Wilbur. On the principle of linear perspective in Euclid's *Optics, Centaurus* **34** (1991), 193–210.

Knorr, Wilbur. Pseudo Euclidean reflection in ancient optics: A re-examination of textual issues pertaining to the Euclidean *Optica* and *Catoptrica, Physis* **31** (1994), 1–45.

Koelblen, Sabine. Un exercice de combinatoire: Les relations issues de la figure sécante de Ptolémée, ou les règles des six quantities en proportion, in *Un Parcours en Histoire des Mathématiques: Travaux et Recherches*, Nantes: Université de Nantes, 1993, 1–21.

Koelblen, Sabine. Une pratique de la composition des raisons dans un exercice de combinatoire, *Revue d'Histoire des Sciences* **47** (1994), 209–247.

Koyré, Alexandre. Mathematics, in René Taton, ed., *History of Science. The Beginnings of Modern Science from 1450 to 1800*, New York: Basic Books, 1964.

Krause, Max. Die Sphärik von Menelaos aux Alexandrien in der Verbesserung von Abū Naṣr Manṣūr b. ʿAlī ibn ʿIrāq, *Abhandlungen der Gesellschaft der Wissenschaften zu Göttingen*, Philologisch-Historische Klasse (3) **17** (1936), 1–254, 1–110 (Arabic numbering).

Kren, Claudia. Astronomy, in David L. Wagner, ed., *The Seven Liberal Arts in the Middle Ages*, Bloomington: Indiana University Press, 1983, 218–247.

Kugler, Franz Xaver. *Die babylonische Mondrechnung*, Freiburg: Herder'sche Verlagshandlung, 1900.

Kunitzsch, Paul; and Lorch, Richard. Abū Naṣr and Ḥabash on *maṭāliʿ al-samt, Zeitschrift für Geschichte der Arabisch-Islamischen Wissenschaften* **9** (1994), 43–82.

Langermann, Y. Tzvi. Peurbach in the Hebrew tradition, *Journal for the History of Astronomy* **29** (1998), 137–150.

Lejeune, Albert. La dioptre d'Archimède, *Annales de la Société Scientifique de Bruxelles* Ser. I, **61** (1947), 27–47.

Leonardo of Pisa. *Scritti di Leonardo Pisano. Matematico del Secolo Decimoterzo*, ed. Baldassarre Boncompagni, 2 vols., Rome: Tipografia delle Scienze Matematiche e Fisiche, 1857/1862.

Leonardo of Pisa. *Leonardo Pisano Fibonacci The Book of Squares: an Annotated Translation into Modern English*, transl. L. E. Sigler. Boston: Academic Press, 1987.

Leonardo of Pisa. *Fibonacci's Liber Abaci*, transl. L. E. Sigler. New York: Springer-Verlag, 2002.

Leonardo of Pisa. *Fibonacci's De practica geometrie*, transl. Barnabas Hughes. New York: Springer, 2008.

Lesley, Mark. Bīrūnī on rising times and daylight lengths, *Centaurus* **5** (1957), 121–141.

Levey, Martin. Abraham bar Ḥiyya, in Charles Coulston Gillispie, ed., *Dictionary of Scientific Biography*, vol. 1, New York: Charles Scribner's Sons, 1970, 22–23.

Levi ben Gerson. *The Astronomy of Levi ben Gerson (1288–1344)*, edition and translation by Bernard R. Goldstein, New York: Springer-Verlag, 1985.

Lévy, Tony. Les débuts de la literature mathématique Hébraïque: la géométrie d'Abraham bar Ḥiyya (XIe–XIIe siècle), *Micrologus* **9** (2001), 35–64.

Linton, Christopher M. *From Eudoxus to Einstein. A History of Mathematical Astronomy*, Cambridge, UK: Cambridge University Press, 2004.

Long, Pamela O.; McGee, David; and Stahl, Alan M. *The Book of Michael of Rhodes: A Fifteenth Century Maritime Manuscript*, Cambridge, MA: MIT Press, to appear.

Lorch, Richard. *Jābir ibn Aflaḥ and his Influence in the West*, doctoral dissertation, University of Manchester, 1971.

Lorch, Richard. The astronomy of Jābir ibn Aflaḥ, *Centaurus* **19** (1975), 85–107. Reprinted in Richard Lorch, *Arabic Mathematical Sciences: Instruments, Texts, Transmission*, Aldershot, UK / Brookfield, VT: Variorum, 1995, article VI.

Lorch, Richard. The *qibla*-table attributed to al-Khāzinī, *Journal for the History of Arabic Science* **4** (1980), 259–264. Reprinted in Richard Lorch, *Arabic Mathematical Sciences: Instruments, Texts, Transmission*, Aldershot, UK / Brookfield, VT: Variorum, 1995, article XIV.

Lorch, Richard. A note on the horary quadrant, *Journal for the History of Arabic Science* **5** (1981), 115–120. Reprinted in Richard Lorch, *Arabic Mathematical Sciences: Instruments, Texts, Transmission*, Aldershot, UK / Brookfield, VT: Variorum, 1995, article XVII.

Lorch, Richard. Jābir ibn Aflaḥ and the establishment of trigonometry in the West, in Richard Lorch, *Arabic Mathematical Sciences: Instruments, Texts, Transmission*, Aldershot, UK / Brookfield, VT: Variorum, 1995, article VIII, 42 pp.

Lorch, Richard. Maslama al-Majrīṭī and Thābit's *al-Shakl al-Qaṭṭāʿ*, in Josep Casulleras and Julio Samsó, eds., *From Baghdad to Barcelona: Studies in the Islamic Exact Sciences in Honour of Prof. Juan Vernet*, Barcelona: Universitat de Barcelona, 1996, vol. 1, 49–57.

Lorch, Richard. Graphical methods in spherical astronomy in treatises by Ḥabash al-Ḥāsib and al-Māhānī, in *Histoire des Mathématiques Arabes. Actes du 3me Colloque Maghrébin sur l'Histoire des Mathématiques Arabes*, Algiers: Association Algérienne d'Histoire des Mathématiques, 1998, vol. 2, 221–226.

Lorch, Richard. Some early applications of the sine quadrant, *Suhayl* **1** (2000), 251–272.

Lorch, Richard. Greek—Arabic—Latin: the transmission of mathematical texts in the middle ages, *Science in Context* **14** (2001), 313–331.

Lorch, Richard. *Thābit ibn Qurra "On the Sector Figure" and Related Texts*, Frankfurt: Institut für Geschichte der Arabisch-Islamischen Wissenschaften, 2001.

Luckey, Paul. Das Analemma von Ptolemäus, *Astronomische Nachrichten* **230** (5498) (1927), cols. 17–46.

Luckey, Paul. T̲habit b. Qurra's Buch über die ebenen Sonnenuhren, *Quellen und Studien zur Geschichte der Mathematik, Astronomie und Physik* **B4** (1937), 95–148.

Luckey, Paul. Zur Entstehung der Kugeldreiecksrechnung, *Deutsche Mathematik* **5** (1941), 405–446.

Luckey, Paul. Der Lehrbrief über den Kreisumfang von Ǧamšīd b. Masʿūd al-Kāšī, *Abhandlungen der Deutschen Akademie der Wissenschaften, Klasse für Mathematik und allgemeine Naturwissenschaften* 1950 (6), Berlin, 1953.

Luckey, Paul. Die Rechenkunst bei Ǧamšīd b. Masʿūd al-Kāšī mit Rückblicken auf die ältere Geschichte des Rechnens, *Abhandlungen für die Kunde des Morgenlandes* **31** (1), Wiesbaden: Franz Steiner, 1951.

Majumdar, Pradip Kumar. The extant *Siddhānta Sārvabhauma*—An error in the sine of one-third part of an angle, *Indian Journal of History of Science* **13** (1978), 6–10.

Malpangotto, Michela. Sul commento di Pappo d'Alessandria alle *Sferiche* di Teodosio, *Bollettino di Storia della Scienze Matematiche* **23** (2003), 121–148.

Mancha, J. L. Heuristic reasoning: approximation procedures in Levi ben Gerson's astronomy, *Archive for History of Exact Sciences* **52** (1998), 13–50.

Maor, Eli. *Trigonometric Delights*, Princeton: Princeton University Press, 1998.

Marar, K. Mukunda; and Rajagopal, C. T. On the Hindu quadrature of the circle, *Journal of the Bombay Branch of the Royal Asiatic Society* (2) **20** (1944), 65–82.

Masiero, Franco. La raxon de marteloio, *Studi Veneziana* (2) **8** (1984), 393–412.

Matvievskaya, G. P. The history of spherics in antiquity and the medieval East [in Russian], in S. K. Sirazhdinov, ed., *Mathematics and Astronomy in the Works of Ibn Sina, his Contemporaries and Successors* [in Russian], Tashkent: Fan, 1981, 95–117.

Matvievskaya, G. P. *An Outline of the History of Trigonometry* [in Russian], Tashkent: Fan, 1990.

Matvievskaya, G. P.; and Tllashev, K. The works of Abū Naṣr ibnʿIrāq on spherics [in Russian], in S. K. Sirazhdinov, ed., *On the History of Medieval Eastern Mathematics and Astronomy* [in Russian], Tashkent: Fan, 1983, 82–171.

May, Kenneth O. *Bibliography and Research Manual of the History of Mathematics*, Toronto: University of Toronto Press, 1973.

McCarthy, Daniel P.; and Byrne, John G. al-Khwārizmī's sine tables and a western table with the Hindu norm of $R = 150$, *Archive for History of Exact Sciences* **57** (2003), 243–266.

Menelaus. *Menelai sphareicorum*, ed. Edmund Halley, Oxford: Sumptibus Academicis, 1758.

Miller, G. A. Archimedes and Trigonometry, *Science* **67** (1928), 555.

Miller, Jeff. *Earliest Known Uses of Some of the Words of Mathematics*, http://members.aol.com/jeff570/mathword.html.

Motzo, Bacchisio R. II compasso da navigare, opera italiana della metà del secolo XIII, *Annali della Facoltà di Lettera e Filosofia della Università di Cagliari* **8** (1947), I–CXXXII, 1–137.

Mundy, John. John of Gmunden, *Isis* **34** (1943), 196–205.

Nadal, Robert; and Brunet, J.-P. Le commentaire d'Hipparque I. La sphere mobile, *Archive for History of Exact Sciences* **29** (1984), 201–236.

Nadal, Robert; and Brunet, J.-P. Le commentaire d'Hipparque II. Position de 78 étoiles, *Archive for History of Exact Sciences* **40** (1989), 305–354.

Nadal, Robert; Taha, Abdelkaddous; and Pinel, Pierre. Le contenu astronomique des *Sphèriques* de Ménélaos, *Archive for History of Exact Sciences* **58** (2004), 381–436.

Naraharayya, S. N. Note on the Hindu table of sines, *Journal of the Indian Mathematical Society* **15** (7) (1924), 105–113.

Naimpally, S. A. Approximate formula for the length of a chord, *Gaṇita-Bhāratī* **9** (1987), 57–58.

Netz, Reviel. The goal of Archimedes' *Sand-Reckoner*, *Apeiron* **36** (2003), 251–290.

Neugebauer, Otto. Zur Entstehung des Sexagesimalsystems, in *Abhandlungen der Gesellschaft der Wissenschaften zu Göttingen*, Mathem.-phys. Kl., Neue Folge **13** (1), 1927.

Neugebauer, Otto. Untersuchungen zur antiken Astronomie III. Die babylonische Theorie der Breitenbewegung des Mondes, *Quellen und Studien zur Geschichte der Mathematik*, Abteilung B: Studien, **4** (1938), 193–346.

Neugebauer, Otto. Archimedes and Aristarchus, *Isis* **34** (1942), 4–6.

Neugebauer, Otto. The transmission of planetary theories in ancient and medieval astronomy, *Scripta Mathematica* **22** (1956), 165–192.

Neugebauer, Otto. *The Exact Sciences in Antiquity*, 2nd ed., Providence: Brown University Press 1957; reprinted New York: Dover, 1969.

Neugebauer, Otto. The equivalence of eccentric and epicyclic motion according to Apollonius, *Scripta Mathematica* **24** (1959), 5–21.

Neugebauer, Otto. On some aspects of early Greek astronomy, *Proceedings of the American Philosophical Society* **116** (1972), 243–251.

Neugebauer, Otto. *A History of Ancient Mathematical Astronomy*, 3 parts, Berlin: Springer, 1975.

Neugebauer, Otto. *Astronomy and History. Selected Essays*, New York: Springer, 1983.

Neugebauer, Otto; and Pingree, David. *The Pañcasiddhāntikā of Varāhamihira*, 2 vols., *Det Kongelige Danske Videnskabernes Selskab Historisk-Filosofiske Skrifter* **6** (1), Copenhagen: Munksgaard, 1970/1971.

Neugebauer, Otto; and Sachs, A. *Mathematical Cuneiform Texts*, New Haven: American Oriental Society / Schools of Oriental Research, 1945.

Neugebauer, Otto; and Schmidt, Olaf. Hindu astronomy at Newminster in 1428, *Annals of Science* **8** (1952), 221–228. Reprinted in Neugebauer, Otto, *Astronomy and History: Selected Essays*, New York: Springer-Verlag, 1983, 425–432.

Newton, Robert R. *The Crime of Claudius Ptolemy*, Baltimore: Johns Hopkins University Press, 1977.

Newton, Robert R. *The Origins of Ptolemy's Astronomical Parameters*, Baltimore: University of Maryland/Johns Hopkins University, 1982.

Newton, Robert R. *The Origins of Ptolemy's Astronomical Tables*, Baltimore: University of Maryland/Johns Hopkins University, 1985.

Nicholas of Cusa. *Die Mathematischen Schriften*, 2nd ed., transl. Josepha Hofmann, Hamburg: Felix Meiner, 1980.

Nīlakaṇṭha. *Tantrasaṅgraha* of Nīlakaṇṭha Somayāji with *Yuktidīpikā* and *Laghuvivṛti* of Śaṅkara, edited by K. V. Sarma, Hoshiarpur: Vishveshvaranand Vishva Bandhu Institute of Sanskrit and Indological Studies, Panjab University, 1977.

Nīlakaṇṭha. *Tantrasaṃgraha* of Nīlakaṇṭha Somayāji, edition by K. V. Sarma, transl. V. S. Narasimhan, *Indian Journal of History of Science* **33** (1998), supplements to issues 1, 2, and 3.

Nissen, H. J.; Damerow, P.; and Englund, R. *Archaic Bookkeeping: Early Writing and Techniques of Economic Administration in the Ancient Near East*, Chicago: University of Chicago Press, 1993.

North, John. *The Fontana History of Astronomy and Cosmology*, London: Fontana Press, 1994.

Ōhashi, Yukio. Early history of the astrolabe in India, *Indian Journal of History of Science* **31** (1997), 199–295.

Ōhashi, Yukio. Remarks on the origin of Indo-Tibetan astronomy, in Helaine Selin, ed., *Astronomy Across Cultures: The History of Non-Western Astronomy*, Dordrecht/Boston/London: Kluwer, 2000, 341–369.

Pedersen, Fritz S. *The Toledan Tables* (4 vols.), Copenhagen: C. A. Reitzel, 2002.

Pedersen, Olaf. Logistics and the theory of functions: an essay in the history of Greek mathematics, *Archives Internationales d'Histoire des Sciences* **24** (1974), 29–50.

Pedersen, Olaf. *A Survey of the Almagest*, Odense: Odense University Press, 1974.

Peet, T. Eric. *The Rhind Mathematical Papyrus*, Liverpool: Liverpool University Press, 1923.

Peurbach, Georg. *Tractatus Georgii Peurbachii super propositiones Ptolemaei de sinubus et chordis*, Nuremberg: apud Iohan. Petreium, 1541.

Pingree, David. Astronomy and astrology in India and Iran, *Isis* **54** (1963), 229–246.

Pingree, David. The *Paitāmahasiddhānta* of the *Viṣṇudharmottarapurāṇa, Adyar Library Bulletin* **31/32** (1967/68), 472–510.

Pingree, David. The fragments of the works of Yaʿqūb ibn Ṭāriq, *Journal of Near Eastern Studies* **27** (1968), 97–125.

Pingree, David. Brahmagupta, in Charles Coulston Gillispie, ed., *Dictionary of Scientific Biography*, vol. 2, New York: Charles Scribner's Sons, 1970, 416–418.

Pingree, David. The fragments of the works of al-Fazārī, *Journal of Near Eastern Studies* **29** (1970), 103–123.

Pingree, David. *Census of the Exact Sciences in Sanskrit*, Philadelphia: American Philosophical Society, vol. A1: 1970, vol. A2: 1971, vol. A3: 1976, vol. A4: 1981, vol. A5: 1994.

Pingree, David. al-Fazārī, Muḥammad ibn Ibrāhīm, in Charles Coulston Gillispie, ed., *Dictionary of Scientific Biography*, vol. 4, New York: Charles Scribner's Sons, 1971, 555–556.

Pingree, David. On the Greek origin of the Indian planetary model employing a double epicycle, *Journal for the History of Astronomy* **2** (1971), 80–85.

Pingree, David. The recovery of early Greek astronomy from India, *Journal for the History of Astronomy* **7** (1976), 109–123.

Pingree, David. History of mathematical astronomy in India, in Charles Coulston Gillispie, ed., *Dictionary of Scientific Biography*, vol. 15, New York: Charles Scribner's Sons, 1978, 533–633.

Pingree, David. Power series in medieval Indian trigonometry, in Peter Gaeffke and David A. Utz, eds., *Science and Technology in South Asia: Proceedings of the South Asia Seminar*, II, 1981/82, Philadelphia: University of Pennsylvania, Dept. of South Asia Regional Studies, 1985, 25–30.

Pingree, David. Āryabhaṭa, the *Paitāmahasiddhānta*, and Greek astronomy, *Studies in History of Medicine and Science* **12** (1993), 69–79.

Pitiscus, Bartholomew. *Trigonometriae*, Augsburg: D. Custodis, 1600.

Playfair, John. Observations on the trigonometrical tables of the Brahmins, *Transactions of the Royal Society of Edinburgh* **4** (1798), 83–106. (Due to a pagination error there are no pages 89–96.)

Plofker, Kim. An example of the secant method of iterative approximation in the fifteenth-century Sanskrit text, *Historia Mathematica* **23** (1996), 246–256.

Plofker, Kim. The "error" in the Indian "Taylor series approximation" to the sine, *Historia Mathematica* **28** (2001), 283–295.

Plofker, Kim. The astrolabe and spherical trigonometry in medieval India, *Journal for the History of Astronomy* **31** (2000), 37–54.

Plofker, Kim. Spherical trigonometry and the astronomy of the medieval Kerala school, in S.M.R. Ansari, ed., *History of Oriental Astronomy*, Dordrecht: Kluwer, 2002, 83–93.

Plofker, Kim. Use and transmission of iterative approximations in India and the Islamic world, in Yvonne Dold-Samplonius, Joseph W. Dauben, Menso Folkerts, and Benno van Dalen, eds., *From China to Paris: 2000 Years Transmission of Mathematical Ideas*, Stuttgart: Franz Steiner Verlag, 2002, 167–186.

Plofker, Kim. The problem of the Sun's corner altitude and convergence of fixed-point iterations in medieval Indian astronomy, in Charles Burnett, Jan Hogendijk, Kim Plofker, and Michio Yano, eds., *Studies in the History of the Exact Sciences in Honour of David Pingree*, Leiden/Boston: Brill, 2004, 552–586.

Plofker, Kim. Relations between approximations to the sine in Kerala mathematics, in G. G. Emch, R. Sridharan, and M. D. Srinivas, *Contributions to the History of Indian Mathematics*, New Delhi: Hindustan Book Agency, 2005, 135–152.

Plofker, Kim. Mathematics in India, in Victor Katz, ed., *The Mathematics of Egypt, Mesopotamia, China, India, and Islam: A Sourcebook*, Princeton: Princeton University Press, 2007, 385–514.

Powell, Marvin A. The antecedents of Old Babylonian place notation and the early history of Babylonian mathematics, *Historia Mathematica* **3** (1976), 417–439.

Prakash, Satya. *A Critical Study of Brahmagupta and his Works*, New Delhi: Indian Institute of Astronomical & Sanskrit Research, 1968.

Ptolemy, Claudius. *Claudii Ptolemaei Pheludiensis Alexandrini Almagestum seu Magnae constructionis mathematicae opus plane diuinum*, ed. George Trebizond, Venice, 1528.

Ptolemy, Claudius. *Geografia cioè descrittione vniuersale della terra*, ed. A. Magini, Venice: Gio. Battista and Giorgio Galignani, 1598.

Ptolemy, Claudius. *The Almagest by Ptolemy*, transl. R. Catesby Taliaferro, Chicago: Encyclopedia Britannica, 1952.

Ptolemy, Claudius. *Ptolemy's Almagest*, transl. Gerald J. Toomer, London: Duckworth / New York: Springer Verlag, 1984; reprinted Princeton: Princeton University Press, 1998.

Ptolemy's Geography: An Annotated Translation of the Theoretical Chapters, transl. J. Lennart Berggren and Alexander Jones, Princeton: Princeton University Press, 2000.

Puig, Roser. *al-Šakkāziyya. Ibn al-Naqqāš al-Zarqālluh*, Barcelona: Universidad de Barcelona, 1986.

Raja, K. Kunjunni. *Astronomy and Mathematics in Kerala*, Madras: Adyar Library and Research Centre, 1995 (reprinted from the *Adyar Library Bulletin*, 1963).

Rajagopal, C. T. A neglected chapter of Hindu mathematics, *Scripta Mathematica* **15** (1949), 201–209.

Rajagopal, C. T.; and Aiyar, T. V. Vedamurthi. On the Hindu proof of Gregory's series, *Scripta Mathematica* **17** (1951), 65–74.

Rajagopal, C. T.; and Aiyar, T. V. Vedamurthi. A Hindu approximation to pi, *Scripta Mathematica* **18** (1952), 25–30.

Rajagopal, C. T.; and Rangachari, M. S. On an untapped source of medieval Keralese mathematics, *Archive for History of Exact Sciences* **18** (1978), 89–102.

Rajagopal, C. T.; and Rangachari, M. S. On medieval Keralese mathematics, *Archive for History of Exact Sciences* **35** (1986), 91–99.

Rajagopal, C. T.; and Venkataraman, A. The sine and cosine power series in Hindu mathematics. *Journal of the Royal Asiatic Society of Bengal. Science* **15** (1949), 1–13.

Rashed, Roshdi. al-Samaw'al, al-Bīrūnī et Brahmagupta: les methods d'interpolation, *Arabic Sciences and Philosophy* **1** (1991), 101–160.

Rawlins, Dennis. An investigation of the ancient star catalog, *Publications of the Astronomical Society of the Pacific* **94** (1982), 359–373.

Rawlins, Dennis. Competence held hostage #2: The Princeton Institute vs. Aubrey Diller, *DIO* **4.2** (1994), 55–57.

Regiomontanus, Johannes. *Epytoma Joannis de Monte Regio in Almagestum Ptolomei*, Venice: Casparis Grossch and Stephani Keomer, 1496.

Regiomontanus, Johannes. *Tabula directionum profectionumque*, Impressum Uenetiis: Ingenio ac impensa Petri Liechtensteyn, 1504.

Regiomontanus, Johannes. *De triangulis omnimodis*, Nuremberg: Io. Petrei, 1533.

Regiomontanus, Johannes. *On Triangles*, transl. Barnabas Hughes, Madison: University of Wisconsin Press, 1967.

Regiomontanus, Johannes. *Joannus Regiomontani Opera collectea*, collected and with an introduction by Felix Schmeidler, Osnabrück: O. Zeller, 1972.

Rheticus, Georg Joachim. *Opus palatinum de triangulis*, Neostadii in Palatinatu: Excudebat Matthaeus Harnisus, 1596.

Richard of Wallingford. *Richard of Wallingford*, edited and translated by John D. North, 3 vols., Oxford: Clarendon Press, 1976.

Richey, Michael. The navigational background to 1492, *Journal of Navigation* **45** (1992), 266–284.

Rius, Mònica. *La Alquibla en al-Andalus y al-Magrib al-Aqṣà*, Barcelona: Universitat de Barcelona, 2000.

Rizvi, Syed Aftab Husain. On trisection of an angle leading to the derivation of a cubic equation and computation of the value of sine, *Indian Journal of History of Science* **19** (1984), 77–85.

Robins, Gay; and Shute, Charles. Mathematical bases of ancient Egyptian architecture and graphic art, *Historia Mathematica* **12** (1985), 107–122.

Robins, Gay; and Shute, Charles. *The Rhind Mathematical Papyrus: An Ancient Egyptian Text*, London: British Museum Publications, 1987.

Robson, Eleanor. *Mesopotamian Mathematics, 2100–1600 BC: Technical Constants in Bureaucracy and Education*, Oxford: Clarendon Press, 1999.

Robson, Eleanor. Neither Sherlock Holmes nor Babylon: a reassessment of Plimpton 322, *Historia Mathematica* **28** (2001), 167–206.

Robson, Eleanor. Words and pictures: new light on Plimpton 322, *American Mathematical Monthly* **109** (2002), 105–120.

Rodet, Léon. Leçons de calcul d'Âryabhaṭa, *Journal Asiatique* (7) **13** (1879), 393–434.

Rome, Adolphe., ed. *Commentaires de Pappus et de Théon d'Alexandrie sur l'Almageste*, 3 vols., Rome: Biblioteca Apostolica Vaticana, 1931/1936/1943.

Rome, Adolphe. Premiers essais de trigonométrie rectiligne chez les Grecs, *Annales de la Société Scientifique de Bruxelles*, Série A **52** (1932), 271–274.

Rome, Adolphe. Premiers essais de trigonométrie rectiligne chez les Grecs, *L'Antiquité Classique* **2** (1933), 177–192.

Rome, Adolphe. Les explications de Théon d'Alexandrie sur le théorème de Ménélas, *Annales de la Société Scientifique de Bruxelles*, Série A **53** (1933b), 39–50.

Rome, Adolphe. Le problème de l'equation de temps chez Ptolémée, *Annales de la Société Scientifique de Bruxelles* **59** (1939), 211–224.

Rose, Paul Lawrence. *The Italian Renaissance of Mathematics*, Geneva: Librairie Droz, 1975.

Rosen, Edward. Regiomontanus, Johannes, in Charles Coulston Gillispie, ed., *Dictionary of Scientific Biography*, vol. 11, New York: Charles Scribner's Sons, 1975, 348–352.

Rosen, Edward. Aristarchus of Samos and Copernicus, *Bulletin of the American Papyrological Society* **15** (1978), 85–93.

Rosenfeld, Boris A. Popytka kvadratichnogo interpolirovaniya u Abu Reikhan al-Biruni, *Istoriko-matematicheskie Issledovaniya* **12** (1959), 421–430.

Rosenfeld, Boris A. "Geometric trigonometry" in treatises of al-Khwārizmī, al-Māhānī and Ibn al-Haytham, in Menso Folkerts and Jan P. Hogendijk, eds., *Vestigia Mathematica: Studies in Medieval and Early Modern Mathematics in Honour of H.L.L. Busard*, Amsterdam/Atlanta: Rodopi, 1993, 305–308.

Rosenfeld, Boris A.; and Hogendijk, Jan P. A mathematical treatise written in the Samarqand observatory of Ulugh Beg, *Zeitschrift für Geschichte der Arabisch-Islamischen Wissenschaften* **15** (2002–2003), 25–65.

Rosenfeld, Boris A.; and Ihsanoğlu, Ekmeleddin. *Mathematicians, Astronomers & Other Scholars of Islamic Civilisation and their Works (7th–19th c.)*, Istanbul: Research Centre for Islamic History, Art and Culture (IRCICA), 2003.

Rosenfeld, Boris A.; and Rožanskaja, M. M. The astronomical work *al-Qānūn al-Masʿūdī* of al-Bīrūnī, *Istoriko-Astronomicheskie Issledovaniya Vyp.* **10** (1969), 63–95.

Rosińska, Grażyna. L'audience de Regiomontanus à Cracovie au XVe et au début du XVIe siècle, in Günther Hamann, ed., *Regiomontanus-Studien*, Vienna: Verlag der Österreichischen Akademie der Wissenschaften, 1980, 315–333.

Rosińska, Grażyna. Giovanni Bianchini—15th-century mathematician and astronomer [in Polish], *Kwartalnik Historii Nauki i Techniki* **26** (1981), 565–577.

Rosińska, Grażyna. Tables trigonométriques de Giovanni Bianchini, *Historia Mathematica* **8** (1981), 46–55.

Rosińska, Grażyna. Don't give to Rheticus what is Regiomontanus's [in Polish], *Kwartalnik Historii Nauki i Techniki* **28** (1983), 615–619.

Rosińska, Grażyna. Tables of decimal trigonometric functions from *ca.* 1450 to *ca.* 1550, in David A. King and George Saliba, eds., *From Deferent to Equant: A Volume of Studies in the History of Science in the Ancient and Medieval Near East in Honor of E. S. Kennedy*, New York: New York Academy of Sciences, 1987, 419–426.

Rossi, Corinna. *Architecture and Mathematics in Ancient Egypt*, Cambridge, UK: Cambridge University Press, 2004.

Rossi, Francesco Saverio. Copernico matematico: La sua trigonometria piana, *Cultura e Scuola* **12** (1973), 317–336.

Roy, Ranjan. The discovery of the series formula for π by Leibniz, Gregory and Nilakantha, *Mathematics Magazine* **63** (1990), 291–306.

Sabra, A. I. The authorship of the *Liber de crepusculis*, an eleventh-century work on atmospheric refraction, *Isis* **58** (1967), 77–85.

Sabra, A. I. Situating Arabic science: locality versus essence, *Isis* **87** (1996), 654–670.

Sachs, Abraham. Babylonian horoscopes, *Journal of Cuneiform Studies* **6** (1952), 49–75.

Saidan, Ahmed. The trigonometry of al-Bīrūnī, in Hakim Mohammed Said, ed., *al-Bīrūnī Commemorative Volume: Proceedings of the International Congress Held in Karachi*, Karachi: Hamdard Academy, 1979, 681–690.

Saidan, Ahmed. al-Bīrūnī on trigonometry, in Ekmeleddin Ihsanoğlu and Feza Günergun, eds., *Science in Islamic Civilisation*, Istanbul: Research Centre for Islamic History, Art and Culture (IRCICA), 2000, 167–178.

Saito, Ken. Mathematical reconstructions out, textual studies in: 30 years in the historiography of Greek mathematics, *Revue d'Histoire des Mathématiques* **4** (1998), 131–142.

Salam, Hala; and Kennedy, Edward S. Solar and lunar tables in early Islamic astronomy, *Journal of the American Oriental Society* **87** (1967), 492–497.

Saliba, George A. The double-argument lunar tables of Cyriacus, *Journal for the History of Aștronomy* **7** (1976), 41–46.

Saliba, George A. The role of the *Almagest* commentaries in medieval Islamic astronomy: A preliminary survey of Ṭūsī's redaction of Ptolemy's *Almagest, Archives Internationales d'Histoire des Sciences* **37** (1987), 3–20.

Samsó, Julio. *Estudios sobre Abū Naṣr Manṣūr b. ʿAlī b. Irāq*, Barcelona: Asociación para la Historia de la Ciencia Española, 1969.

Samsó, Julio. Levi ben Gerson, in Charles Coulston Gillispie, ed., *Dictionary of Scientific Biography*, vol. 8, New York: Charles Scribner's Sons, 1973, 279–282.

Samsó, Julio. Notas sobre la trigonometría esférica de Ibn Muʿād̲, *Awrāq* **3** (1980), 60–68. Reprinted as article VII in Julio Samsó, *Islamic Astronomy and Medieval Spain*, Aldershot, UK / Brooksfield, VT: Variorum, 1994.

Samsó, Julio. al-Bīrūnī in al-Andalus, in Josep Casulleras and Julio Samsó, eds., *From Baghdad to Barcelona: Studies in the Islamic Exact Sciences in Honour of Prof. Juan Vernet*, Barcelona: Universitat de Barcelona, 1996, vol. II, 583–612.

Samsó, Julio; and Mielgo, Honorino. Ibn Isḥāq al-Tūnisī and Ibn Muʿādh al-Jayyānī on the *qibla*, in Julio Samsó, *Islamic Astronomy and Medieval Spain*, Aldershot, UK / Brooksfield, VT: Variorum, 1994, article VI, 1–25.

Saraswathi, T. A. The development of mathematical series in India after Bhāskara II, *Bulletin of the National Institute of Sciences of India* **21** (1963), 320–343.

Sarma, K. V.; and Hariharan, S. *Yuktibhāṣā* of Jyeṣṭhadeva: a book of rationales in Indian mathematics and astronomy. An analytical appraisal, *Indian Journal of History of Science* **26** (1991), 185–207.

Saud, Muhammad. A part of al-Bīrūnī's *Istikhrāj al-autār fi al-dā'irah*, in Hakim Mohammed Said, ed., *al-Bīrūnī Commemorative Volume: Proceedings of the International Congress Held in Karachi*, Karachi: Hamdard Academy, 1979, 691–705.

Schmalzl, Peter. *Zur Geschichte des Quadranten bei den Arabern*, München: Salesianischen Offizin, 1929.

Schmidt, Wilhelm. Sind die Heronischen Vielecksformeln trigonometrisch? *Bibliotheca Mathematica* (3) **1** (1900), 319–320.

Schmidt, Olaf H. The computation of the length of daylight in Hindu astronomy, *Isis* **35** (1944), 205–211.

Schneider, Ivo. *Archimedes. Ingenieur, Naturwissenschaftler und Mathematiker*, Darmstadt: Wissenschaftliche Buchgesellschaft, 1979.

Schoy, Karl. Arabische Gnomonik, *Archiv der Deutschen Seewarte* **36** (1913), 3–42. Reprinted in Karl Schoy, *Beiträge zur Arabisch-Islamischen Mathematik und Astronomie*, Fuat Sezgin, ed., Frankfurt: Institut für Geschichte der Arabisch-Islamischen Wissenschaften, 1988, vol. I, 49–88.

Schoy, Karl. Abhandlung des al-Ḥasan ibn al-Haiṯam (Alhazen) über die Bestimmung der Richtung der Qibla, *Zeitschrift der Deutschen Morgenländischen Gesellschaft* **75** (1921), 242–253. Reprinted in Karl Schoy, *Beiträge zur*

Arabisch-Islamischen Mathematik und Astronomie, Frankfurt: Institut für Geschichte der Arabisch-Islamischen Wissenschaften, 1988, vol. 1, 230–241.

Schoy, Karl. Abhandlung von al-Faḍl b. Hâtim an-Nairîzî: Über die Richtung der Qibla Arab. Hdschr. Nr. 2457, 17° der Bibl. nat. in Paris, *Sitzungsberichte der Bayerischen Akademie der Wissenschaften*, Math.-Phys. Klasse 1922, 55–68. Reprinted in Karl Schoy, *Beiträge zur Arabisch-Islamischen Mathematik und Astronomie*, Frankfurt: Institut für Geschichte der Arabisch-Islamischen Wissenschaften, 1988, vol. 1, 252–265.

Schoy, Karl. Beiträge zur arabischen trigonometrie, *Isis* **5** (1923), 364–399. Reprinted in Karl Schoy, *Beiträge zur Arabisch-Islamischen Mathematik und Astronomie*, Frankfurt: Institut für Geschichte der Arabisch-Islamischen Wissenschaften, 1988, vol. 2, 448–483.

Schoy, Karl. *Die Gnomonik der Araber*, Band I, Lieferung F of Ernst von Bassermann-Jordan, ed., *Die Geschichte der Zeitmessung und der Uhren*, Berlin/Leipzig: Walter de Gruyter, 1923, 1–95. Reprinted in Karl Schoy, *Beiträge zur Arabisch-Islamischen Mathematik und Astronomie*, Fuat Sezgin, ed., Frankfurt: Institut für Geschichte der Arabisch-Islamischen Wissenschaften, 1988, vol. II, 351–447.

Schoy, Karl. The geography of the Moslems of the Middle Ages, *Geographical Review* **14** (1924), 257–269. Reprinted in Karl Schoy, *Beiträge zur Arabisch-Islamischen Mathematik und Astronomie*, Frankfurt: Institut für Geschichte der Arabisch-Islamischen Wissenschaften, 1988, vol. 2, 513–525.

Schoy, Karl. al-Bîrûnî's method of approximation of chord 40°, *American Mathematical Monthly* **33** (1926), 95–96.

Schoy, Karl. *Die trigonometrischen Lehren des Persischen Astronomen Abu'l-Raiḥān Muḥ. ibn Aḥmad al-Bīrūnī*, Hannover: Orient-Buchhandlung Heinz Lafaire K. G., 1927.

Sédillot, L. A. *Prolégomènes des Tables Astronomiques d'Oloug-Beg*, 2 vols., Paris: L'Institute de France, 1847/1853.

Sédillot, L. A. L'algèbre chez les Arabes, *Journal Asiatique* (5) **2** (1853), 323–356.

Sen, S. N. Astronomy, in D. M. Bose, S. N. Sen, and B. V. Subbarayappa, *A Concise History of Science in India*, New Delhi: Indian National Science Academy, 1971, 58–135.

Sen, S. N. Epicyclic eccentric planetary theories in ancient and medieval Indian astronomy, *Indian Journal of History of Science* **9** (1974), 107–121.

Sen, S. N. Planetary theories in Sanskrit astronomical texts, in G. Swarup, A. K. Bag, and K. S. Shukla, eds., *History of Oriental Astronomy*, Cambridge: Cambridge University Press, 1987, 113–124.

Sengupta, Prabodh Chandra. Brahmagupta on interpolation, *Bulletin of the Calcutta Mathematical Society* **23** (1931), 125–128.

Sengupta, Prabodh Chandra. Greek and Hindu methods in spherical astronomy, *Journal of the Department of Letters, Calcutta University* **21** (1931), paper 4, 1–25.

Sengupta, Prabodh Chandra. Infinitesimal calculus in Indian mathematics, *Journal of the Department of Letters, Calcutta University* **22** (1932), paper 5, 1–17.

Shank, Michael. Academic consulting in fifteenth-century Vienna: the case of astrology, in Edith Sylla and Michael McVaugh, eds., *Texts and Contexts in Ancient and Medieval Science: Studies on the Occasion of John E. Murdoch's Seventieth Birthday*, Leiden/New York/Köln: Brill, 1997, 245–270.

Shapiro, Alan E. Archimedes's measurement of the Sun's apparent diameter, *Journal for the History of Astronomy* **6** (1975), 75–83.

Shelby, Lon R. The geometrical knowledge of the mediaeval master masons, *Speculum* **47** (1972), 395–421.

Shelby, Lon R. Geometry, in David L. Wagner, ed., *The Seven Liberal Arts in the Middle Ages*, Bloomington: Indiana University Press, 1983, 196–217.

Shukla, Kripa Shankar, ed. *The Sūrya-Siddhānta with the Commentary of Parameśvara*, Lucknow: Lucknow University Department of Mathematics and Astronomy, 1957.

Shukla, Kripa Shankar. Early Hindu methods in spherical astronomy, *Gaṇita* **19** (1968), 49–72.

Shukla, Kripa Shankar. Hindu mathematics in the seventh century as found in Bhāskara I's commentary on the Āryabhaṭīya (four parts), *Gaṇita* **22** (1971), 115–130; **22** (1971), 61–78; **23** (1972), 57–79; and **23** (1972), 41–50.

Shukla, Kripa Shankar. The *Pañca-Siddhāntikā* of Varāhamihira (1), *Gaṇita* **24** (1973), 59–73.

Shukla, Kripa Shankar. The *Pañca-Siddhāntikā* of Varāhamihira (2), *Gaṇita* **28** (1977), 99–116.

Sidersky, D. La division de la circonférence en 360 parties, *Revue d'Assyriologie* **26** (1929), 31–32.

Sidoli, Nathan. Hipparchus and the ancient metrical methods on the sphere, *Journal for the History of Astronomy* **35** (2004), 71–84.

Sidoli, Nathan. Heron's *Dioptra* 35 and analemma methods: an astronomical determination of the distance between two cities, *Centaurus* **47** (2005), 236–258.

Sidoli, Nathan. The sector theorem attributed to Menelaus, *SCIAMVS* **7** (2006), 43–79.

Sidoli, Nathan; and Berggren, J. Lennart. The Arabic version of Ptolemy's *Planisphere* or *Flattening of the Surface of the Sphere*: Text, translation, commentary, *SCIAMVS* **8** (2008), 37–140.

Simi, Annalisa; and Toti Rigatelli, L. Some 14th and 15th century texts on practical geometry, in Menso Folkerts and Jan P. Hogendijk, eds., *Vestigia Mathematica*, Amsterdam/Atlanta: Rodopi, 1993, 453–470.

Simi, Annalisa. L'eredità della *Practica geometriae* di Leonardo Pisano nella geometria del basso medioevo e del primo rinascimento, *Bollettino di Storia delle Scienze Matematiche* **24** (2004), 9–41.

Smith, A. Mark. Ptolemy's search for a law of refraction: A case-study in the classical methodology of "saving the appearances" and its limitations, *Archive for History of Exact Sciences* **26** (1982), 221–240.

Stahl, William H. Aristarchus of Samos, in Charles Coulston Gillispie, ed., *Dictio-*

nary of Scientific Biography, vol. 1, New York: Charles Scribner's Sons, 1970, 246–250.

Steinschneider, Moritz. Die arabischen Uebersetzungen aus dem Griechischen, *Zeitschrift der Deutschen Morgenländischen Gesellschaft* **50** (1896), 161–219.

Strabo. *The Geography of Strabo*, 8 vols., Loeb Classical Library, ed. and transl. Horace Leonard Jones, Cambridge, MA: Harvard University Press, 1917.

Suter, Heinrich. Zur Trigonometrie der Araber, *Bibliotheca Mathematica* (3) **10** (1910), 156–160.

Suter, Heinrich. Das Buch der Auffindung der Sehnen im Kreise von Abū'l-Raiḥān Muḥ. el-Bīrūnī, *Bibliotheca Mathematica* (3) **11** (1910–1911), 11–78.

Suzuki, Takanori. A solution of the *qibla*-problem by Abu 'l-Qāsim Aḥmad ibn Muḥammad al-Ghandajānī, *Zeitschrift für Geschichte der Arabisch-Islamischen Wissenschaften* **4** (1987/88), 139–149.

Swerdlow, Noel M. Hipparchus on the distance of the Sun, *Centaurus* **14** (1969), 287–305.

Swerdlow, Noel M. Ptolemy on trial, *American Scholar* **48** (1979), 523–531.

Swerdlow, Noel M. Review of [Toomer 1973], *Mathematical Reviews* **58**, **#26714** (1979).

Swerdlow, Noel M. *The Babylonian Theory of the Planets*, Princeton: Princeton University Press, 1998.

Swerdlow, Noel M. The empirical foundations of Ptolemy's planetary theory, *Journal for the History of Astronomy* **35** (2004), 249–271.

Swerdlow, Noel M.; and Neugebauer, Otto. *Mathematical Astronomy in Copernicus's De Revolutionibus*, 2 parts, New York/Berlin/Heidelberg/Tokyo: Springer-Verlag, 1984.

Szabó, Árpád. Eudoxus und das Problem der Sehnentafeln, in *Aristoteles. Werk und Wirkung, Band 1* (Berlin: de Gruyter, 1985), 499–517.

Szabó, Árpád. Strabon und Pytheas—die geographische Breite von Marseille. Zur Frühgeschichte der mathematischen Geographie, *Historia Scientiarum* **29** (1985), 3–15.

Talbot, C. H. Simon Bredon (c. 1300–1372). Physician, mathematician and astronomer, *British Journal for History of Science* **1** (1962–63), 19–30.

Tannery, Paul. L'arithmétique des Grecs dans Héron d'Alexandrie, *Mémoires de la Société des Sciences Physiques et Naturelles de Bordeaux* (2) **4** (1882), 161–194.

Tannery, Paul. Aristarque de Samos, *Mémoires de la Société des Sciences Physiques et Naturelles de Bordeaux* (2) **5** (1883), 237–258.

Tannery, Paul. *Recherches sur l'Histoire de l'Astronomie Ancienne*, Paris: Gauthier-Villars & Fils, 1893; reprinted New York: Arno Press, 1976.

Tannery, Paul. L'inscription astronomique de Keskinto, *Revue des Études Grecques* **8** (1895), 49–58.

Tannery, Paul. La coudée astronomique et les anciennes divisions du cercle, reprinted in Paul Tannery, *Mémoires Scientifiques*, Toulouse: Édouard Privat / Paris: Gauthier-Villars, 1912, vol. 2, 256–268.

Taylor, Eva G. R. Five centuries of dead reckoning, *Journal of Navigation* **3** (1950), 280–285.

Taylor, Eva G. R. *The Haven-Finding Art: A History of Navigation from Odysseus to Captain Cook*, New York: Abelard-Schuman, 1957.

Taylor, Eva G. R. Mathematics and the navigator in the thirteenth century, *Journal of Navigation* **13** (1960), 3–14.

Theodosius. *Theodosii Sphaericorum elementorum*, ed. Francisco Maurolico, Impressit Petrus Spira, 1558.

Theodosius. *Die Sphärik des Theodosius*, trans. E. Nizze, Stralsund, 1826.

Theodosius. Theodosii *De habitationibus liber De diebus et noctibus libri duo*, *Abhandlungen der Gesellschaft der Wissenschaften zu Göttingen*, Philologisch-Historische Klasse (2) **19** (1927), edited and translated into Latin by R. Fecht.

Theodosius. Theodosius Tripolites Sphaerica, *Abhandlungen der Gesellschaft der Wissenschaften zu Göttingen*, Philologisch-Historische Klasse (2) **19** (3) (1927), edited by J. L. Heiberg.

Theodosius. *Les Sphériques de Théodose de Tripoli*, transl. Paul ver Eecke, Bruges: Desclée de Brouwer, 1927.

Thomas, Ivor. *Selections Illustrating the History of Greek Mathematics*, 2 vols., Loeb Classical Library, Cambridge, MA: Harvard University Press, 1941.

Thoren, Victor E. Prosthaphaerisis revisited, *Historia Mathematica* **15** (1988), 32–39.

Thureau-Dangin, F. Numération et Métrologie Sumériennes, *Revue d'Assyriologie* **18** (1921), 123–142.

Thureau-Dangin, F. L'origine du système sexagesimal, *Revue d'Assyriologie* **25** (1928), 115–118.

Thureau-Dangin, F. La division du cercle, *Revue d'Assyriologie* **25** (1928), 187–188.

Thureau-Dangin, F. Encore un mot sur la division du cercle, *Revue d'Assyriologie* **27** (1930), 53–54.

Thureau-Dangin, F. Mesures de temps et measures angulaires dans l'astronomie babylonienne, *Revue d'Assyriologie* **28** (1931), 111–114.

Thureau-Dangin, F. History of the sexagesimal system, *Osiris* **7** (1939), 95–141.

Thurston, Hugh. Greek and Indian planetary longitudes, *Archive for History of Exact Sciences* **44** (1992), 191–195.

Thurston, Hugh. Greek mathematical astronomy reconsidered, *Isis* **93** (2002), 58–69.

Tichenor, Mark J. Late medieval two-argument tables for planetary longitudes, *Journal of Near Eastern Studies* **26** (1967), 126–128.

Toomer, Gerald J. The chord table of Hipparchus and the early history of Greek trigonometry, *Centaurus* **18** (1973), 6–28.

Toomer, Gerald J. al-Khwārizmī, in Charles Coulston Gillispie, ed., *Dictionary of Scientific Biography*, vol. 7, New York: Charles Scribner's Sons, 1973, 358–365.

Toomer, Gerald J. Hipparchus on the distances of the Sun and Moon, *Archive for History of Exact Sciences* **14** (1974), 126–142.

Toomer, Gerald J. Ptolemy, in Charles Coulston Gillispie, ed., *Dictionary of Scientific Biography*, vol. 11, New York: Charles Scribner's Sons, 1975, 186–206.

Toomer, Gerald J. Hipparchus, in Charles Coulston Gillispie, ed., *Dictionary of Scientific Biography*, vol. 15, New York: Charles Scribner's Sons, 1978, 207–224.

Toomer, Gerald J. Hipparchus and Babylonian astronomy, in Erle Leichty, Maria de J. Ellis, and Pamela Gerardi, eds., *A Scientific Humanist: Studies in Memory of Abraham Sachs*, Philadelphia: The University Museum, 1988, 353–362.

Tropfke, Johannes. Archimedes und die Trigonometrie, *Archiv für Geschichte der Mathematik, der Naturwissenschaften und der Technik* **10** (1928), 432–463.

al-Ṭūsī, Naṣīr al-Dīn. *Traité du Quadrilatère Attribué à Nassiruddin-el-Toussy*, translated by Alexandre Pacha Caratheodory, Constantinople: Typographie et Lithographie Osmanié, 1891. Reprinted as vol. 47 of *Islamic Mathematics and Astronomy*, Frankfurt: Institut für Geschichte der Arabisch-Islamischen Wissenschaften, 1998.

al-Ṭūsī, Naṣīr al-Dīn. *Traktat o Polnom Chetyrehstoronnike*, translated by G. D. Mamedbeyli, S. P. Riznichenko and Boris A. Rosenfeld, Baku: Inst. Fiz. I Mat. Akad. Nauk Azerbaĭdžan. SSR, 1952.

al-Ṭūsī, Naṣīr al-Dīn. *Naṣīr al-Dīn al-Ṭūsī's Memoir on Astronomy (al-Tadhkira fī ʿilm al-hay'a)*, edited and translated by F. Jamil Ragep, 2 vols., New York: Springer-Verlag, 1993.

Uiblein, Paul. Die Wiener Universität, ihre Magister und Studenten zur Zeit Regiomontanus, in Günther Hamann, ed., *Regiomontanus-Studien*, Vienna: Verlag der Österreichischen Akademie der Wissenschaften, 1980, 395–432.

Unguru, Sabetai. On the need to rewrite the history of Greek mathematics, *Archive for History of Exact Sciences* **15** (1975/76), 67–114.

Van Brummelen, Glen. The numerical structure of al-Khālīlī's auxiliary tables, *Physis* **28** (1991), 667–698.

Van Brummelen, Glen. *Mathematical Tables in Ptolemy's Almagest*, doctoral dissertation, Simon Fraser University, 1993.

Van Brummelen, Glen. Lunar and planetary interpolation tables in Ptolemy's *Almagest, Journal for the History of Astronomy* **25** (1994), 297–311.

Van Brummelen, Glen. Jamshīd al-Kāshī: Calculating genius, *Mathematics in School* **27** (1998), 40–44; reprinted in Chris Pritchard, ed., *The Changing Shape of Geometry*, Cambridge, UK: Cambridge University Press, 2003, 130–135.

Van Brummelen, Glen. Mathematical methods in the tables of planetary motion in Kūshyār ibn Labbān's *Jāmiʿ Zīj, Historia Mathematica* **25** (1998), 265–280.

Van Brummelen, Glen. Taking latitude with Ptolemy: Jamshīd al-Kāshī's novel geometric model of the motions of the inferior planets, *Archive for History of Exact Sciences* **60** (2006), 353–377.

Van Brummelen, Glen; and Butler, Kenneth. Determining the interdependence of historical astronomical tables, *Journal of the American Statistical Association* **92** (1997), 41–48.

Van Brummelen, Glen; and Kerai, Yousuf. al-Samaw'al's curious approach to trigonometry, to appear.

Van Dalen, Benno. *Ancient and Mediaeval Astronomical Tables: Mathematical Structure and Parameter Values,* doctoral dissertation, Utrecht: University of Utrecht, 1993.

Van Dalen, Benno. al-Khwārizmī's astronomical tables revisited: analysis of the equation of time, in Josep Casulleras and Julio Samsó, eds., *From Baghdad to Barcelona: Studies in the Islamic Exact Sciences in Honour of Prof. Juan Vernet*, Barcelona: Universitat de Barcelona, 1996, vol. I, 195–252.

Van Dalen, Benno. Tables of planetary latitude in the *Huihui li* (II), in *Current Perspectives in the History of Science in East Asia*, Seoul: Seoul National University Press, 1999, 316–329.

Van Dalen, Benno. Islamic and Chinese astronomy under the Mongols: a little-known case of transmission, in Yvonne Dold-Samplonius, Joseph W. Dauben, Menso Folkerts, and Benno van Dalen, eds., *From China to Paris: 2000 Years Transmission of Mathematical Ideas*, Stuttgart: Franz Steiner, 2002, 327–356.

Van der Waerden, Bartel L. History of the zodiac, *Archiv für Orientforschung* **16** (1953), 216–230.

Van der Waerden, Bartel L. Ausgleichpunkt, 'methode der perser', und indische planetenrechnung, *Archive for History of Exact Sciences* **1** (1961), 107–121.

Van der Waerden, Bartel L. Das heliozentrische System in der griechischen, persischen un indischen Astronomie, *Neujahrsblatt herausgegeben von der Naturforschenden Gesellschaft in Zürich* **172**, Zürich: Kommissionsverlag Leemann AG, 1970.

Van der Waerden, Bartel L. *Science Awakening II. The Birth of Astronomy*, Leyden: Noordhoff / New York: Oxford University Press, 1974.

Van der Waerden, Bartel L. On Greek and Hindu trigonometry, *Bulletin de la Société Mathématique de Belgique* **38** (1986), 397–407.

Van der Waerden, Bartel L. Reconstruction of a Greek table of chords, *Archive for History of Exact Sciences* **38** (1988), 23–38.

Varāhamihira. *The Pañcasiddhāntikā of Varāhamihira*, edited and with commentary by Otto Neugebauer and David Pingree, *Det Kongelige Danske Videnskabernes Selskab Historisk-Filosofiske Skrifter* **6** (1) (1970) part I, 1–206, and **6** (1) (1971) part II, 1–154.

Vernet, Juan; and Samsó, Julio. The development of Arabic science in Andalusia, in Roshdi Rashed, ed., *Encyclopedia of the History of Arabic Science*, London/New York: Routledge, 1996, vol. 1, 243–275.

Vetter, Q. Poznámka k. t. zv. trigonometrii Ahmōseově a k rozměrům pyramidy Chufuovy, *Časopis pro Pěstování Matematiky a Fisiky* **54** (1925), 281–283.

Vettius Valens. *Vettii Valentis Antiocheni Anthologiarum Libri Novem*, ed. David Pingree, Leipzig: Teubner, 1986.

Victor, Stephen K. *Practical Geometry in the High Middle Ages: Artis cuiuslibet consummatio and the Pratike de Geometrie*, Philadelphia: American Philosophical Society, 1979.

Villuendas, M. V. *La Trigonometria Europea en el Siglo XI. Estudio de la Obra de Ibn Muʿāḏ El Kitāb Maŷhūlāt*, Barcelona: Instituto de Historia de la Ciencia de la Real Academia de Buenas Letra, 1979.

Vitruvius. *The Ten Books on Architecture*, transl. Morris Hicky Morgan, Cambridge, MA: Harvard University Press, 1914.

Vitruvius. *On Architecture*, 2 vols., transl. Frank Granger, Loeb Classical Library, Cambridge, MA: Harvard University Press, 1934.

Vogel, Kurt. *Beiträge zur griechischen Logistik*, Munich: Verlag der Bayerischen Akademie der Wissenschaften in Kommission der C. H. Beck'schen Verlagsbuchhandlung, 1936.

Vogel, Kurt. *Vorgriechische Mathematik. Teil I. Vorgeschichte und Ägypten*, Hannover: Hermann Schroedel Verlag / Paderborn: Verlag Ferdinand Schöningh, 1959.

Vogel, Kurt. *Der Donaurum, die Wiege mathematischer Studien in Deutschland*, München: Werner Fritsch, 1973.

Vogel, Kurt. John of Gmunden, in Charles Coulston Gillispie, ed., *Dictionary of Scientific Biography*, vol. 7, New York: Charles Scribner's Sons, 1973, 117–122.

Von Braunmühl, Anton. Beitrag zur Gechichte der prosthaphäretischen Methode in der Trigonometrie, *Bibliotheca Mathematica* (2) **10** (1896), 105–108.

Von Braunmühl, Anton. Beiträge zur Geschichte der Trigonometrie, *Abhandlungen der Kaiserlichen Leopoldinisch-Carolinischen Deutschen Akademie der Naturforscher* **71** (1897), 3–30.

Von Braunmühl, Anton. Nassir eddin Tusi und Regiomontan, *Abhandlungen der Kaiserlichen Leopoldinisch-Carolinischen Deutschen Akademie der Naturforscher* **71** (1897), 33–67.

Von Braunmühl, Anton. Zur Geschichte der prosthaphaeretischen Methode in der Trigonometrie, *Zeitschrift für Mathematik und Physik* **44** (1899), supplement, Festschrift zum siebstigen Geburtstage Moritz Cantors, 15–29.

Von Braunmühl, Anton. *Vorlesungen über Geschichte der Trigonometrie*, 2 vols., Leipzig: Teubner, 1900/1903.

Von Erhardt, Rudolf; and von Erhardt-Siebold, Erika. Archimedes' *Sand-Reckoner*: Aristarchos and Copernicus, *Isis* **33** (1942), 578–602.

Von Erhardt, Rudolf; and von Erhardt, Erika. Reply to [Neugebauer 1942], *Isis* **34** (1943), 214–215.

Waerden, van der. *See* van der Waerden.

Wall, Byron Emerson. Anatomy of a precursor: the historiography of Aristarchos of Samos, *Studies in History and Philosophy of Science* **6** (1975), 201–228.

Werner, John. *Ionannis Verneri De triangulis sphaericis libri quatuor*, ed. Axel Anthon Björnbo, *Abhandlungen zur Geschichte der Mathematischen Wissenschaften mit Einschluss ihrer Anwendungen Begründet von Moritz Cantor* **24**, part I, Leipzig: Teubner, 1907.

Werner, John. *Ionannis Verneri De meteoroscopis libri sex*, ed. Axel Anthon Björnbo, *Abhandlungen zur Geschichte der Mathematischen Wissenschaften mit Einschluss ihrer Anwendungen Begründet von Moritz Cantor* **24**, part II, Leipzig: Teubner, 1913.

Whish, Charles M. On the Hindu quadrature of the circle, and the infinite series of the proportion of the circumference to the diameter exhibited in the four Śāṣtras, the *Tantrasaṃgraham, Yuktibhāṣā, Caraṇapaddhati* and *Sadratnamālā, Journal of the Royal Asiatic Society* **3** (1835), 509–523.

Wilkinson, Lancelot. *Hindu Astronomy II. The Siddhánta Śiromani*, Calcutta: Asiatic Society of Bengal, 1861.

Wilson, Curtis. Hipparchus and spherical trigonometry, *DIO* **7.1** (1997), 14–15.

Woepcke, Franz. Discussion de deux methods arabes pour determiner une valeur approchée de sin 1°, *Journal de Mathématiques Pures et Appliquées* **19** (1854), 153–176; 301–303. Reprinted in Franz Woepcke, *Études sur les Mathématiques Arabo-Islamiques*, Frankfurt: Institut für Geschichte der Arabisch-Islamischen Wissenschaften, 1986, vol. 1, 614–640.

Woepcke, Franz. Sur le mot kardaja et sur une méthode Indienne pour calculer les sinus, *Nouvelles Annales de Mathématiques* **13** (1854), 386–395.

Woepcke, Franz. Recherches sur l'histoire des sciences mathématiques chez les orientaux, d'après des traits inédits arabes et persans, troisième article: Sur une mesure de la circonférence du cercle, due aux astronomes arabes, et fondée sur un calcul d'Aboûl Wafâ, *Journal Asiatique* (5) **15** (1860), 281–320. Reprinted in Franz Woepcke, *Études sur les Mathématiques Arabo-Islamiques*, Frankfurt: Institut für Geschichte der Arabisch-Islamischen Wissenschaften, 1986, vol. 1, 573–612.

Yano, Michio. Three types of Hindu sine tables, *Indian Journal of History of Science* **12** (1977), 83–89.

Yano, Michio. Trigonometry in India, in Helaine Selin, ed., *Encyclopaedia of the History of Science, Technology, and Medicine in Non-Western Cultures*, Dordrecht/Boston/London: Kluwer, 1997, 987–990.

Zeller, Mary Claudia. *The Development of Trigonometry from Regiomontanus to Pitiscus*, Ann Arbor: University of Michigan doctoral dissertation, 1944.

Zeuthen, Hieronymus Georg. Note sur la trigonométrie de l'antiquité, *Bibliotheca Mathematica* (3) **1** (1900), 20–27.

Zhitomirskiĭ, S. V. The heliocentric hypothesis of Aristarchos of Samos and ancient cosmology [in Russian], *Istoriko-Astronomicheskie Issledovaniya* **18** (1986), 151–160.

Zinner, Ernst. *Regiomontanus: His Life and Work*, transl. Ezra Brown, Amsterdam/New York/Oxford/Tokyo: North-Holland, 1990.

Index

Arabic names are alphabetized after the initial "al-" if it appears. Thus, for instance, "al-Bīrūnī" appears in the Bs. Historical works are listed under their author's name, if known.